19.50

Polymers in Nature

Polymers in Nature

E. A. MacGregor
Department of Chemistry, University of Manitoba, Canada

and

C. T. Greenwood
Department of Chemistry, University of the South Pacific, Fiji

JOHN WILEY & SONS
Chichester · New York · Brisbane · Toronto

British Library Cataloguing in Publication Data:

MacGregor, E. A.
Polymers in nature.
1. Macromolecules
I. Title II. Greenwood, Charles Trevor
574.1′924 QP801.P64 79-41787

ISBN 0 471 27762 2

Typeset by Preface Ltd., Salisbury, Wiltshire and printed in the United States of America

To S. and D.

Contents

Preface

In Nature's infinite book of secrecy
A little I can read

(Shakespeare—*All's Well That Ends Well*)

The wide variety of polymers in Nature impinges on every aspect of our existence, and currently exciting advances are being made in all areas of the field.

It is against this background that we have outlined the present state of knowledge of the structure, properties, and functions of these fascinating substances. We bring together in this book the whole spectrum of different natural macromolecules—both organic and inorganic—with all their inherent variations.

The range of these natural polymeric materials is so wide that a complete coverage is not possible, but examples have been chosen from each different class of macromolecules to illustrate the types of problems which are encountered in any particular field. Necessarily some groups of polymers have not been included because they have not yet been investigated in sufficient detail, for example, the polylipids of plants.

The book is intended as a text for undergraduates in chemistry, biochemistry, agricultural chemistry, and related subjects, and also as supplementary material for graduate courses on polymers. Additional reading material is included in each chapter for further study.

Finally, we are grateful to our former colleague, Dr. I. G. Jones, for his helpful commentary on the chapter dealing with nucleic acids.

E.A.M.
C.T.G.

Chapter 1

Introduction

1.1. POLYMERS IN NATURE

Natural high polymers are chemical substances which are basic to the whole scheme of Nature. This group of macromolecules includes the *proteins*, extremely important constituents of all organisms where they carry out an infinite variety of functions; the *nucleic acids*, constituting the material from which chromosomes—the very core of the life process itself—are made; the *polysaccharides*, which play not only an important structural rôle in plants and animals, but also provide an energy source for many life-forms; *lignin*, the plant cement; and the unusual elastomer, *rubber*. All these materials are *organic* in their constitution, but there is another very important group of natural *inorganic* high polymers. This latter group includes the silicates, which form the major part of the earth's crust.

The characteristics of a natural high polymer are outlined below. Although many of these recent concepts have been developed from studies of *synthetic* high polymers, where the properties of the resultant polymer molecule can be controlled by the mode of synthesis, such concepts are equally valid for natural polymers. Indeed, it must not be forgotten that the basic concept of a high polymer was developed from classical studies carried out in the 1930s on materials such as proteins, rubber, and cellulose.

The majority of natural high polymers occur in admixture, and before any structural examination or physicochemical measurements can be made, these materials have to be isolated without modification from their natural source. Problems of extraction and purification are outlined in general terms in Chapter 2, and then after considering in Chapter 3 the techniques for determining the size and shape of a macromolecule, the following chapters discuss the structure, rôle, and—where appropriate—the biological function of the major classes of natural high polymers.

1.2. THE CONCEPT OF A HIGH POLYMER

A high polymer is simply a chemical substance composed of giant molecules, each consisting of a large number of structural, or repeating, units joined by covalent bonds, i.e.

$$\cdots\!-\!A\!-\!A\!-\!A\!-\!A\!-\!A\!-\!A\!-\!A\!-\!A\!-\!\cdots,$$

where A is the structural unit, and — is the covalent bond.

Organic polymers are synthesized from monomers, and depending on the type of monomer, the polymerization reaction can occur by either a condensation or an addition mechanism.

1.2.1. Condensation and addition polymers

Linear condensation polymers are formed from monomers containing two reactive groups, and a condensation reaction occurs with the elimination of some small molecule such as water or ammonia. For example, the polymerization of the amino acid, $H_2N{-}R{-}COOH$, occurs as

$$n\,H_2N{-}R{-}C{\overset{\displaystyle O}{\underset{\displaystyle OH}{\lessgtr}}} \longrightarrow H_2N{+}R{-}\overset{O}{\overset{\|}{C}}{-}NH{+}_{n-1}R{-}C{\overset{\displaystyle O}{\underset{\displaystyle OH}{\lessgtr}}} + (n-1)H_2O.$$

Amino acid — Polyamino acid

Most natural polymers can be classified as belonging to the condensation type, although biosynthesis from the monomer is much more complex than outlined above, and involves other natural polymers, the enzymes, which are biological catalysts.

Linear addition polymers are formed from unsaturated monomers under specific conditions when the olefinic double bond becomes reactive, and the monomers join together in a head-to-tail arrangement. For example, the diolefin, butadiene, the monomer of synthetic rubber, will react as

$$n\;CH_2{=}CH{-}CH{=}CH_2 \longrightarrow {-}\!\!\left(CH_2{-}CH{=}CH{-}CH_2\right)\!\!{-}_n$$

Butadiene — Polybutadiene

In an addition polymer, the structural unit contains the same proportion of atoms as does the original monomer.

It should be noted that both condensation and addition polymers can be depolymerized, for example, by enzymes or acids. This reaction may occur readily, and very often it is a problem to avoid inadvertent depolymerization during isolation.

1.3. ARRANGEMENT OF MONOMER UNITS IN POLYMERS

Very few natural polymers are simple linear molecules, and more complicated structures exist. It is necessary, therefore, to consider in more detail the various ways in which monomers may join together. Indeed, only if the monomer units are identical and contain no asymmetric centres can the polymer structure be uniquely defined.

1.3.1. Linear polymers

Even in a simple linear polymer, the presence of asymmetric centres in the repeating units gives rise to different chain configurations, or different tacticities, of the polymer. These can be illustrated by considering the polymer, $-(CH_2-CHX)_n-$, where X is any group. As shown in Figure 1.1,

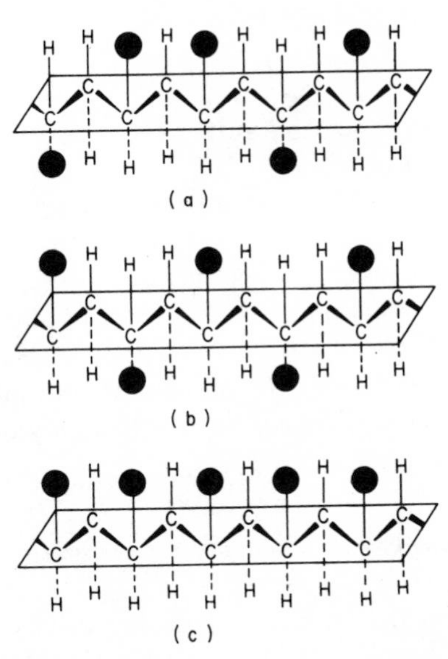

FIGURE 1.1. The various stereospecific forms of the polymer $-(CH_2-CHX)_n-$: (a) atactic; (b) syndiotactic; (c) isotactic. Carbon—carbon bonds are in a plane at right-angles to the paper

in the atactic form the X groups on the alternate carbon atoms are arranged at random on either side of the plane of the —C—C— backbone; in the syndiotactic polymer the X groups alternate; and in the isotactic form they are on the same side.

These configurations are the simplest possible ones, and with more complicated monomers, the stereoregular forms of the polymer are even more complex. With the widespread occurrence of stereospecific biosynthetic mechanisms, such stereoregular macromolecules are far more common in Nature than amongst the man-made polymers.

Different geometric configurations of the polymer chain also occur with diolefin monomers, such as the butadiene mentioned above. In this case, after polymerization, the bonds on either side of the double bond in the repeating unit can be in either the *cis* or *trans* configuration:

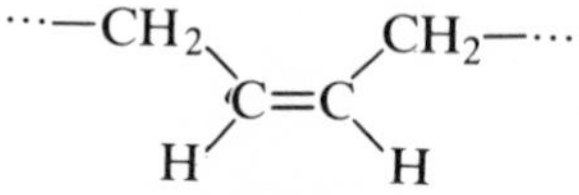

$$\begin{array}{ccc} \cdots\!-\!CH_2 & & CH_2\!-\!\cdots \\ & C\!=\!C & \\ H & & H \end{array}$$

Cis form

or

$$\begin{array}{ccc} \cdots\!-\!CH_2 & & H \\ & C\!=\!C & \\ H & & CH_2\!-\!\cdots \end{array}$$

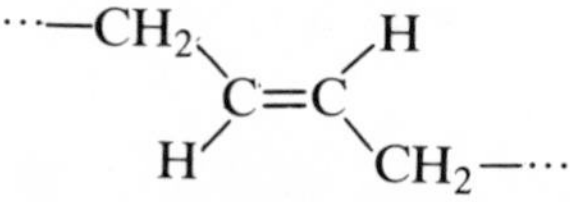

Trans form

We shall see later that rubber and gutta percha are natural polymers where this type of isomerism arises.

If the polymer contains two or more types of repeating unit—and there are many such polymers in Nature—it is known as a copolymer. Even for only two monomers, these may be arranged in different ways, either randomly to form a random copolymer; or alternately to form an alternating copolymer; or in groups or blocks of the same monomer, to form a block copolymer. Taking the monomers arabinose (Ara) and xylose (Xyl) as an example, these different copolymers may be represented as:

···— Ara-Ara-Xyl-Ara-Xyl-Xyl-Ara-Xyl —··· random copolymer

···— Ara-Xyl-Ara-Xyl-Ara-Xyl-Ara-Xyl —··· alternating copolymer

···— Ara-Ara-Ara-Ara-Xyl-Xyl-Xyl-Xyl —··· block copolymer

With more monomers, these theoretical structural possibilities increase, but in the very complex copolymers which occur in Nature (e.g. in the case of the proteins the number of monomers is twenty-three), we shall see that the enzymic mode of synthesis is such that each polymer molecule is synthesized in a unique pattern.

1.3.2. Non-linear polymers

With non-linear polymers, structures of many degrees of complexity exist. A monomer with more than two functional groups (e.g. glucose, Glc, see Chapter 6) can polymerize to form a branched homopolymer molecule as in Figure 1.2(a). The branched chains may be regular or irregular in length, whilst the branch-points may be also situated at regular, or irregular, intervals. Furthermore, depending on the stereospecific

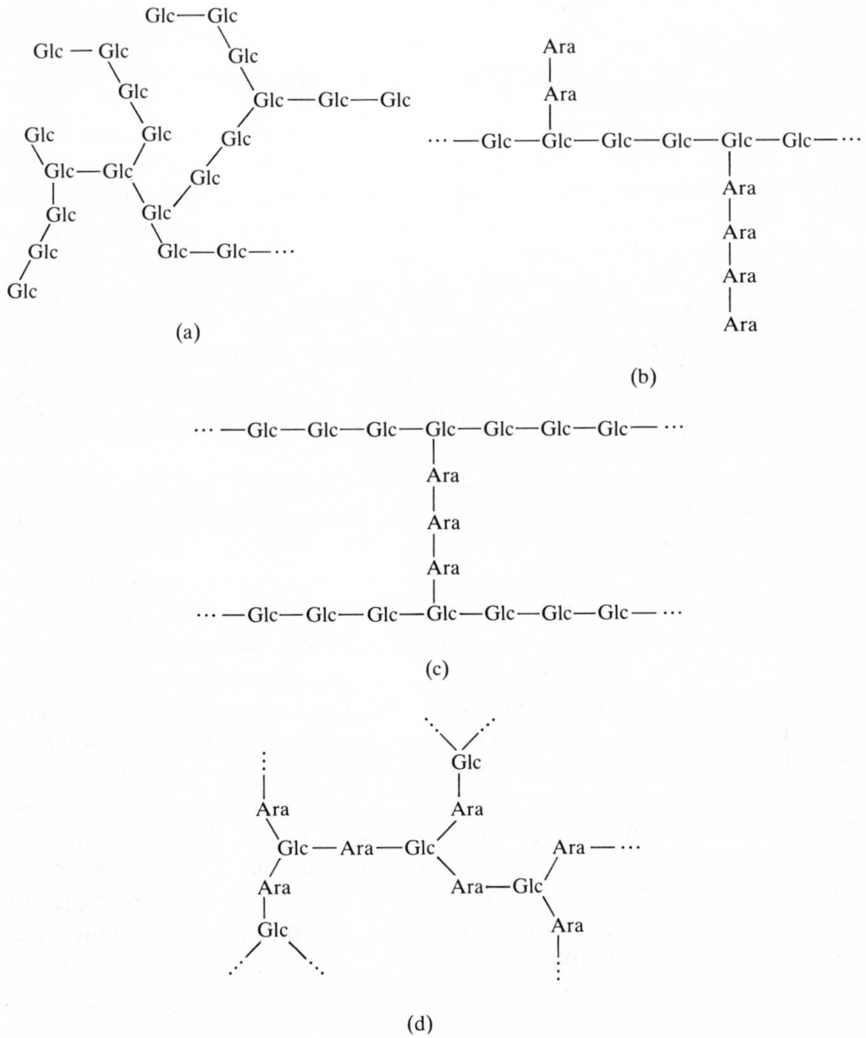

FIGURE 1.2. Schematic forms of non-linear polymers: (a) branched homopolymer; (b) block copolymer; (c) cross-linked copolymer; (d) copolymer network

requirements in enzyme synthesis, the polymer may be produced in either a two-dimensional or a three-dimensional structure.

Copolymers formed from multifunctional units can also take various forms. For example, in a hypothetical arabinoglucan (see Chapter 6), we can have arabinose side-chains of the same, or different lengths on a glucan backbone, i.e. a block copolymer as in Figure 1.2(b); or a cross-linked copolymer of arabinose chains joining glucose chains of the same or varying lengths, as in Figure 1.2(c); or an arabinoglucan network, which is either regular as in Figure 1.2(d), or irregular in the number of the units between branch-points.

1.4. CHARACTERISTICS OF POLYMERS

The unique properties of macromolecules arise primarily from their high molecular weights. The size of the polymer chain is specified by the number of structural units it contains, i.e. the degree of polymerization (DP). The corresponding molecular weight is simply the sum of the molecular weights of the structural units.

The majority of natural macromolecules, however, do not possess a unique molecular weight, but consist of a mixture of molecules with different molecular weights varying around some mean value. The polymer is said to possess a molecular weight distribution, and the significance of this distribution and the corresponding average molecular weight values will be discussed in Chapter 3.

Not only may the molecular size of the individual polymer molecules vary, but as has been mentioned above, their structure may differ, and no two may be exactly the same. For example, in a branched three-dimensional macromolecule, the degree of branching can be different, and experimental results can define only an average structure.

In solution, depending on its chemical structure and detailed size and shape, the macromolecule may exhibit varying conformations (see Chapter 3), but in the solid state, intermolecular attractions may be very important, and sufficiently strong, for the polymer to exhibit crystallinity as shown by X-ray diffraction measurements. Indeed X-ray photographs from cellulose fibres and stretched rubber gave the first information regarding the geometry of these macromolecules, i.e. the repeating distance along the chain. The importance of strong hydrogen bonding between the polar groupings of proteins is one example of this effect, but many examples of intermolecular forces effectively controlling the polymer architecture will be found throughout the following chapters.

Chapter 2

Extraction and Purification

2.1. INTRODUCTION

Most natural polymers are found in close association, or in admixture, with other related biopolymers. For example, a cell may contain hundreds of different proteins, and many nucleic acids. If the cell is of plant origin, the cell-wall itself will consist of an assembly of polymers, such as lignin and several polysaccharides. Hence, in order to characterize one type of biopolymer completely, that substance has to be extracted from its source and purified until it is free of contaminants, whether large or small molecules. This task is, of course, extremely difficult.

Nowadays, the majority of studies on natural macromolecules, particularly those of biological origin, are carried out on the polymer in solution, and so the first requirement is to extract the required material in a soluble form. Usually the physical structure of the tissue containing the desired polymer must be destroyed, e.g. cell-walls or intercellular matrix must be disrupted, but some polymers are either insoluble, or are chemically bonded to insoluble structures. To obtain such a polymer in solution without degradation presents many problems.

Once in solution, a polymer must be purified before any structural study can be carried out. Usually, no one procedure suffices to give a pure product, and instead a combination of techniques must be used. We shall in this chapter discuss some of the most important of these procedures, and indicate their applicability to the different types of natural polymers. In general, the extraction techniques used should be mild enough so that no chemical modification or large physical change in the structure of the desired substance takes place. Obviously, for industrial exploitation of a polymer, these rigid criteria need not be met, but they are of vital importance to any fundamental investigation.

It should be mentioned at this stage that many of the procedures used to purify natural polymers can also be employed for the separation of polymer fragments, and hence are of great importance in the detailed structural studies of large molecules.

During a purification procedure, a sensitive and quantitative method must be available to detect and follow the substance under investigation. Where a polymer is biologically active, assays of activity may be used; in other cases, physical properties, such as the absorption of ultraviolet light, form the basis of a detection system. Nowadays, many of the procedures for the purification and detection of a polymer based on such properties are automated so that the presence of the polymer may be indicated by a peak on a recorder chart, or by numbers on a digital 'print-out'.

The successful outcome of a detailed structural investigation of a natural polymer depends on the purity of the biopolymer. This aspect of a study has often been ignored, and much published work is of doubtful validity because of the questionable purity of the substance involved. This problem has been particularly acute in the field of complex polysaccharides, and remains so even today.

We shall, therefore, discuss methods for extracting and purifying natural polymers, and outline some of the existing difficulties.

2.2 EXTRACTION

For most biological polymers, extraction must first be preceded by disruption of the tissue in which the polymer of interest is found. Animal cell-walls can usually be broken by mechanical blending or homogenization, but many plant and bacterial cells require more drastic treatment such as milling, grinding, sand fragmentation under high pressure, or the use of ultrasonics. In cases where the required polymer is not modified by, and is easily separated from, certain biological catalysts, the enzymes (see Section 4.5.2, p. 114), these substances may be added to aid in the degradation of tissue structure. Extraction may then be accomplished by shaking or blending the fragmented tissue material with a suitable aqueous solution, or in the case of lignin or the polyisoprenes with a non-aqueous solvent.

Where the polymer under study is localized within an animal, plant, or microbial cell in one type of subcellular particle these particles may be separated by differential centrifugation (see Section 2.3.3, p. 27). The cell fraction may then be used as the starting point for extraction and purification, thus affecting a substantial degree of purification at the outset.

The process of extraction can, in itself, constitute a step in the purification procedure. Thus a careful choice of conditions and solvent may bring the polymer of interest into solution whilst leaving many contaminants undissolved, e.g. some bacterial polysaccharides can be extracted in cold diethylene glycol which has limited dissolving power for many of the cell proteins and nucleic acids.

For insoluble polymers, prior extractions may remove unwanted soluble macromolecules leaving the insoluble substance relatively pure in the solid state. Severe problems may then be encountered: by its very nature, an insoluble polymer can be dissolved only by changing the substance in some way, and at best, these changes must be kept to a minimum. As will be seen later, lignin is a prime example of this type of polymer, for only a small fraction of the lignin in a wood sample can be brought into solution without drastic modification.

The *proteins* form such a heterogeneous group of natural polymers that no general method for their extraction can be given. It should be noted, however, that the three-dimensional structure of most proteins is extremely labile, and care must be taken during extraction and subsequent purification to attempt to maintain that structure intact. For this purpose, proteins are often extracted into a dilute salt solution, where the ionic strength, temperature, pH, and the nature of the ions present are carefully controlled. In some cases detergents may be added to the extraction medium to disrupt certain subcellular structures and facilitate solution of protein.

The solubility of most proteins decreases when the original three-dimensional structure of the molecules is destroyed, and so a preferential extraction of a particularly stable protein may sometimes be carried out. For example, an acid-stable protein may be extracted at low pH where many other proteins lose their original structures, and hence become insoluble. Protein solubility often depends on ionic strength, and again preferential extractions may be carried out using media of different ionic strengths.

Many tissue-extracts contain enzymes—themselves proteins—capable of breaking down the desired protein molecule into small fragments. These catalysts must quickly either be separated from the protein under study, or have their activity inhibited by the addition of specific chemical compounds.

With care, however, it is now a relatively simple matter to extract a soluble protein and maintain its natural structure. For insoluble proteins this is not necessarily the case, and more drastic extraction procedures must be used. These might involve higher temperatures or relatively strong acids or bases, and in the process chemical bonds in the original polymer are often broken. Studies of such solutions tell us, therefore, about the soluble form of the protein, but cannot always give details of the polymer structure in its original, insoluble state.

Some *polysaccharides* can be extracted with little difficulty, e.g. gums from plant exudates, or food storage polysaccharides from plants or animals. The former, however, are obtained in a complex mixture, which is exceedingly difficult to resolve into single polysaccharides, whilst the latter are easily purified.

On the other hand, extraction of certain plant cell-wall polysaccharides without concomitant degradation is almost impossible to achieve. For example, many polysaccharides of wood cannot be extracted without prior delignification, a drastic process involving the use of chlorine or sulphur dioxide and sulphites. These processes inevitably produce oxidation and/or degradation of the polysaccharides.

Differing solubilities of plant cell-wall polysaccharides allow preferential extraction of certain polymers without delignification, but again the solvents used—dilute potassium hydroxide solution, or sodium hydroxide solution containing borate—can lead to depolymerization. Cellulose is usually left as the insoluble material of cell-walls, and can itself be dissolved in cuprammonium hydroxide, but has often undergone degradation during delignification or solution of other cell-wall polysaccharides.

Structural polysaccharides of animal origin often occur covalently linked to protein. In order to extract a protein-free polysaccharide, the tissue is treated first with protein-degrading enzymes or with alkali. Again, care must be taken to minimize the effect of the alkali on the polysaccharide. Sometimes, with the appropriate choice of tissue as starting material, an extract containing only one animal polysaccharide can be obtained directly.

Nucleic acids occur in all living cells and in viruses, and are often found as nucleoprotein complexes. (Some viruses are simply a nucleic acid with a protein coat.) After cell disruption, certain subcellular fractions are normally prepared by differential centrifugation (see Section 2.3.3, p. 27), and nucleoprotein complexes may then be extracted into salt solution from the appropriate fractions. These complexes can subsequently be dissociated by changes in ionic strength and ion content of the solution. It has proved very difficult to keep intact the largest nucleic acid molecules during extraction, for the forces required to disrupt the appropriate cells are often sufficient to fragment these substances. Because simple viruses contain one nucleic acid only, such polymers are readily purified. Many studies of the function of nucleic acids have been carried out using viral preparations.

Rubber and *gutta percha* are both readily extracted in almost pure form. The rubber in latex can be coagulated by heat or acid, and the coagulate can be washed in alkali to remove soluble polymers such as proteins. After further washing, the rubber may be dissolved in a non-polar solvent such as benzene. Gutta percha can be extracted directly from a suitable source into a solvent such as benzene, after washing the starting material with water and acetone to dissolve soluble contaminants.

Lignin occurs in plant cell-walls in close association with several polysaccharides. It is an inert and extremely insoluble polymer, and it is not possible to dissolve all the lignin of, say, a wood sample without using drastic chemical conditions; these are assumed to cause extensive degradation of the lignin. A small proportion of wood lignin may be extracted in dioxane or ethanol, often after the ground-up wood has been treated with alkali to remove some of the associated polysaccharides. It is this substance which is taken as the model for lignin in its 'native' state. Conditions used in commercial pulping, which involve strong alkali or bisulphites, dissolve all the lignin of a sample at the expense of substantial chemical modification. Many other methods of extraction also give solutions of lignin derivatives, not of lignin itself.

Extraction and purification of *inorganic polymers* present completely different problems. These substances are normally studied and used (if industrially important) in the solid state. Thus extraction on a commercial scale would involve location of a rock stratum or structure, rich in the mineral of interest, and mining of that rock, followed by crushing and grinding to particles of the correct size for subsequent refining procedures.

2.3. METHODS OF PURIFICATION

2.3.1. Chromatography

The Russian scientist, Tswett, developed an adsorption method for separating plant pigments into coloured bands, and introduced the name chromatography (from the Greek word 'Chroma' meaning colour). Subsequently, his methods were extended, and new separation procedures were developed for colourless as well as coloured substances, but the term *chromatography* was retained. There are now so many different techniques in use that it is convenient to discuss them under five subheadings.

2.3.1.1. Adsorption chromatography

Many natural polymers can be adsorbed on columns of certain finely divided solids, where the ratio of surface area to particle size is high. Such solids include activated charcoal, silica gel, alumina, and hydroxyapatite (the calcium phosphate mineral found in bone). The nature of the forces between the adsorbent and polymer is not always well understood, but it seems likely that with hydroxyapatite they are highly polar in nature, i.e. the Ca^{2+} sites binding negative charges of an adsorbed polymer and the phosphate sites binding positive charges.

Adsorbed molecules are washed from the adsorbent—a process known as elution—using a suitable solvent. As biopolymers differ in their strength of binding to an adsorbent, some separation of a polymer mixture can occur on elution, the polymer most weakly bound to the adsorbent emerging first from the column.

Hydroxyapatite has been widely used for separations of both proteins and nucleic acids. Both types of macromolecules can be eluted using phosphate solutions; often a solution of continuously increasing phosphate concentration is used as eluant, removing first the weakly bound protein or nucleic acid, and later the more strongly held polymers. This procedure is known as *gradient elution.*

Charcoal columns are often used to separate oligosaccharides (see Section 6.2.2, p. 254), a technique which is useful in the determination of polysaccharide structure.

Advances in the theory and technique of adsorption chromatography have resulted in the construction of more sophisticated apparatus, where conditions for separation are optimized and high pressures (several hundred atmospheres) may be used, with finer particles of adsorbent. Good separations are obtained in very short times, but so far this method, *high-pressure liquid chromatography*, is used on the analytical rather than preparative scale.

For preparative work, the adsorbent is most often used in columns, but it may be spread also in a thin layer on an inert support such as a sheet of glass. A mixture may then be applied to the adsorbent near the bottom of the plate, and the eluting solvent allowed to move up the plate by capillary attraction (in an atmosphere saturated in solvent to minimize evaporation). This variation of adsorption chromatography is one type of *thin-layer chromatography* (TLC), and its main advantage over column work is a greatly increased rate of separation. However, only small quantities of mixtures can be used, so that this method is more frequently employed for qualitative analysis and investigation of purity than for preparation.

2.3.1.2. *Partition chromatography*

This type of chromatography depends on the use of two immiscible phases, of which one is held stationary, and the other is moving. Separation of constituents of a mixture then depends on differences in solubility between the two phases.

For one substance we can define a partition coefficient, a:

$$a = \frac{\text{concentration in phase A (say, the stationary phase)}}{\text{concentration in phase B (say, the moving phase)}},$$

where a is constant at any temperature and pressure. During partition chromatography, molecules of solute (i.e. a constituent of a mixture) may be thought of as diffusing between the stationary and moving phases. The time spent in each phase is a function of the solubility in that phase, i.e. depends on the partition coefficient. Obviously, the greater the solubility of a substance in the moving phase, the further that substance will diffuse. Many natural polymers have little or no solubility in the moving phase of partition chromatography, hence scarcely move from the point of application. Therefore, these types of chromatography are used more often for the separation of polymer fragments during structural studies than for the purification of macromolecules. An exception to this general rule is the use of reversed-phase chromatography in the purification of some types of ribonucleic acid (see later in this section).

Partition chromatography can be carried out using columns of materials such as cellulose, starch, or silica gel as an inert support. All of these substances function as a support for partition chromatography only when they contain water: a water–cellulose, water–starch, or water–silica complex is thought to act as the stationary phase, immiscible with the moving phase liquid, which may itself be an organic liquid, or organic liquid/water mixture. When a polymer mixture is applied to the top of the column and the constituents are eluted, that member of the mixture having the highest solubility in the eluting liquid emerges from the column first, whilst the substance of lowest solubility is retained longest on the column.

Materials such as cellulose or silica gel can also be used spread thinly on plates. This technique gives rise to a second type of thin-layer chromatography where the separation is by partition rather than by adsorption. In practice, the distinction may not be sharp—on silica gel both adsorption and partition may operate simultaneously to separate mixtures. However, only small quantities of sample can be used and again this thin-layer chromatography (TLC) is an analytical rather than a large-scale preparative method. TLC is frequently used to separate breakdown products during investigations of the structure of a protein or nucleic acid.

A great deal of partition chromatography is carried out using paper as the inert support, and this procedure is always called *paper*, rather than partition, *chromatography*. Again, the stationary phase is thought to consist of a cellulose–water complex, this water being derived from the moving phase which is normally an organic liquid/water mixture. Paper chromatography is frequently used during structural studies for analysis of the degradation products from natural high polymers, e.g. peptides from proteins (see Section 4.4.1, p. 87), mono- and oligo-saccharides from polysaccharides (see Section 6.4.1, p. 261), and nucleosides from nucleic acids (see Section 5.4.1, p. 180).

After preliminary calibration work, a substance can often be identified by the distance it moves relative to the distance travelled by the moving-phase solvent. For example, the R_f factor can be defined as:

$$R_f = \frac{\text{distance moved by substance}}{\text{distance moved by solvent front}}.$$

If a substance travels distance x and the moving-phase solvent travels distance $(x + y)$, then x is proportional to the solubility of the substance in the moving phase, and y is proportional to its solubility in the stationary phase.

Thus y/x is proportional to a, the partition coefficient, and $y/x = a/k$, where k is a constant depending on the cross-sectional areas occupied by the moving and stationary phases.

Now

$$R_f = \frac{x}{x+y}, \quad \text{or} \quad \frac{1}{R_f} = \frac{x+y}{x} = 1 + \frac{y}{x} = 1 + \frac{a}{k},$$

and hence the relative distance moved by a substance depends on its partition coefficient between the stationary and moving phases.

Paper chromatography can be used as a preparative technique as well as for analysis. For preparative work, large sheets of thick paper are employed, and the mixture to be analysed is applied in solution as a streak near one end of the paper. In contrast, qualitative analysis can best be carried out on thin paper and the sample is applied in solution as a small spot.

So far we have discussed a water-containing stationary phase used in conjunction with an essentially organic moving phase. It is also possible to coat an inert support with an organic substance to provide the stationary phase, and then employ an aqueous moving phase. Such a technique is known as *reversed-phase chromatography*, and has proved extremely useful for the separation of transfer ribonucleic acids (see Section 5.5.5, p. 213), and also of ribosomal ribonucleic acids (see Section 5.5.5, p. 211).

One other type of partition chromatography should briefly be mentioned here—*gas–liquid chromatography*. The technique is essentially analytical and is not used for purification. The stationary phase is a liquid on an inert solid support, and the moving phase is a gas. Gas–liquid chromatography is being used increasingly to separate and investigate the structural units of polysaccharides.

We shall now describe several of the most important techniques currently employed in natural high polymer purification. (Adsorption chromatography was exceedingly important in the past, but has now been superseded by more modern methods, whilst partition chromatography is used more in the separation of small, rather than large, molecules.

2.3.1.3. Gel-permeation (molecular sieve) chromatography

In gel permeation chromatography (also widely known as gel-filtration chromatography) separation of polymers depends on molecular size. The gels on which this type of chromatography is carried out are themselves polymers—inert, cross-linked, and highly hydrated, and usually available commercially in the form of small beads. On the preparative scale, a column packed with beads of a suitable gel is used. The mixture of biopolymers to be analysed is applied to the top of the column, and the

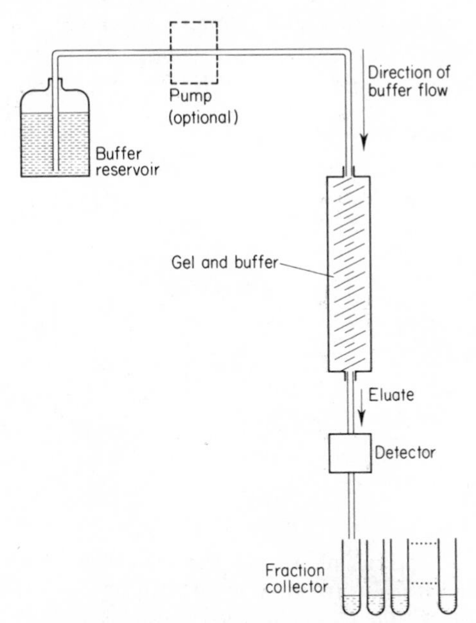

FIGURE 2.1. Apparatus for column chromatography (schematic)

constituents are eluted through the column by a suitable buffer solution, the eluant, the gel having previously been equilibrated with this buffer. With some gels, organic liquids can be used. Fractions of liquid emerging from the column (the eluate) are collected and examined for the polymer under study (see Figure 2.1).

It has been found that the largest molecules of a mixture emerge first from the column, whilst the smallest emerge last. The column may be considered to consist of porous gel particles containing, and surrounded by, buffer. If the pores of the gel beads are relatively small, then large polymer molecules cannot penetrate them and are free to move only in the surrounding buffer; these molecules can be eluted using a small volume of buffer. Small molecules, on the other hand, can readily diffuse to the interior of the gel particles, where they are retarded and thus move more slowly down the column. These small molecules therefore require larger volumes of eluting buffer to emerge from the column (see Figure 2.2).

For any particular gel, molecules above a certain minimum molecular size cannot penetrate the gel pores, i.e. they are excluded from the gel, and are eluted from the column by a volume of eluant equal to the volume of buffer surrounding the gel beads. This volume is known as the *void volume* (V_0) of the gel column. Although the behaviour of a polymer on a gel depends on the actual molecular size and shape, these are frequently difficult to measure (cf. Chapter 3). It is usually easier to find the molecular weight of the substance, and so behaviour on gels is frequently discussed in terms of molecular weight. For one class of polymers, e.g. proteins, the minimum molecular weight of molecules which cannot penetrate a gel is said to be the *exclusion limit* of that gel. (It should be noted that the same gel would have a different exclusion limit for another

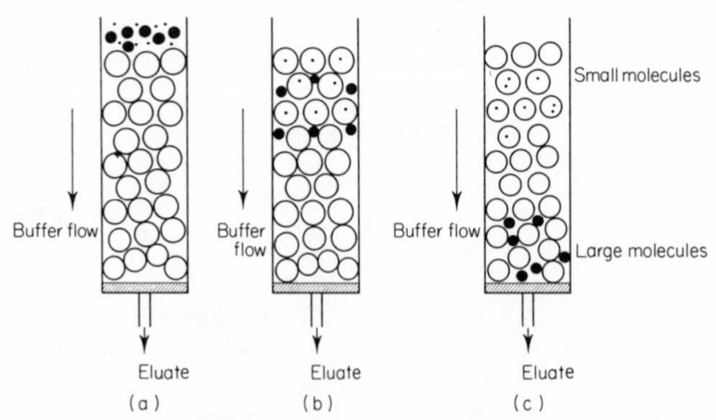

FIGURE 2.2. Principle of gel-permeation chromatography (schematic) (a) Mixture of large (●) and small (·) molecules is applied to top of column gel particles (○); (b) small molecules are retarded within the gel beads; (c) large molecules may be completely separated from small ones

class of polymers, such as polysaccharides, because of differences in molecular shape.)

If a small molecule does not interact with gel particles other than by penetrating the pores and being washed out, then this substance should be eluted by one bed volume of eluant, i.e. a volume equal to the volume of buffer surrounding and within the gel particles. A molecule of intermediate size may penetrate the gel particles to some extent, and will eluted in a volume between those required for the small and the large molecules. This *elution volume* (V_e) is the volume of buffer which passes through the column between the application of the substance to the top of the column and its emergence from the bottom. A substance may be characterized by a constant, K_{av}, such that

$$K_{av} = \frac{V_e - V_0}{V_t - V_0},$$

where V_t is the total volume of the column.

This constant depends only on the nature of the gel, and is independent of the geometry and packing density of the column. For a given polymer/gel system, V_e, and hence K_{av}, depend only on the molecular size of that substance, and are independent of the flow rate of buffer through the gel and the polymer concentration in a mixture. In fact, if the polymer does not interact with the other constituents of the mixture, K_{av} and V_e are independent of the nature of the other constituents present, and the behaviour of the desired substance will be very reproducible.

In current practice, three main types of gel are available. All can be manufactured with different pore sizes, and hence can be used to separate molecules of different ranges of molecular weight. If, say, a protein of molecular weight 50,000 is contained in admixture with high and low molecular weight substances, it may first be chromatographed on a gel effective for separating proteins in the molecular weight range 1000–30,000. The protein would emerge from the column at the void volume along with the other substances of weight above 30,000. Subsequent use of a gel column effective for the weight range 3000–70,000 would separate out all molecules of molecular weight over 70,000 (which would emerge at the void volume), whilst the protein of interest would be retarded and eluted later. Thus, at worst, the protein would now be in a mixture of molecules of weight range 30,000–70,000, but with the use of a long second column, this range could be narrowed considerably.

For separation of substances of molecular weight up to approximately one million, two types of gel are available: one, ‘Sephadex’, is a cross-linked dextran (a glucose polysaccharide, see Sections 6.5.1 and 6.8, pp. 288 & 327), whilst the other, ‘Biogel’, is a polyacrylamide. A third type of gel, ‘Agarose’ (which is also a polysaccharide), has a very open-pore structure, and is used for the separation of biopolymers of large molecular weights of more than one million.

Gel-permeation (or molecular sieve) chromatography can be carried out under very mild conditions, e.g. at low temperature and controlled pH, which may often be necessary to preserve the original three-dimensional structure of a natural polymer. It also requires relatively simple equipment and procedures, and can be carried out in short times. With care, the recovery of product is good, and may be as high as 100% in good separations. However, the technique of gel-permeation chromatography alone rarely gives a pure polymer, and it must be used in conjunction with other preparative procedures.

Gel-permeation chromatography can be used whenever mixtures contain molecules of different sizes, e.g. for desalting polymer solutions, in the separations of polymer fragments during structural determinations, and in the purification of various types of natural polymers such as proteins, polysaccharides, nucleic acids, and lignins. Care must be taken, however, when using 'Sephadex' in polysaccharide preparation, as dextran from the 'Sephadex' itself may contaminate the polysaccharide sample. In fact, it is advisable to use 'Biogel', rather than 'Sephadex' for polysaccharide studies.

Difficulties can be encountered when a small molecule, or ion, loosely bound to a natural polymer, is essential for biological activity or maintenance of the structure of the polymer. Gel-permeation chromatography usually separates the small molecule from the large, thus leading to loss of activity or structure, unless the eluting buffer contains a sufficient concentration of the same small molecule or ion.

Gel-permeation chromatography can also be carried out as a thin-layer process, using a layer of gel on an inert support, but this is a fast analytical procedure rather than a preparative one.

2.3.1.4. Ion-exchange chromatography

Separations by this technique depend on differences in acid–base behaviour, i.e. on differences in electric charge on molecules. Again on the preparative scale, a column of 'ion-exchanger' would be used, but ion-exchange thin-layer chromatography can be employed as a fast analytical technqiue, and indeed small samples can be prepared by such TLC.

For ion exchange, the column is filled with a synthetic resin (often of the polystyrene type), or, very commonly, a cellulose derivative containing charged groups. There are two major types of ion-exchange resin: cation exchangers, which are themselves negatively charged, and anion exchangers, which are positively charged.

A cation exchanger is most often used in the sodium form, after washing with NaOH or NaCl to saturate the resin with Na^+ ions, or in the hydrogen form after washing with acid. This resin can then best be used to purify a positively-charged natural polymer. The mixture containing the polymer is applied to the top of the column, and washed through the resin with an eluting buffer of carefully controlled pH and ionic strength. All positively-charged components of the mixture will bind to the column, either strongly or weakly depending on their degree of ionization. Changes in the pH and/or salt concentration of the eluting buffer cause different components of the mixture to elute in different volumes; the components emerging at lowest pH or salt concentration are those with the weakest positive charge, whilst components with strong positive charges elute only at high pH or in high salt concentration. Frequently, the change in buffer pH or salt concentration is made continuously, rather than in a step-wise manner, i.e. gradient elution is used.

Some of the most commonly used ion-exchange resins are sulphonated polystyrenes, cation-exchange resins used for the separation of amino acids during the study of the amino acid composition of a protein (see Section 4.3.2, p. 85). Other widely used ion exchangers are cellulose derivatives such as the anion exchanger, diethylaminoethylcellulose (DEAE-cellulose), which contains positive groups at neutral pH, and the cation exchangers, carboxymethylcellulose (CM-cellulose) (see Figure 2.3), and phosphocellulose (P-cellulose), which contain negative groups at neutral pH.

These resins are commonly used for the purification of proteins. Changes in pH have an important effect on the purification of a protein by ion-exchange chromatography. The amphoteric nature of proteins (i.e. carrying both anionic and cationic groups) enables the net charge on a protein molecule to be changed from, say, positive to negative by increasing the pH. Thus a protein which would bind to a cation-exchange resin at low pH may not do so at high pH. Separations can, therefore, be considerably improved by judicious choice of pH.

Anion exchangers have also been employed in the preparation of negatively-charged polysaccharides, and can even be used for neutral

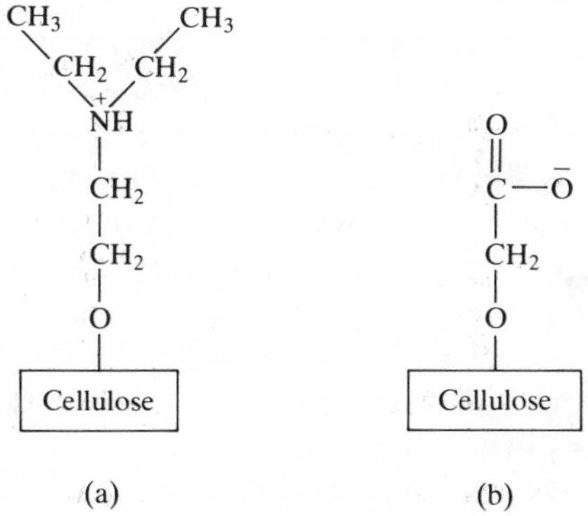

(a) $(CH_3CH_2)_2\overset{+}{N}H-CH_2-CH_2-O-$Cellulose

(b) $^{-}O-C(=O)-CH_2-O-$Cellulose

FIGURE 2.3. Examples of modified cellulose ion exchangers: (a) DEAE-cellulose; (b) CM-cellulose (see text)

polysaccharides in borate buffer (where a negative polysaccharide–borate complex is formed). A benzoylated derivative of DEAE-cellulose (BD-cellulose) has an increased affinity for aromatic groups, and is widely used in the separation of certain ribonucleic acids. In addition, fragments of natural high polymers, obtained during studies of macromolecular structure, are frequently separated by ion-exchange chromatography.

The advantages of ion-exchange chromatography are largely those of gel-permeation chromatography—it is a fast, simple technique with high levels of recovery. Sometimes, however, a protein may lose activity at a pH required to give the protein sufficient charge to bind to an ion exchanger; also if a polymer requires, say, a loosely bound cation for activity, then use of a cation-exchanger resin may separate the ion from the polymer, and activity can be lost.

Because the two types of chromatography, gel-permeation and ion-exchange, depend on completely different molecular properties for separation, i.e. on size and charge, respectively, consecutive use of these two methods frequently gives good purification of a natural polymer. When dextran gels containing charged groups (e.g. DEAE-'Sephadex', or sulphopropyl-'Sephadex'), are used, polymer mixtures are resolved on the basis of charge, but size also can have an effect.

2.3.1.5. Affinity chromatography

Certain natural polymers can be isolated from very complex mixtures, sometimes even in a single step, by affinity chromatography. This method depends on the ability of the desired polymer to bind tightly to a specific molecule, which may be small or relatively large. The specific molecule must first be attached covalently to the surface of the hydrated porous particles of an inert material such as 'Agarose'. A column of the modified 'Agarose' is prepared, a polymer mixture is applied to the column, and all polymers can easily be eluted from the column except those which bind to the specific molecule now attached to the 'Agarose' (see Figure 2.4). The bound polymer, or polymers, can often be eluted from the column using a solution of the free specific molecule as eluant. Buffers at high or low pH, or high salt concentrations, can also be effective in removing the bound molecules from the column. Thus a high degree of purification may be obtained in a single step.

The advantage of affinity chromatography is obvious—excellent purification in one operation. However, it is frequently difficult to find a 'specific molecule' for each natural polymer, and often even more difficult to attach such a specific molecule to 'Agarose' in such a way that it can still bind the polymer.

The method has, nevertheless, been used in the purification of many proteins, particularly enzymes (see Section 4.5.2, p. 114), where the specific molecule may be related in structure to a coenzyme or substrate.

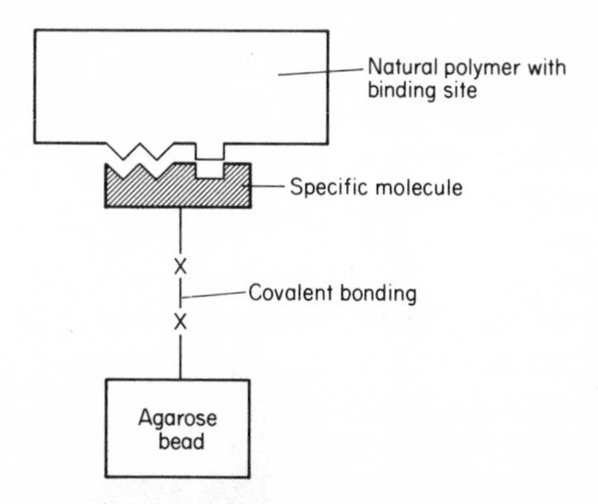

FIGURE 2.4. Affinity chromatography.

Affinity chromatography has also proved useful in the separation of animal messenger-ribonucleic acid (see Section 5.5.4, p. 207) from other nucleic acids by virtue of the polyadenylic acid 'tail' possessed by these acids (see Section 5.4.1, p. 192), which binds readily to polyuridylic acid–cellulose.

2.3.2 Electrophoresis

Electrophoretic techniques all involve the molecular movement of charged polymers under the influence of an electric field, relative to the stationary solvent. Thus separation of the components of a mixture again depends on electric charge. Once more, the variety of methods now available makes it convenient to discuss the more important ones under separate subheadings.

2.3.2.1. Free (moving) boundary electrophoresis

This technique developed by the Swedish scientist, Tiselius (who received the Nobel Prize in 1948 for this work) is the prototype of all modern electrophoretic procedures, but was more important as an analytical tool rather than a preparative method. Essentially, buffer is carefully layered on top of a solution of a biopolymer mixture (usually a protein mixture) in a U-tube, and an electric current is passed through the tube (see Figure 2.5a). Negatively-charged molecules move to the anode, whilst those carrying positive charges move to the cathode; in practice an attempt is made to choose the pH of the buffer so that most of the proteins of the mixture move in one direction. The protein of highest mobility migrates first into the buffer above the mixture and forms a sharp boundary. Slower moving proteins do not necessarily separate completely from the first, but each one gives a sharp boundary in the solution (see Figure 2.5b).

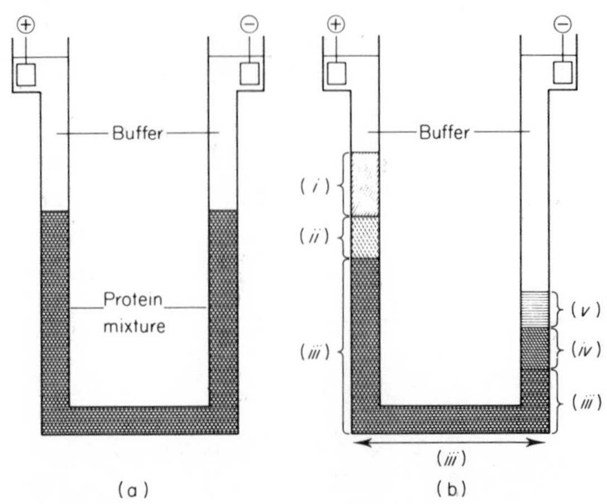

FIGURE 2.5. Moving boundary electrophoresis of mixture of three negatively-charged proteins (schematic): /// protein 1 of highest mobility, \\\ protein 2 of intermediate mobility, ≡ protein 3 of lowest mobility. (a) Initial situation; (b) distribution of proteins after passage of current for some time: (i) pure protein 1, (ii) proteins 1 + 2, (iii) proteins 1 + 2 + 3, (iv) proteins 2 + 3, (v) protein 3

It can be seen that a small quantity of the proteins of the highest and lowest mobility can be obtained from the leading and trailing zones, respectively, but in general free boundary electrophoresis is not a useful preparative technique. Optical systems have been developed (see Section 3.3.3, p. 47) which allow each boundary to be identified and the amount of each component of a mixture to be estimated, so that the procedure has some usefulness in, for example, the examination of the purity of a polymer preparation. Furthermore, the Tiselius apparatus is costly and bulky, and requires relatively large amounts of charged polymers, so that this method has been superseded by other techniques.

2.3.2.2. Zone electrophoresis

Zone electrophoresis is carried out on an inert support and has a greater resolving power than moving boundary electrophoresis; charged molecules are separated into distinct zones, hence the general name for a number of different procedures. The inert support may be paper or cellulose acetate strips (see Figure 2.6). For preparative work, the mixture under study is applied as a streak across the support (near the centre if the direction of migration of the required component is unknown), current is passed for some time, and some method of detecting the required polymer is used on strips cut from the sides of the support. The section of the support

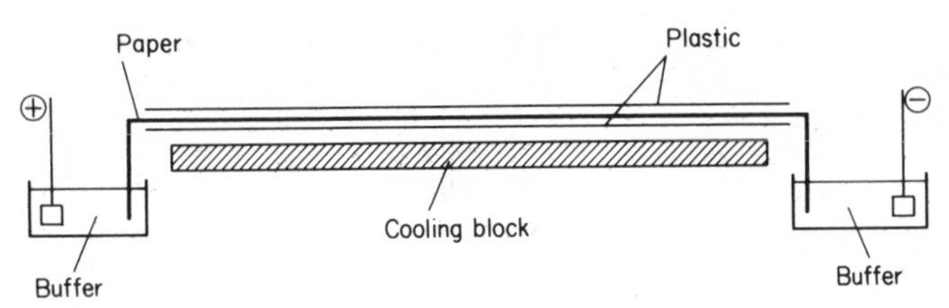

FIGURE 2.6. Typical apparatus for high-voltage paper electrophoresis: the paper sheet is moistened with buffer after the sample has been applied to the paper

containing the desired compound may then be cut out, and the compound obtained by elution with a suitable solvent.

Increased resolution can be obtained if a *porous gel* is used as the 'inert' support. The gels can retard, or exclude, polymer molecules on the basis of molecular size, and so separations can be made which depend simultaneously on differences in both electric charge and size. (In this technique the porosity of the gel physically controls the movement of the macromolecules, i.e. large polymers move more slowly—in contrast to their behaviour in gel-permeation.) Gels of starch, polyacrylamide, or 'Agarose' are used, sometimes in columns, but more often in the form of flat slabs, or thin layers. These techniques can be used analytically on very small quantities of polymer if the purity of a particular preparation is to be examined. After electrophoresis on, say, a gel, the number of components present is usually investigated by applying to the gel a chemical which gives a coloured stain with the type of polymer being studied. Gel electrophoresis can also be used preparatively by 'scaling-up' the analytical technique.

Although some electrophoretic techniques are used for polysaccharides and nucleic acids, they are most frequently employed on proteins. Care must be taken when working with a protein that the pH of the buffer used does not destroy the three-dimensional structure of that protein.

In a useful variation of an analytical type of zone electrophoresis, the polymers (usually proteins) are subjected to an electric field in a gel support divided into two sections of different porosity and buffered at different pH values. As a protein migrates from the more porous to the less porous gel, and simultaneously into a medium at different pH, the protein molecules become concentrated in a very narrow band; thus much better resolution is achieved than in a continuous buffer. The technique is known as *disc gel electrophoresis*.

2.3.2.3. Isoelectric focusing

In 1966, two Scandinavian scientists, Versterberg and Svensson, showed that a stable pH-gradient could be set up in a column by passing an electric current through a solution of mixed amphoteric substances with continuous

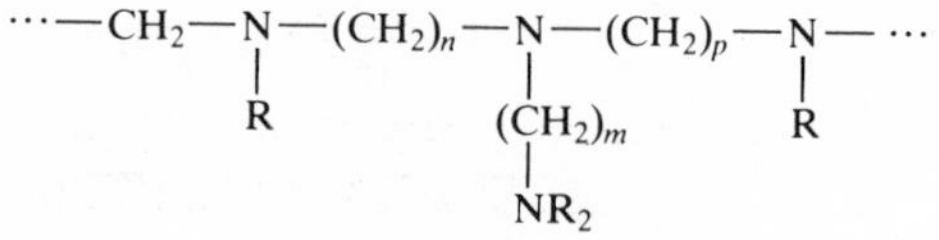

$$\cdots-CH_2-\underset{\displaystyle R}{\underset{|}{N}}-(CH_2)_n-\underset{\underset{\displaystyle NR_2}{\underset{|}{(CH_2)_m}}}{\underset{|}{N}}-(CH_2)_p-\underset{\displaystyle R}{\underset{|}{N}}-\cdots$$

$$R = H \text{ or } (CH_2)_q-COOH$$

$$m,\ n,\ p,\ q < 5$$

FIGURE 2.7. Ampholyte for isoelectric focusing

isoelectric points. (The isoelectric point of an amphoteric substance is the pH at which the substance has zero electrophoretic mobility.) An example of the type of amphoteric substance, i.e. ampholyte, used originally is given in Figure 2.7. For isoelectric focusing in solution, the pH gradient is stabilized by the presence of sucrose or glycerol throughout the solution, but not at a constant concentration; the sucrose or glycerol solution varies in density from top to bottom of the apparatus, being densest at the bottom. Thus the apparatus contains both a pH and a density gradient. A protein in solution in this apparatus (see Figure 2.8) migrates in the electric field until the molecules reach the pH at which they are isoelectric, i.e. have zero mobility, and there form a narrow band. As proteins usually differ in isoelectric point, a good separation of components of a mixture

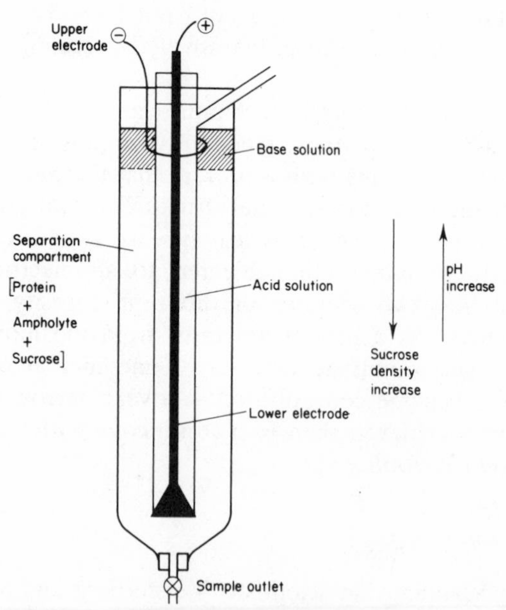

FIGURE 2.8. Apparatus for column isoelectric focusing

may be obtained. Unfortunately, it may take several days to focus the proteins of a mixture, and prolonged exposure to a pH at, or near, the isoelectric point may cause a protein to lose its biological activity and original three-dimensional structure. Also, a protein is least soluble at its isoelectric point, and so can precipitate from solution, causing experimental problems.

Smaller quantities of proteins may be separated much more quickly by isoelectric focusing, not in free solution, but on a thin layer of polyacrylamide gel. Here the gel is not intended to act as a molecular sieve, but rather as an inert support and a stabilizer of the pH gradient. The technique is most often used for analytical studies, but may be used for small-scale preparations, in which case the portion of the gel containing the required protein is cut out and the protein subsequently eluted.

To date, the technique of isolectric focusing has been used for the study of proteins rather than other types of natural high polymer.

2.3.2.4. Isotachophoresis

This term, derived from the Greek, means electrophoresis where all zones acquire the same speed. It is a useful analytical tool which employs a discontinuous buffer system, and the procedure may be carried out in free solution or on a gel support. For free solution work, a long narrow tube connected to two electrode compartments is used. If, say, anionic samples are to be separated, the anode compartment and capillary tube are filled with an electrolyte solution containing an anion of high mobility (higher than that of any anion in the sample mixture) and a cation with buffering capacity—this is known as the leading electrolyte. The cathode compartment is filled with the 'terminating' electrolyte, which has an anion of mobility lower than any anion in the sample. Semipermeable membranes may be fitted between the electrode compartments and the capillary tube to prevent hydrodynamic flow. The sample is fed into the tube between the leading and terminating electrolyte (Figure 2.9a) and a current is passed. The different anions separate out into sharp zones along the tube, depending on their mobility. The speed at which an ion moves along a tube depends on its mobility, concentration, and the potential gradient it experiences. A steady state is reached within the tube where all zones move with the same speed, i.e. the concentration of one species of ion becomes adjusted so that the speed of that zone of ions equals the speed of the neighbouring zones (Figure 2.9b). Each zone acquires a constant potential gradient, different from that of every other zone. The anions move along the tube towards the anode in order of effective mobility, in zones of decreasing (in the direction of the anode) potential gradient and increasing concentration, i.e. the most mobile ion first with the lowest potential gradient and highest concentration (see Figure 2.9).

The zone boundaries are exceedingly sharp because of self-sharpening.

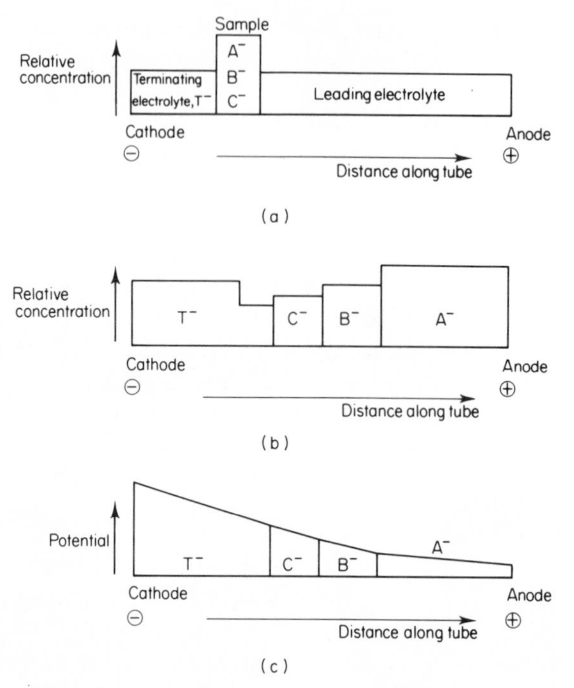

FIGURE 2.9. Separation by isotachophoresis: (a) Initial concentration and position of polyanions; (b) concentration and position of polyanions at steady state; (c) variation of potential along tube at steady state

For example, if a very mobile ion A falls back into zone B (see Figure 2.9c), the increased potential gradient speeds A up until it reaches the main A zone once again. Increased resolution can be obtained if a gentle flow of electrolyte is applied counter to the direction of anion migration; this flow slows down the forward movement of the zones and increases the effective length of the capillary tube for separation.

This type of apparatus provides an extremely sensitive analytical technique, particularly for polymer fragments such as peptides (see Section 4.4.1, p. 91). Improved protein separation is obtained on a gel support if low molecular weight ampholytes are added to the system. These ampholytes dilute and space the protein zones at the steady state, hence giving the possibility of better purification.

2.3.3. Preparative ultracentrifugation

Ultracentrifugation is simply centrifugation using high rotor speeds (e.g. 60,000 rpm) to generate correspondingly very large centrifugal forces (e.g. 250,000 **g**). These cause the dissolved macromolecules to sediment in the direction of the force field; the larger the polymer molecule, the faster it moves to the bottom of the centrifuge tube.

Crude extracts of biological origin often contain subcellular structures as well as dissolved molecules. Different structures can be obtained from the pellets formed on centrifugation at a series of increasing rotor speeds, whilst cell-sap molecules are left in solution. This technique is known as *differential centrifugation*.

Food-storage polysaccharides can also be readily purified by differential centrifugation.

Nowadays, however, elegant separations of soluble macromolecules are carried out by centrifuging a sample in a dense solution, where there is a concentration gradient of a low molecular weight solute. We can distinguish two main types of procedure involving density gradients:

2.3.3.1. Zonal density gradient centrifugation

In this technique, a density gradient of a substance such as sucrose or glycerol, in aqueous buffer, is set up in an ultracentrifuge tube prior to centrifugation. A continuous gradient is prepared using a device which mixes water and sucrose solution in an increasing ratio as the tube is filled, so that the density of the liquid is greatest at the bottom. The sample mixture is then carefully layered on top of the tube, and, during centrifugation, macromolecules move down the tubes in separate bands at rates determined by their size, shape, and density (see Figure 2.10). Usually centrifugation is stopped before equilibrium is reached. The density gradient protects against mechanical and convective disturbances.

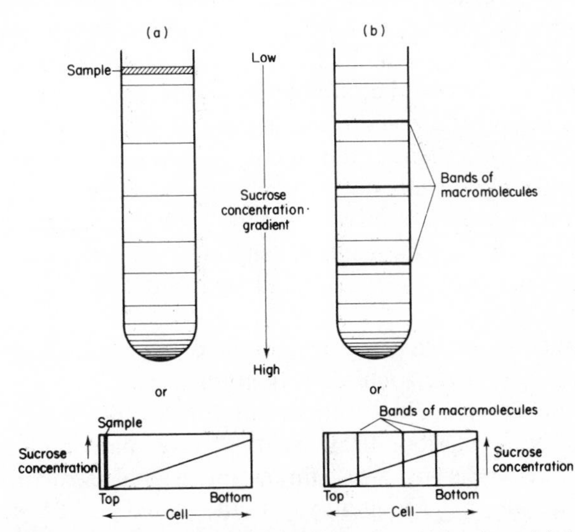

FIGURE 2.10. Zonal density gradient centrifugation: (a) before centrifugation; (b) after centrifugation

Separated macromolecules can be recovered from the tube in a number of ways: the bottom of the tube can be pierced and the liquid, which drips out slowly, collected in small samples and analysed for the desired substance; or a dense solution can be pumped carefully in at the bottom of the tube, forcing the sucrose solution with the separated polymer bands out at the top. This method is used for the purification of some proteins, particularly lipoproteins, and is widely used in the preparation of nucleic acids.

If the macromolecules under study are unaffected by 'pelleting', then good purification can be obtained after two zonal centrifugation runs. The sample solution may first be centrifuged in a fairly concentrated sucrose gradient until the required polymer nears the bottom of the tube; if the band of polymer is kept while the pellet is discarded, all contaminants of density and sedimentation coefficient (see Section 3.3.3, p. 48) greater than the desired molecule are removed. The polymer solution can be recentrifuged in a less dense gradient until the required molecules form a pellet at the tube bottom. This pellet should contain only substances with very similar sedimentation characteristics.

The use of continuous flow rotors, where most of the supernatant is continuously discarded, allows large volumes of solution to be processed in relatively short times. Obviously these rotors are used to purify molecules which are removed from the supernatant on centrifugation, and are retained in the rotor.

2.3.3.2. Density gradient sedimentation equilibrium (isopycnic) centrifugation

Here the density gradient is provided by an inorganic salt, most often caesium chloride, in aqueous solution. The gradient is not preformed, but is set up during centrifugation. In this method, centrifugation is continued until equilibrium is reached, i.e. a macromolecule sediments down the tube until it reaches the position where the density of the salt solution equals its own buoyant density (hence the name, isopycnic centrifugation); the macromolecule then proceeds no further down the tube (except for limited diffusional motion). Thus a mixture of polymers separates into discrete bands depending on the buoyant density of each species; the buoyant density, in turn, depends on the size, shape, and chemical nature of the molecules.

Separated macromolecules may be obtained after centrifugation, by the same techniques used after zonal density gradient runs.

This technique is exceedingly important for the purification of nucleic acids, particularly deoxyribonucleic acids. These macromolecules can be separated on the basis of size, three-dimensional structure, and also chemical composition. Frequently, both zonal and sedimentation equilibrium procedures are used in the preparation of one nucleic acid, to give excellent purification.

These techniques are simple and can give good separations, but also may give large dilution of a sample. Also, such procedures can only be used on macromolecules which retain their three-dimensional structure when exposed to strong sucrose or salt solutions, often for prolonged periods of time—several days' centrifugation may be necessary to reach equilibrium in isopycnic centrifugation. In some cases, low yields of product may be obtained. For example, some nucleic acids adsorb strongly onto the walls of centrifuge tubes, although this phenomenon may be lessened by the addition of detergent.

2.3.4. Other techniques

Here we shall discuss a number of widely differing techniques, including those specifically designed for inorganic polymers.

2.3.4.1. Dialysis and ultrafiltration

Dialysis involves the use of membranes which are permeable to molecules of a certain size—usually solvent and small contaminants—but which are impermeable to polymers. Several types of such membrane are available commercially in the form of tubes, hollow fibres, or simply horizontal plates.

Frequently, however, this process can be speeded-up by the application of high pressure to the side of the membrane containing the macromolecules; this forces the smaller molecules much more quickly through the membrane pores. Where the small molecules, forced through the membrane, are continuously removed, the process is known as *ultrafiltration*. This method is useful for concentrating polymer solutions, as the solvent is also forced through the filter.

Such techniques are useful for removing small molecules and ions from polymer solutions, i.e. they can be used for desalting. Care must be taken, however, that ions or small molecules required for biological activity or maintenance of structure are not removed from a polymer solution. Another problem can be irreversible adsorption of polymer molecules on the membrane itself, and this is particularly serious in high-pressure work.

2.3.4.2. Fractional precipitation

Polymers present in admixture usually differ in their solubilities in a variety of solvents, or in aqueous solutions of varying salt concentration. For example, some polymers may be precipitated at a particular salt concentration, whilst others remain in solution, and so a partial purification can be achieved by separating the precipitate from the solution.

Organic liquids such as ethanol or acetone can be used to bring about fractional precipitation of polymers. Ethanol is widely used as a precipitant in polysaccharide and nucleic acid purification, whilst rubber and gutta percha can be purified by acetone precipitation from benzene. Preparations of 'soluble' lignin can be improved by precipitation of the lignin from dioxane solution by the addition of water. Precipitants such as ethanol or acetone were formerly used in protein purification; however, the yields of active product were frequently low because the organic liquid tended to alter the three-dimensional structure of protein molecules, i.e. to denature the protein (see Section 4.4.3, p. 101).

Protein may, however, be fractionally precipitated with greater recovery if salts are used to alter the ionic strength of aqueous solutions, and ammonium sulphate is most often the salt of choice. At low ionic strengths, most proteins become more soluble as salt concentration is increased. Qualitatively, the small ions can be considered to be shielding the protein molecules from each other, by coming between the charges of opposite sign on different polymer molecules. Thus the attractive forces between protein molecules are decreased, and precipitation is less likely. This process is known as 'salting-in'. At high ionic strengths (usually greater than 1), protein solubility decreases as salt is added. This gives 'salting-out' of proteins, and may be due to a 'dehydration' of the polymer molecules—the solvent becomes organized about the salt ions to such an extent that the normal organization round the protein molecules is decreased. The protein molecules can then associate, and thus precipitate, more readily.

This method may also be used for nucleic acids and charged polysaccharides. Copper salts, barium hydroxide, and quarternary ammonium salts are frequently employed as precipitants in the purification of polysaccharides.

Proteins may be precipitated by changes in pH of a solution, for a protein is least soluble at its isoelectric point, but again the risk of altering three-dimensional structure is high.

Fractional precipitation is a very simple technique, but it is relatively inefficient and often does not give good separation of polymers. Usually, it must be used in conjunction with other, more sophisticated, procedures.

Sometimes a specific precipitant can be found, which separates one or two polymers from a mixture. Some enzymes may be precipitated by substances related to their substrates (see Section 4.5.2, p. 114), whilst certain polysaccharides precipitate with proteins having a high specificity for particular polysaccharides structures. Obviously, a specific precipitation is much more useful than a general one, but specific precipitants have been developed for very few natural polymers.

2.3.4.3. Countercurrent distribution

The principle on which this technique depends is basically the same one which makes partition chromatography so successful—a mixture of

Consider 100 parts of one polymer with a partition coefficient of 2,

$$\text{i.e.} \quad \frac{\text{solubility in upper (light) phase}}{\text{solubility in lower (heavy) phase}} = 2.$$

After each distribution shown schematically below, the light phase is moved to the right, and a new light phase is added to the tube on the extreme left. We will consider five such distributions. The numbers alongside show the amount of polymer in each tube after each cycle of operations.

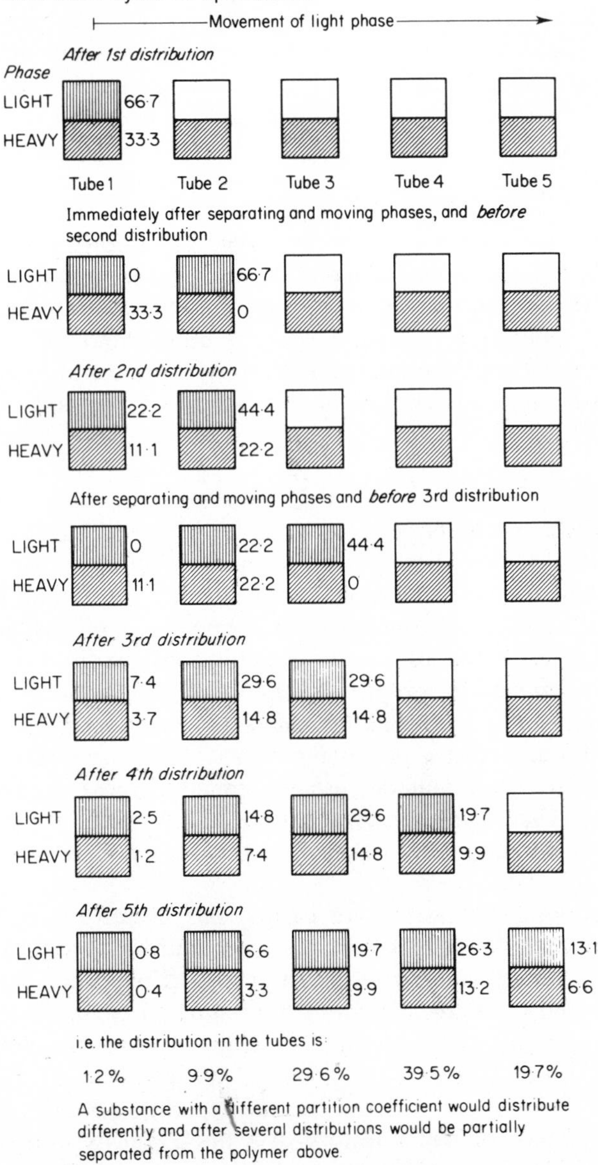

FIGURE 2.11. Principle of counter current distribution

polymers is shaken with two immiscible phases, and different polymers distribute themselves in the two phases according to the partition coefficient. The two phases are separated, shaken again with a fresh sample of the respective immiscible liquid, whereupon the polymers are redistributed between the two phases. This process can be repeated automatically many times, and partial, or complete, separation of polymers can result (see Figure 2.11).

The technique is time-consuming and requires bulky apparatus, so that it has largely been replaced by more modern chromatographic procedures. There is increased interest in the technique at the present time, however, with the use of polyethylene glycol–buffer systems which minimize degradation or denaturation of biopolymers.

2.3.4.4. Use of biological catalysts

As each enzyme catalyses a specific chemical reaction (see Section 4.5.2, p. 114), it is possible, for example, to use an enzyme to catalyse the breakdown of, say, protein, in a protein–polysaccharide mixture, leaving the polysaccharide intact. This procedure is useful for freeing one type of natural polymer from another unrelated in chemical structure, but the desired macromolecule must then be separated from the added enzyme

2.3.4.5. Methods for inorganic polymers

The procedures described so far are, in general, useful for macromolecules of biological origin. Completely different techniques have been developed for inorganic polymers which are frequently prepared on the industrial scale. Rock samples, after mining, are usually crushed or powdered, and the minerals may be separated by methods depending on, for example, magnetic properties, hardness, or density. Thus ores containing iron can be held back by magnets, whilst non-magnetic materials would pass through such a processing device. With proper adjustments to milling machinery, hard minerals usually remain as relatively large particles after grinding. These can then be separated out from the remainder of the powdered rock by screening. Asbestos fibres, for example, are obtained in this way.

Many important separations are carried out, however, on the basis of mineral density. In the simplest of these procedures, powdered rock would be added to a liquid which may be water or a denser medium such as an oil. On agitation, the lighter minerals float and can be skimmed off, whilst the heavier minerals sink and can be removed from the bottom of the tank. In the process known as *flotation*, use is made of differences in surface properties, as well as density, to aid separation of minerals. Water is added to a mixture of minerals to give a thick aqueous suspension, then specific chemicals are added in an attempt to give a hydrophobic coating to, say, one mineral, whilst other chemicals are used to emphasize the hydrophilic nature of the remaining compounds of the mixture. The suspension is then agitated by a stream of air bubbles. Hydrophobic particles associate with the air bubbles and rise to the surface to form a froth, which can then be skimmed off. The substances which endow floatability are called collectors, and are often detergents, i.e. ionizable, organic chemicals, which are capable of forming an hydrophobic ion. The chemical binds to the surface of a mineral and gives a monolayer of hydrocarbon groups projecting out from the mineral particles. This monolayer then aids association of the mineral particles with the air bubbles, causing the mineral to float to the surface. Quartz has no natural hydrophobicity, and this method can be used, for example, to separate quartz from feldspar.

2.4. EVALUATION OF PURITY

As mentioned earlier, the purity of a polymer sample is critical if a detailed investigation of its structure and function is to be carried out. For some types of macromolecules, it is a relatively easy task to ensure that the sample is, say, at least 99% pure; for others this situation is very difficult to achieve.

In general, the molecules of one *protein* all have the same molecular weight and chemical composition, and many of the techniques described in this chapter can be used to verify the purity of a protein sample—for example, gel-permeation chromatography, isotachophoresis, disc gel electrophoresis, and isolectric focusing. In particular, if one protein band is obtained after the use of the latter two techniques, this can be taken as very strong evidence that the substance under study is pure. There is always the possibility that traces of contaminants are present in quantities below the limit of the detection procedure being used, but these traces are extremely unlikely to interfere with studies of the structure of the main component. In unfortunate cases, however, where trace materials are biologically active, such substances could have serious effects on the investigation of the function of a biological polymer.

Formerly, free-boundary electrophoresis and the analytical ultracentrifuge were used to study protein purity, but these techniques have been superseded by more sensitive chromatographic and electrophoretic procedures.

A *nucleic acid*, like a protein, usually consists of molecules of the same molecular weight and chemical composition and again many of the procedures already described can be used to investigate purity. Of these, the more important would be gel-permeation chromatography, thin-layer chromatography (both ion-exchange and partition), particularly for smaller nucleic acids, and density gradient sedimentation equilibrium for larger nucleic acids. In earlier investigations of large nucleic acids, a source containing one nucleic acid only, e.g. a simple virus, was chosen. Thus once a 'pure' strain of virus was obtained and the protein removed, the nucleic acid could be assumed to be pure.

When studying *polysaccharides*, the problem of determining purity becomes much more complex. Chemical tests can show that only carbohydrate is present in a sample, but it can be extremely difficult to show that only *one* polysaccharide is present. In many cases this problem stems from a difficulty in defining precisely the polymer under study. Unlike proteins and nucleic acids, a polysaccharide sample does not consist of a collection of identical molecules—the molecules may differ in molecular size and, in complex polysaccharides, also in chemical composition. A polysaccharide sample made up of, say, three different monomers joined together, may have molecules differing in the amount of each of the three monomers and the distribution of these three along the polymer chains. At what point do we say two molecules are closely enough related to constitute one polymer, and at what point would they differ sufficiently to be said to belong to two different polymers? At the moment there is no definite answer to this question. Thus when examining polysaccharide purity, we can only call 'contaminants' those polysaccharides which are demonstrably quite different from the one under study.

Procedures described in this chapter are generally used to investigate polysaccharide purity, e.g. gel-permeation chromatography and zone electrophoresis (the latter can be used directly for charged polysaccharides, or for borate complexes of neutral polysaccharides). The analytical ultracentrifuge can also give information on polysaccharide purity, as can certain immunochemical techniques. In one example of the latter, a protein antibody (see Section 4.5.4, p. 138), which reacts specifically with a certain type of pure polysaccharide, must first be prepared. If an unknown polysaccharide sample has a structure related to that of the pure polysaccharide, then the whole sample, or a fraction of it, will form a complex with the antibody. If, further, the polysaccharide in the complex has a chemical composition different to that of the original sample, then that sample must have contained a mixture of polysaccharides.

Rubber and *gutta percha* are rarely found together in the same plant, and so there is not often the problem of freeing one from the other. Thus, in the preparation of rubber, or gutta percha, once low molecular weight contaminants and polymers such as nucleic acid and protein are removed, the hydrocarbon polymer can be assumed to be pure.

The problems encountered in dealing with *lignin* are even greater than those with polysaccharides. Lignin is so insoluble that it must be altered and degraded to be obtained in solution—the details of the polymer in its original state are largely unknown. Thus it is not known with certainty whether the substance called lignin is actually one or several polymers, and hence whether the fragments obtained in solution come from one or a number of substances. At best, the dissolved fragments can be separated and their purity examined by chromatographic techniques.

With the *inorganic polymers* also, there is the difficulty of defining one polymer, at least for several of the silicates. The molecules of one substance can have a wide range of molecular weight, and here a single crystal may consist of one molecule. In many silicates, one cation may readily be replaced by another, whilst the silicon itself can be replaced by aluminium, to give a wide range of possible chemical compositions. In many cases, the permitted range of composition recognized as a particular mineral has already been defined. Properties such as density, refractive index, chemical composition, and three-dimensional structure (determined using X-rays) would be used as criteria of purity.

2.5. ADDITIONAL READING

New Techniques in Amino Acid, Peptide, and Protein Analysis (Eds. A. Niederweiser and G. Pataki), Ann Arbor Science Publisher Inc., Ann Arbor, Michigan, U.S.A., 1971.

Liquid Column Chromatography (Eds. Z. Deyl, K. Macek, and J. Janek), Elsevier Scientific Publishing Company, Amsterdam, London and New York, 1975.

Methods in Enzymology (Eds. S. P. Colowick and N. D. Kaplan), Academic Press, London and New York; in particular Vol. 22 (1971), Vol. 27 (1973), and Vol. 34 (1974).

Advances in Carbohydrate Chemistry and Biochemistry (Eds. R. S. Tipson and D. Horton), Academic Press, London and New York; in particular Vol. 32 (1976).

Density Gradient Centrifugation R. Hinton and M. Dokata, North Holland Publishing Co., Amsterdam, Oxford, and New York, 1976 [part of a series *Laboratory Techniques in Biochemistry and Molecular Biology* (Eds. T. S. Work and E. Work)].

Separation Methods in Biochemistry C. J. O. R. Morris and P. Morris, Pitman Publishing Ltd., London and New York, 1976.

Methods of Protein Separation (Ed. N. Catsimpoolas), Plenum Press, New York, 1975.

Isoelectric Focussing (Ed. N. Catsimpoolas), Academic Press, London and New York, 1976.

Chapter 3

Molecular Size and Shape

3.1. INTRODUCTION

In order to characterize a polymer completely, its *molecular weight* must be determined. Furthermore, if the polymer molecules are not identical with respect to size, the *distribution of molecular weights* within the sample must also be investigated.

We shall discuss in this chapter a number of the more important methods for determining the weight and shape of polymer molecules.

Most techniques for investigating molecular size and shape involve studying the polymer in solution, and hence a suitable solvent must be found. Solvents may be defined as 'good' when they promote contact between polymer and solvent molecules; conversely in 'bad' solvents, polymer–polymer contacts are preferred. In good solvents, even at low solute concentrations, the volume occupied by a polymer molecule is so large that considerable molecular interaction takes place. Thus polymer solutions usually behave non-ideally (i.e. deviate from Raoult's law) even at low concentrations. This type of behaviour is observed, but is less marked, in poor solvents. The theories on which most determinations of molecular weight are based have been developed for ideal solutions, and so when dealing with polymer solutions of finite concentration, it is often necessary to take a series of measurements at different polymer concentrations, and then to extrapolate to infinite dilution.

In addition to problems caused by non-ideality, other complications may arise because of self-association of polymer molecules, or because of dissociation into subunits. Unless it is wished to investigate these particular phenomena, it is preferable to choose a solvent where such effects are minimized as far as is possible. In some cases, it may be necessary to prepare a derivative of the polymer in order to surmount problems of this nature. Again, if a polymer is unstable in one solvent, another solvent must be found so that the macromolecules are stable at least for the duration of the measurements.

The situation becomes even more complex when charged polymers are involved—*polyelectrolyte solutions* deviate markedly from ideality, and the theoretical bases for measurements may no longer hold. Thus it is necessary to investigate such substances in solutions of high ionic strength—which depress the ionization of the macromolecules—and then extrapolate the measurements to zero ionic strength.

Each of the methods of molecular weight determination described in this chapter is only suitable for a certain range of molecular weights. Indeed, most methods are *not* universally applicable, and the size of the molecule to be investigated dictates the procedures which are available.

Where a sample has a molecular weight distribution, each procedure measures only an *average* value, and the nature of this average depends on the method used. Thus different values of average molecular weight of the sample can be obtained, and it is important to realise how they may be interpreted.

3.2. MOLECULAR WEIGHT AVERAGES

When the molecules in a polymer sample are not all of the same size, we say the sample has a *distribution of molecular weights*. Suppose that the sample consists of n_1 molecules of molecular weight M_1, n_2 molecules of molecular weight $M_2, \ldots, n_i$ molecules of molecular weight M_i, etc., then a number of *different* average molecular weights may be defined.

The arithmetic mean of the molecular weights of the individual molecules is called the *number-average molecular weight*, $\bar{M}_n$, and is given by

$$\bar{M}_{\mathrm{n}} = \frac{n_1M_1 + n_2M_2 + n_3M_3 + \ldots n_iM_i}{n_1 + n_2 + n_3 + \ldots n_i} = \frac{\Sigma\, n_iM_i}{\Sigma n_i}. \tag{3.1}$$

This average is obtained using methods of determination which depend on the number of molecules present, and is equally affected by light and heavy molecules.

In other methods of molecular weight determination, each particle contributes to the average according to its mass. Then a *weight-average molecular weight*, $\bar{M}_{\mathrm{w}}$, is obtained, which is given by:

$$\bar{M}_{\mathrm{w}} = \frac{n_1M_1^2 + n_2M_2^2 + n_3M_3^2 + \ldots n_iM_i^2}{n_1M_1 + n_2M_2 + n_3M_3 + \ldots n_iM_i} = \frac{\Sigma\, n_iM_i^2}{\Sigma n_iM_i}\,. \tag{3.2}$$

In this case, the average is influenced more by the heavier molecules present than the lighter ones. Consider some examples:

Example A. For a polymer sample where 50% of the molecules have a molecular weight of 10,000 and 50% of the molecules have a molecular weight of 100,000 then

$$\bar{M}_{\mathrm{n}} = \frac{(0.5 \times 10^3) + (0.5 \times 10^5)}{0.5 + 0.5} = 55{,}000,$$

and

$$\bar{M}_{\mathrm{w}} = \frac{[0.5 \times (10^3)^2] + [0.5 \times (10^5)^2]}{(0.5 \times 10^3) + (0.5 \times 10^5)} = 92{,}000.$$

We see immediately that $\bar{M}_{\mathrm{w}}$ is greater than $\bar{M}_{\mathrm{n}}$. In general, for a polymer sample with a molecular weight distribution, $\bar{M}_{\mathrm{w}}$ is greater than $\bar{M}_{\mathrm{n}}$, and the sample is said to be *polydisperse*. If all the molecules of a polymer have the same molecular weight, then $\bar{M}_{\mathrm{w}} = \bar{M}_{\mathrm{n}}$, and the sample is said to be *monodisperse*. The ratio $\bar{M}_{\mathrm{w}}/\bar{M}_{\mathrm{n}}$ can be used to give an indication of the polydispersity of a polymer. If $\bar{M}_{\mathrm{w}}/\bar{M}_{\mathrm{n}} = 1$, the sample is monodisperse, whilst if $\bar{M}_{\mathrm{w}}/\bar{M}_{\mathrm{n}} > 1$, the sample is polydisperse and has a molecular weight distribution. In general, the larger is the ratio $\bar{M}_{\mathrm{w}}/\bar{M}_{\mathrm{n}}$, the greater is the spread of molecular weights in the sample.

Example B. For a polymer where numerically 40% of the molecules have molecular weight 10,000, 50% of the molecules have molecular weight 100,000, and 10% of the molecules have molecular weight 1,000,000, then

$$\bar{M}_{\mathrm{n}} = \frac{(0.4 \times 10^4) + (0.5 \times 10^5) + (0.1 \times 10^6)}{0.4 + 0.5 + 0.1} = 154{,}000,$$

and

$$\bar{M}_{\mathrm{w}} = \frac{[0.4 \times (10^4)^2] + [0.5 \times (10^5)^2] + [0.1 \times (10^6)^2]}{(0.4 \times 10^4) + (0.5 \times 10^5) + (0.1 \times 10^6)} = 682{,}000.$$

In *Example A*, $\bar{M}_w/\bar{M}_n = 1.67$, in *Example B*, $\bar{M}_w/\bar{M}_n = 4.43$. In the second example, $\bar{M}_w$ has been influenced more by the large molecules than $\bar{M}_n$, so the ratio $\bar{M}_w/\bar{M}_n$ has increased, indicating a greater spread of molecular weight than in the first example.

It should be remembered, therefore, that if two methods of molecular weight determination (one yielding $\bar{M}_n$ and the other $\bar{M}_w$) give very different values for a given polymer, this does not indicate an error in the procedures, but rather is a measure of the polydispersity of the sample.

Another complication is the fact that *different* molecular weight distributions may give the *same* values of $\bar{M}_w$ and $\bar{M}_n$:

Example C. For a polymer where numerically 56.7% of the molecules have molecular weight 20,000, 11.9% have molecular weight 50,000, and 31.4% have weight 120,000 then

$$\bar{M}_n = \frac{(0.567 \times 2 \times 10^4) + (0.119 \times 5 \times 10^4) + (0.314 \times 1.2 \times 10^5)}{0.567 + 0.119 + 0.314}$$

$$= 55{,}000,$$

and

$$\bar{M}_w = \frac{[0.567 \times (2 \times 10^4)^2] + [0.119 \times (5 \times 10^4)^2] + [0.314 \times (1.2 \times 10^5)^2]}{(0.567 \times 2 \times 10^4) + (0.119 \times 5 \times 10^4) + (0.314 \times 1.2 \times 10^5)}$$

$$= 92{,}000.$$

This distribution, different from that in *Example A*, gives $\bar{M}_w$ and $\bar{M}_n$ very close to those in *Example A*. (It must be noted that these examples are artificial and have been chosen only to illustrate the properties of molecular weight distributions: such discontinuous distributions never exist in natural polymers.)

In practice, it is not sufficient only to find $\bar{M}_w$ and $\bar{M}_n$ to characterize a polydisperse polymer; the complete molecular weight distribution should be investigated.

Two other molecular weight averages may be encountered when dealing with polymers. One is obtained by viscosity measurements (see Section 3.3.4), and the *viscosity-average* may be defined as

$$\bar{M}_v = [(\Sigma\, n_i M_i^{1+\alpha})/(\Sigma\, n_i M_i)]^{1/\alpha}, \tag{3.3}$$

where α is a constant which depends on the shape of the polymer molecules in solution. In cases when $\alpha = 1$, then $\bar{M}_v = \bar{M}_w$.

Another average, the *Z-average molecular weight*, $\bar{M}_Z$, emphasizes the contribution of high molecular weight. This average, which can be obtained from some ultracentrifuge measurements (see Section 3.3.3, p. 46), is given by

$$\bar{M}_Z = (\Sigma n_i M_i^3)/(\Sigma\, n_i M_i^2). \tag{3.4}$$

3.3. METHODS OF DETERMINING MOLECULAR WEIGHT

3.3.1. Methods based on colligative properties

Colligative properties of a substance are those properties which depend on the *number* of molecules of solute in a solution and not on the nature of the solute (under ideal conditions). Hence *number-average molecular weights* may be obtained by measuring these properties.

The molecular weights of oligomers and small polymers (molecular weight less than 20,000) may be determined using a *vapour-pressure osmometer*. The method depends on the lowering of the vapour pressure of a solvent by the presence of solute molecules.

The essential part of the apparatus (commercial equipment is available) is a carefully thermostated box containing two thermistors (temperature-sensitive resistances) which form part of a Wheatstone bridge; the box is saturated with solvent vapour. A drop of solvent is placed on each thermistor, and the Wheatstone bridge is balanced. One of the solvent drops is then replaced by a drop of solution. Because the solution has a lower vapour pressure, solvent distils into the solution drop and as the solvent condenses, the heat of vaporization is released and warms up the drop of solution. The temperature increases until the vapour pressure of the warm solution equals the vapour pressure of the solvent in the apparatus. The change in resistance of the thermistor in contact with the solution is measured.

For dilute ideal solutions, the change in resistance is proportional to the molar concentration of the solute. This concentration cannot, in practice, be calculated directly from the resistance change because of heat losses in the apparatus. The system is, therefore, usually calibrated with a substance of known molecular weight, so that the resistance change produced by a known molar concentration can be measured. From this information, the molar concentration of an unknown substance can then be found.

Ideally, the resistance change, Δr, may be related to the molecular weight of the solute M by

$$\Delta r/c = k/M, \tag{3.5}$$

where k is the calibration constant, and c is the solute concentration. However, polymer solutions rarely behave ideally, and the dependence of Δr on concentration is given more correctly by

$$\Delta r/c = k[(1/M) + Bc)], \tag{3.6}$$

Thus a series of measurements of $\Delta r/c$ should be made for different values of c, and a graph drawn showing the variation of $\Delta r/c$ with c. The extrapolation of $\Delta r/c$ to $c = 0$ gives a value $(\Delta r/c)_0$ from which M can be determined.

Larger molecular weights, up to 1,000,000, may be determined by measuring the *osmotic pressure* of the polymer solution. If a semipermeable membrane (i.e. a membrane through which solvent, but not large solute molecules, can pass) separates a polymer solution from pure solvent, there is a tendency for solvent molecules to flow through the membrane into the solution. This flow can be prevented by the application of pressure to the solution, and the applied pressure can be used as a measure of the osmotic pressure developed by the system. Alternatively, the system may be allowed to come to equilibrium without an external applied pressure; solvent flowing into the solution causes a hydrostatic pressure head to develop above the solution, which, at equilibrium, prevents the ingress of further solvent (see Figure 3.1). If there has been no appreciable dilution of the solution, the hydrostatic pressure head may be used as a measure of the osmotic pressure, Π. This second method takes longer, but tends to be more accurate.

The whole success of osmometery depends on the availability of efficient semipermeable membranes. In theory, they should be completely impermeable to the polymer molecules, whilst allowing free movement of solvent molecules. Films of cellulose and cellulose derivatives are often used as membranes. With the correct choice of membrane, measurements may be made on both aqueous and non-aqueous solutions. Because of the difficulty of obtaining membranes which are impermeable to small polymer

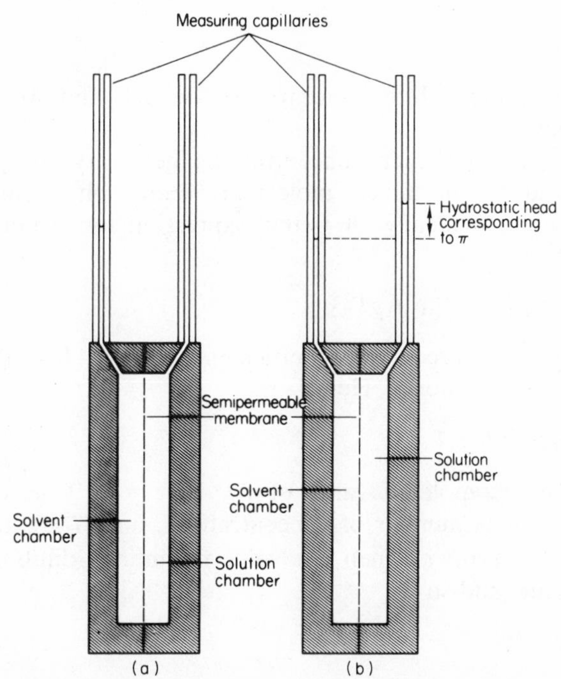

FIGURE 3.1. Schematic representation of an osmometer: (a) initially; (b) at osmotic equilibrium

molecules, the method is most suitable for macromolecules of molecular weight greater than 20,000.

The osmotic pressure depends on the number of solute molecules in solution, i.e. on the molar concentration, and hence yields a *number-average molecular weight* for polydisperse samples. For very large polymers, it is often difficult to obtain, or even to dissolve, sufficient material to give a solution which would yield a measurable osmotic pressure (concentrations of the order of 10^{-5} M may be required), and so the larger the molecular weight of a substance, the more difficult it is to measure. Thus the upper limit of molecular weight which can be determined by osmometry is in the region of 1,000,000.

Care must be taken, when studying a very polydisperse polymer, that the smallest or largest molecules do not exceed the limits given above; otherwise an erroneous average molecular weight would be obtained from osmotic pressure measurements.

For ideal solutions the osmotic pressure, Π, is given by

$$\Pi = -(RT/V_0) \ln (1 - x), \tag{3.7}$$

where R = ideal gas constant, T = temperature (°K), V_0 = molar volume of the solvent, and x = mole-fraction of solute. For very dilute solutions, this reduces to the van't Hoff equation

$$\Pi = RTc/M,$$

$$\text{or } \Pi/c = RT/M \tag{3.8}$$

where c is now the solute concentration in g/l, and M is the solute molecular weight.

However, at the polymer concentrations necessary for good osmotic pressure measurements, much molecular interaction occurs, and most solutions behave non-ideally. A better expression for osmotic pressure is then given by:

$$\Pi/c = (\Pi/c)_0 \left[1 + \Gamma_2 c + g\Gamma_2^2 c^2\right], \tag{3.9}$$

where Γ_2 and $g\Gamma_2^2$ are virial coefficients, and at low concentrations (<0.3%), $g\Gamma_2^2$ can be ignored. Hence

$$\Pi/c = (\Pi/c/_0)[1 + \Gamma_2 c]. \tag{3.10}$$

$(\Pi/c)_0$ is the extrapolated value of Π/c at $c = 0$. Thus measurements should be made at a number of concentrations, and $(\Pi/c)_0$ obtained from the graph of Π/c *versus* c. Then this value for infinite dilution is set equal to the ideal value, and so

$$(\Pi/c)_0 = RT/M. \tag{3.11}$$

Additional complications arise when charge polymers, i.e. polyelectrolytes, are being studied. Because of the so-called Donnan effects, at equilibrium an unequal distribution of small ions across the membrane results. If the solution contains, say, a positively charged polymer, there will be fewer small positive ions and more small negative ions on the solution side of the membrane than on the solvent side. This increases markedly the dependence of Π/c on c. However, the effects may be minimized if measurements are carried out at high ionic strength to suppress polymer ionization.

Osmometry has been used to measure the molecular weights of many natural polymers, notably proteins, polysaccharides, and rubber, and commercial equipment is available. The latter often uses transducers to measure and record the observed osmotic pressure.

3.3.2. Light-scattering measurements

Macromolecules in solution scatter a beam of incident light according to their size, shape, and concentration. Indeed, measurements of the intensities of the light scattered at various angles from a polymer solution (see Figure 3.2) can be interpreted to yield information about the molecular weight and shape. Because large molecules scatter more light than small ones, measurements become easier as the size of a polymer increases: molecular weights of 10^8 can be easily measured by light scattering. For a polydisperse sample, the results give the *weight-average molecular weight.*

The technique may be carried out using aqueous or non-aqueous solutions, but a major problem is posed by the necessity of preparing dust-free solutions. The presence of dust gives large errors in the intensity of scattered light, and attempts are usually made to remove dust by subjecting the polymer solution to filtration and ultracentrifugation.

The light scattered from particles at an angle, θ, to the incident beam (see Figure 3.2) may be conveniently expressed in the terms of the quantity, R_θ, where

$$R_\theta = I_\theta \, r^2/I_0 \, (1 + \cos^2\theta). \tag{3.12}$$

Here, I_θ is the intensity of scattered light, I_0 is the intensity of the incident beam, and r is the distance of the measuring device from the solution.

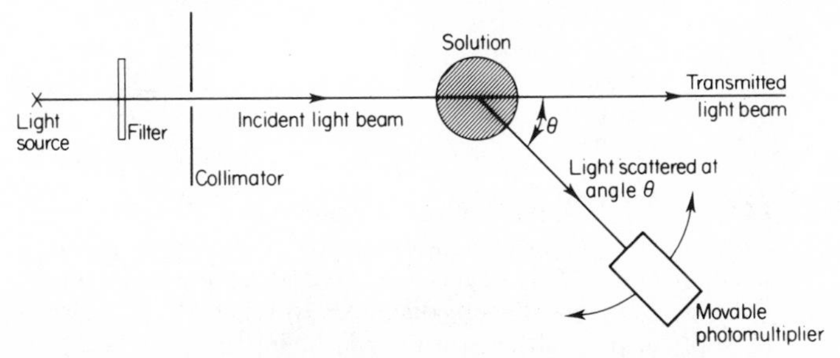

FIGURE 3.2. Diagram of light-scattering equipment

Now for an ideal solution of small molecules,

$$Kc/R_\theta = 1/M, \tag{3.13}$$

where K is a constant, and c is the concentration of the solute of molecular weight M. [$K = \{2\pi^2 n_0^2 (dn/dc)^2\}/N_0\lambda_0^4$, where n_0 is the refractive index of the solvent, using light of wavelength λ_0; N_0 is Avogadro's number, and dn/dc is the change in refractive index of the solution with concentration of solute.]

Unfortunately, polymer solutions do not behave ideally and c/R_θ is found to depend on concentration. In addition, when polymer molecules are large (approaching the order of magnitude of the wavelength of light) different parts of one molecule scatter light differently; the light scattered from one end of the macromolecule may interfere with light scattered from the other end, thus reducing the intensity of scattered light (see Figure 3.3). This decrease depends on the size and shape of the polymer molecules, and so we must introduce a particle scattering factor, P_θ, where

$$P_\theta = \frac{\text{observed intensity at angle } \theta}{\text{intensity in absence of interference}}.$$

P_θ is a function of $\sin^2(\theta/2)$. Then for non-ideal polymer solutions at low concentrations, we have

$$Kc/R_\theta = 1/MP_\theta + 2Bc, \tag{3.14}$$

where B is the virial coefficient.

Thus extrapolation of Kc/R_θ to zero concentration would give a value of $1/MP_\theta$. If the shape of the polymer molecules is known, P_θ can be calculated and M found. In most cases, however, the exact shape of the macromolecules is unknown. Fortunately for $\theta = 0°$, $P_\theta = 1$ for all shapes of molecules, and thus the value of M can be found from a double

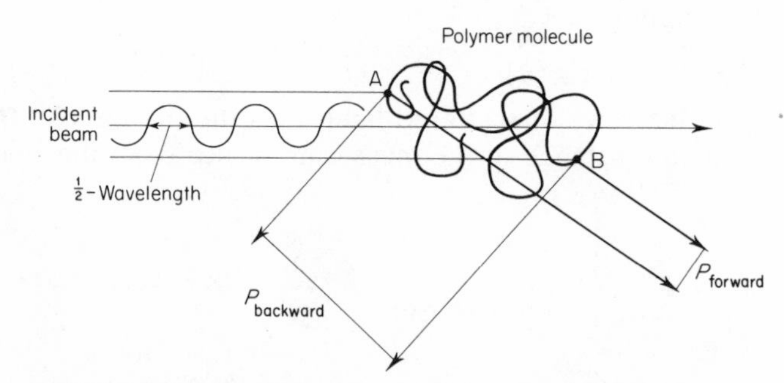

FIGURE 3.3. Interference of light when the polymer molecule is comparable in size to the wavelength of the incident beam. The beams scattered from points A and B have to travel different distances to reach P_{forward}; also the optical path difference for the beam to reach P_{forward} and P_{backward} is different

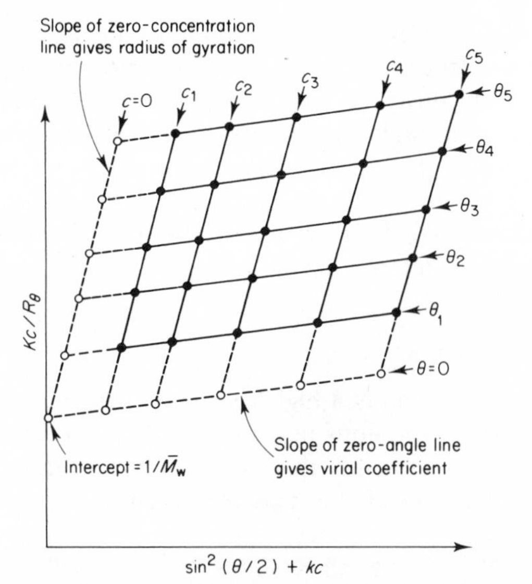

FIGURE 3.4. Double extrapolation of light-scattering data to obtain molecular weight, a Zimm graph: ● experimental results; ○ extrapolated values. k is an arbitrary constant chosen so that $\sin^2 (\theta/2) \approx kc$; Kc/R_θ is defined in the text

extrapolation to zero concentration and zero scattering-angle. This type of extrapolation is known as a Zimm graph, and is shown in Figure 3.4. From the intercept $\theta = 0°$, $c = 0$, M (or $\bar{M}_w$ for a polydisperse polymer) can be obtained. In addition, an estimate can be made of the size of the molecule in solution; the slope of the zero concentration line is related to the mean square radius of gyration of the polymer molecules. [The slope = $(16\pi^2/3\lambda_0^2 n_0^2)(1/M)R_G^2$, where R_G^2 is the mean square radius of gyration (see also Section 3.5, p. 62.]

The molecular weights of many natural polymers including proteins, polysaccharides, lignin, and nucleic acids have been determined using this technique. Initially, problems were encountered with large deoxyribonucleic acids (molecular weights greater than 6×10^6) because Kc/R_θ was not a linear function of $\sin^2 (\theta/2)$, and linear extrapolation gave molecular weights which were too low. With the development of apparatus capable of giving measurements at low scattering angles, these difficulties are now less serious.

Light-scattering methods are also useful in studying changes in molecular weight with time: thus continuous monitoring of scattered intensities can give information of the time-course of polymer degradation, dissociation, or association.

Currently, light-scattering equipment is available with a laser light source. This refinement enables measurements to be made on very small volumes of solution (thus less polymer sample is required and solutions are easier to free from dust) at low scattering angles, so that only small extrapolations to zero angle are necessary. In addition, with a laser light source, the scattered light can be analysed for variations produced by local fluctuations in the concentration of polymer molecules, and from this analysis the translational diffusion coefficient (see Section 3.3.3.1, p. 48) of the polymer can be determined.

3.3.3. Ultracentrifugation

An ultracentrifuge is simply a high-speed centrifuge capable of speeds of, for example, 60,000 revolutions per minute, which generate force fields of 250,000*g*. In force fields of this size, polymer molecules move, or sediment, in the direction of the field (see Figure 3.5a, b). The behaviour

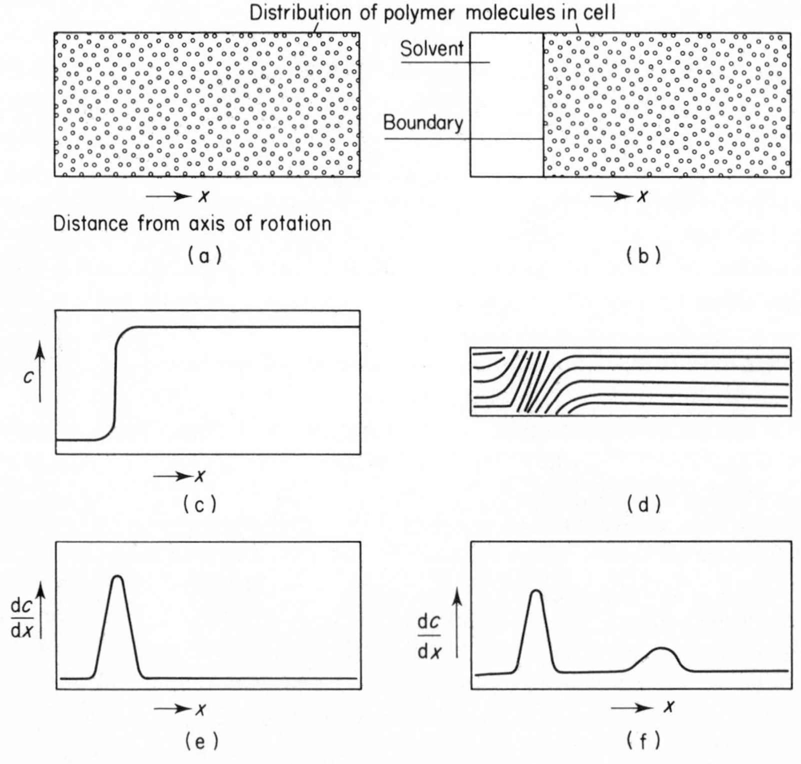

FIGURE 3.5. Effects of high force fields on polymer molecules in an ultracentrifuge cell (schematic): (a) initially; (b) after some time in centrifugal field; (c) concentration as shown by absorption optics at time corresponding to (b); (d) pattern of interference fringes corresponding to (b) and (c); (e) Schlieren pattern corresponding to (b), (c), and (d); (f) Schlieren pattern obtained for a two-component polymer system

of macromolecules in such fields can give information on polymer homogeneity, molecular weight, and molecular weight distribution. In general, ultracentrifugal methods can be used to determine molecular weights from a few thousand to several million.

The analytical ultracentrifuge was developed by a Swedish scientist, The Svedberg. He was awarded the Nobel Prize for Chemistry in recognition of his work in establishing the polymer-nature and homogeneity of molecular weight of proteins, using this technique.

A schematic diagram of a typical, commercially-available, analytical ultracentrifuge is shown in Figure 3.6. Light passing through the solution can be used to monitor changes in solute concentration throughout the cell. Three different optical systems are commonly used in the ultracentrifuge: absorption optics, interference optics, and Schlieren optics. With *absorption optics*, an ultraviolet light source is employed, and a photoelectric scanner measures the amount of light absorbed by the solution compared to the absorption by pure solvent. The scanner may be attached to a recorder which traces out directly the changes in concentration along the cell (see Figure 3.5c). This method is extremely sensitive for detecting nucleic acids and proteins at very low concentrations (a few μg of polymer per ml of solution may be used). Visible light is usually used for *interference optics*, and the light beams passing through the solvent and solution are allowed to interfere to produce interference fringes (see Figure 3.5d); sensitivity is less than for absorption optics, but quantitative measurements tend to be more accurate. In this method, the number of non-horizontal fringes is proportional to the concentration of solute. *Schlieren optics* give a pattern related to the change in concentration with distance along the cell, i.e. the derivative of the concentration versus distance curve (see Figure 3.5e). Again this method is less sensitive than absorption optics, but is extremely

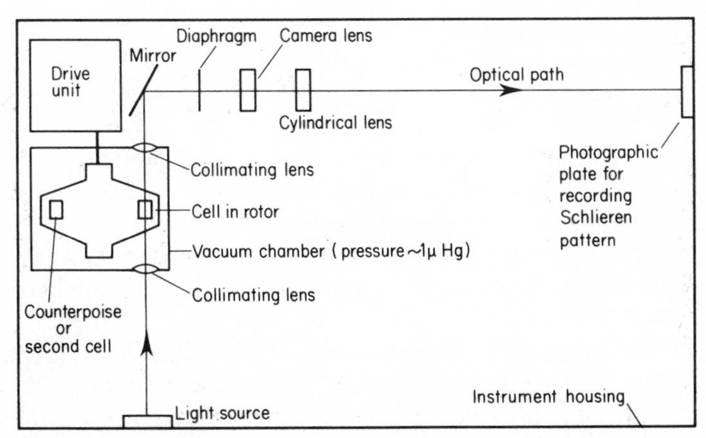

FIGURE 3.6. Schematic diagram of an electrically driven ultracentrifuge fitted with Schlieren optics

useful for the measurement of sedimentation velocities (see below), and for the investigation of homogeneity (see Figure 3.5f).

Nowadays, the ultracentrifuge is used principally in two ways to obtain molecular weights—either the rate of movement, *sedimentation velocity*, of solute in high force fields is studied, or the distribution of solute at *sedimentation equilibrium* is investigated.

3.3.3.1. Sedimentation velocity

When polymer molecules sediment down a cell, pure solvent is left above the solution and a sharp boundary forms between solvent and solution (see Figure 3.5). The rate of movement of this boundary can be measured from photographic plates, or photoelectric scans, taken at various times during the ultracentrifuge run, and this sedimentation rate can be related to the molecular weight of the polymer.

The sedimentation coefficient, S, of a polymer is defined as the velocity, $\mathrm{d}x/\mathrm{d}t$, of net movement of the molecules in the direction of the force field, i.e. the velocity of the solvent–solution boundary, in a unit force field; x is the distance from the axis of rotation. Thus

$$S = (\mathrm{d}x/\mathrm{d}t)/x\omega^2,$$
$$\text{or} \quad S = (1/\omega^2)\mathrm{d}\ln x/\mathrm{d}t, \tag{3.15}$$

where ω is the angular velocity of the rotor.

Sedimentation coefficients fall in the range $1\text{–}200 \times 10^{-13}$ sec, and 10^{-13} sec has been called one Svedberg unit (S) in honour of the originator of the technique. Thus sedimentation coefficients are given as 1–200 S.

Sedimentation coefficients may be related to molecular weight (M) by the Svedberg equation

$$M = RTS/(1 - \bar{V}\rho)D, \tag{3.16}$$

where R is the gas constant, T is the temperature (°K), D is the diffusion coefficient of the polymer, $\bar{V}$ is the partial specific volume of the polymer, and ρ is the density of the solvent. [Molecules diffuse across a sharp solvent–solution boundary at a rate proportional to the concentration gradient ($\mathrm{d}c/\mathrm{d}x$) and the cross-sectional area of the container (A). If the mass of molecules transferred per unit time is $\mathrm{d}m/\mathrm{d}t$, then the diffusion coefficient is defined by Fick's law: $\mathrm{d}m/\mathrm{d}t = -DA\mathrm{d}c/\mathrm{d}x$.]

Unless very low solute concentrations are used (which is now possible with absorption optics), the sedimentation coefficient is found to depend on polymer concentration. In general, as a polymer solution is diluted, there is less intermolecular entanglement and the sedimentation coefficient increases. The value of S to be used in the Svedberg equation is that for infinite dilution, S_0. This value is usually obtained by graphing $1/S$ against c; the straight line which results can be extrapolated to $c = 0$, to give $1/S_0$. (However, using absorption optics and low concentrations of protein or nucleic acid, a value close to S_0 may be obtained directly for these biopolymers.)

Sedimentation coefficients may be obtained from zonal density gradient centrifugation (see Section 2.3.3.1, p. 27), but corrections must be made for the viscosity and density of the solvent and changes in partial specific volume of the polymer. In general, sedimentation coefficients are given for 'standard' conditions, i.e. water solvent at 20°C.

For a polydisperse polymer, the average molecular weight obtained in this way is not a well-defined average, neither $\bar{M}_n$ nor $\bar{M}_w$, but depends on molecular shape.

However, before a molecular weight can be determined from a sedimentation coefficient, both $\bar{V}$ and D must be known. Difficulties in the determination of the diffusion coefficient have been a major drawback in the use of the Svedberg equation.

The partial specific volume of a polymer can be obtained, if the densities of solutions of known concentrations of this polymer can be measured very exactly. Solution densities may be determined, for example, from the position at which a drop of, say, aqueous solution settles in a calibrated density gradient of organic water-immiscible liquids. [Then $\bar{V} = (1/\rho)\{1 - (d\rho_s/dc_s)_0\}$, where ρ is the solvent density; $(d\rho_s/dc_s)_0$ is the change of solution density with change in polymer concentration, extrapolated to infinite dilution.] For proteins, the partial specific volume can be calculated if the amino acid composition (see Section 4.3.2, p. 85) is known.

A number of procedures are now available for measuring D, the diffusion coefficient of a polymer. With the advent of laser light-scattering photometers, diffusion coefficients can be measured using very small quantities of solution. The 'classical' technique for measuring D involved the establishment of a sharp boundary between solvent and solution and following changes in concentration with time. For such a system $dc/dx = -[c_0/(4\pi Dt)^{\frac{1}{2}}] \exp(-x^2/4Dt)$, where c_0 is the original concentration of the solution. Now the maximum value of $dc/dx = c_0/(4\pi Dt)^{1/2}$, so that D may be determined from c_0 and the maximum value of dc/dx at any time t. The value of D to be used in the Svedberg equation should be obtained by measuring the diffusion coefficient at a number of concentrations and extrapolating to infinite dilution.

Sedimentation velocity measurements have been used to obtain the molecular weights of proteins, polysaccharides, and nucleic acids, and indeed sedimentation coefficients are used to distinguish one type of ribonucleic acid from another (see Chapter 5).

3.3.3.2. Sedimentation equilibrium

A lower force field than that for sedimentation velocity studies is used here. A rotor speed is chosen so that an equilibrium is reached, where the movement of polymer molecules in the direction of the force field is exactly balanced by movement in the opposite direction due to diffusion.

Then, at one initial concentration of solution:

$$M_{\text{apparent}} = \frac{2\,RT}{(1 - \bar{V}\rho)\omega^2} \cdot \frac{\mathrm{d} \ln c}{\mathrm{d}x^2}. \tag{3.17}$$

Thus if the variation of c with x^2 within a cell can be measured, and $\bar{V}$ and ρ are known, M can be found. One advantage of the method is that the diffusion coefficient need not be determined in order to find M. However, the procedure can be a lengthy one, as it may take several days for the solution to reach sedimentation equilibrium.

As with the other techniques described so far, M should be measured at a variety of initial solute concentrations, and the 'true' value for the molecular weight found by extrapolation to infinite dilution. For polydisperse samples, the weight-average molecular weight is obtained by this method.

Results may be obtained using less polymer if only a small volume of solution is used in an ultracentrifuge cell. The optical system then shows a short column of liquid in the cell, a few millimetres in length. Then equation (3.17) can be used in the integrated form:

$$M_{\text{apparent}} = \frac{2\,RT(\ln c_{\mathrm{b}} - \ln c_{\mathrm{m}})}{(1 - \bar{V}\rho)\omega^2(x_{\mathrm{b}}^2 - x_{\mathrm{m}}^2)}, \tag{3.18}$$

where the subscripts b and m refer to the base of the cell and the meniscus of the solution, respectively. Here c_{m} is not zero, and there is a gradient of solute concentration throughout the cell.

For polydisperse samples, this technique gives the apparent weight-average molecular weight for the whole cell. From equation (3.17), apparent molecular weights may be found at particular positions in the cell, and the apparent Z-average molecular weight (see Section 3.2) can be obtained from:

$$\bar{M}_{\mathrm{Z}} = (c_{\mathrm{b}} M_{\mathrm{w,b}} - c_{\mathrm{m}} M_{\mathrm{w,m}})/)c_{\mathrm{b}} - c_{\mathrm{m}}), \tag{3.19}$$

where M_{w} is the apparent molecular weight obtained from equation (3.17), and b and m refer to the cell base and solution meniscus, respectively.

Frequently, it is difficult to determine c_{m} accurately, and so Yphantis introduced a modification of the procedure. The rotor speed is increased sufficiently so that the solute concentration at the meniscus, and for a fraction of a millimetre below the meniscus, is effectively zero. Then it becomes easier using, say, interference fringes to determine the concentration at any point in the cell (beyond the point where $c = 0$). By use of the *Yphantis method* on a polymer with a very narrow range of molecular weights, the number-average, weight-average, and Z-average molecular weight at any point in the cell (beyond the point where $c = 0$) can be found by considering different functions of the variation of c with x.

In addition, the weight-average and Z-average molecular weight over the whole cell can be calculated.

Again, values should be determined at a variety of different initial solute concentrations, and extrapolated to infinite dilution. For best results, the measurements should be repeated at a number of different rotor speeds, and extrapolated to zero speed. The effect of speed is most marked with self-associating polymers where the system is influenced by changes in pressure; obviously the greater the speed of the rotor, the greater is the pressure developed at the base of the cell.

The Yphantis method is now widely used for the determination of the molecular weights of proteins and nucleic acids. Care must be taken, however, as the values obtained are very sensitive to the presence of small molecular weight contaminants at the meniscus.

Molecular weights of polymers can also be measured by density gradient sedimentation equilibrium centrifugation in salts such as caesium chloride (see Section 2.3.3.2, p. 28). At equilibrium, a band of polymer molecules form where the buoyant density of the macromolecules equals the density of the salt solution. Then for a monodisperse polymer, the apparent molecular weight is related to the concentration distribution within the band by

$$\frac{\mathrm{d}\ln c}{\mathrm{d}(x - x_0)} = -\frac{M_{\mathrm{app}}\omega^2\bar{V}(\mathrm{d}\rho/\mathrm{d}x)}{RT} \,.\, x_0(x - x_0) \qquad (3.20)$$

where x_0 is the distance of the centre of the band from the axis of rotation, and $\mathrm{d}\rho/\mathrm{d}x$ is the density gradient within the cell. Again a series of values of M_{app} at different concentrations should be obtained and extrapolated to infinite dilution.

This method has been used for nucleic acids, but difficulties have been encountered because of band-spreading caused by small variations in chemical composition of the molecules of a nucleic acid preparation. (The buoyant density of a nucleic acid is critically dependent on the chemical composition of the molecules.) This band-spreading, due to heterogeneity, gives an erroneous value for the molecular weight. In addition, problems arise with very large nucleic acid molecules, as these may be degraded by the shearing forces within the ultracentrifuge cell.

A third method of employing the ultracentrifuge for measuring molecular weights was used in the past, but is not now of great importance in view of the relatively large amounts of polymer required. This technique was the *pseudo-equilibrium method* developed by Archibald, which did not require the establishment of a true equilibrium. It was realized that the rate of transport of molecules across the meniscus and the base of the cell must at all times be zero. (In other words, at the meniscus and cell base, diffusion must balance movement due to the centrifugal force.) From this situation, equations were developed to give the molecular weight from measurements near the meniscus and the base of the cell. This technique has been used to determine the molecular weights of proteins and polysaccharides.

In all cases where the ultracentrifuge is used, the salt concentration in polymer solutions should be high enough to suppress polymer ionization; if the polymer itself is charged, the theories for molecular weight determination become extremely complicated.

3.3.4. Viscosity measurements

Measurements of the viscosity of polymer solutions provide an easy and fast method for determining molecular weights in the range 10^4–10^6. However, the technique does not provide a direct measure of molecular weight, and *must first be calibrated* using polymer fractions very similar in nature to the polymer being studied, and having known molecular weights. This requirement for calibration is an obvious disadvantage of the technique; another is that relatively large amounts of polymer sample are needed (often several hundred milligrams).

The method depends on the fact that the addition of a polymer to a solvent increases the viscosity of the liquid, and this increase is related to the size and shape of the macromolecules.

If η_0 is the viscosity of the solvent at one particular temperature, and η is the viscosity of a polymer solution of concentration c (in g/ml) at the same temperature, then a quantity known as the specific viscosity, η_{sp}, is given by

$$\eta_{sp} = (\eta - \eta_0)/\eta_0. \tag{3.21}$$

In general, the specific viscosity varies with concentration according to the relation

$$\eta_{sp}/c = [\eta] + k'\,[\eta]^2 c, \tag{3.22}$$

where k' is a constant, and $[\eta]$ is known as the *limiting viscosity number*. This value can be obtained by measuring η_{sp}/c at a variety of concentrations and extrapolating to zero concentration.

For a series of fractions (of one type of polymer) which differ in molecular weight, the relation holds that

$$[\eta] = KM^{\alpha}, \tag{3.23}$$

where K and α are both constants which depend on the solvent, the temperature, and the nature of the polymer; α is also related to the shape of the polymer in solution. Thus for a series of fractions of known molecular weight, a graph of $\log[\eta]$ against $\log M$ should be linear, and K and α can be determined from the intercept and slope, respectively. The molecular weights of the fractions must first be determined by an independent method such as osmometry or light scattering, but once K and α are known, the molecular weight of an *unknown* sample can be obtained directly from its limiting viscosity number $[\eta]$.

If the sample is polydisperse, a viscosity-average molecular weight, $\bar{M}_v$, is obtained (see Section 3.2). The value of α is usually between 0 and 1, and so the viscosity-average normally lies between the number-average, $\bar{M}_n$, and weight-average, $\bar{M}_w$, molecular weights; when $\alpha = 1$, $\bar{M}_v = \bar{M}_w$. For many globular proteins, $\alpha = 0$, and the limiting viscosity number is almost independent of M. Thus viscosity measurements cannot be used to determine the molecular weight of these types of protein, unless the original three-dimensional structure of the protein is first destroyed.

Viscosities can be studied by comparing the time of flow of a polymer solution through a capillary with the time taken by the pure solvent. The time, t, taken for the liquid to flow between two marks can be measured (see Figure 3.7). It then follows that

$$\eta = A'' \rho t + B'' \rho/t, \tag{3.24a}$$

where A'' and B'' are constants which can be determined for one viscometer by calibration with a liquid of known viscosity, and ρ is the solution density. By careful viscometer design, the second term $B'' \rho/t$ can be made negligible. (This factor corrects for energy losses when the liquid changes flow characteristics on entering and leaving the capillary.) Equation (3.24a) then reduces to

$$\eta = A''\rho t, \tag{3.24b}$$

and the specific viscosity, from equation (3.21), becomes

$$\begin{aligned} \eta_{sp} &= (\eta - \eta_0)/\eta_0 \\ &= (\rho t - \rho_0 t_0)/\rho_0 t_0, \end{aligned} \tag{3.25a}$$

where ρ_0 is the solvent density, and t_0 the solvent flow-time.

For dilute polymer solutions $\rho \approx \rho_0$, and so we have

$$\eta_{sp} = (t - t_0)/t_0. \tag{3.25b}$$

Thus specific viscosities can be calculated directly from flow-times. In the viscometer shown in Figure 3.7, dilutions can be made *directly in the viscometer*, so that η_{sp}/c can be measured at a number of concentrations, as is required for extrapolation of η_{sp}/c to infinite dilution.

When the limiting viscosity number, $[\eta]$, is very large, polymer solutions may show variations in viscosity with applied shear forces. In such cases, viscosities should be measured at different shear forces and extrapolated to zero shear. Different shear forces can be obtained by using viscometers with different capillary diameters, or a special concentric cylinder viscometer.

Polymers which are polyelectrolytes should be studied in solutions of fairly high salt concentration to suppress ionization of the macromolecules, otherwise complications arise with possible changes in shape of the polymer molecules on dilution of the solution.

The molecular weights of polysaccharides, rubber, and lignin have been

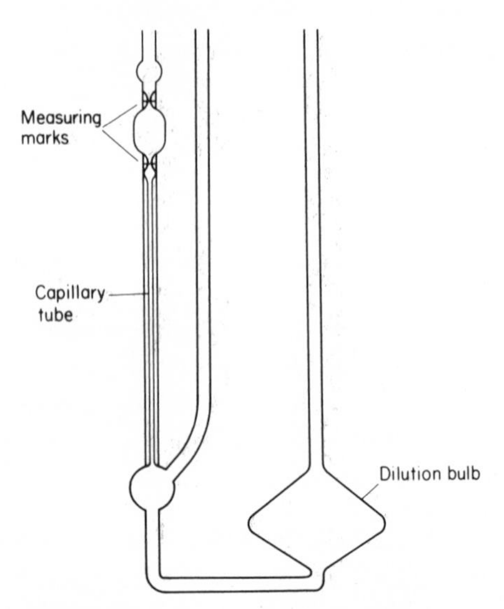

FIGURE 3.7. A modified Ubbelohde viscometer

investigated in this way. Proteins may also be studied, provided that the native three-dimensional structure of globular proteins is first destroyed. The sizes of ribonucleic and deoxyribonucleic acids have been determined by viscosity measurements, but great care must be taken with large deoxyribonucleic acids, because the shear forces developed in a viscometer may be sufficient to degrade the macromolecules.

Molecular weights of some polymers, e.g. nucleic acids, have been estimated using a combination of sedimentation velocity and viscosity measurements. [A combination of sedimentation coefficient and limiting viscosity number was used to avoid the need to know the diffusion coefficient. The equation used was $M = S_0[\eta]^{\frac{1}{3}}\eta_0/\{\beta'(1 - \bar{V}\rho_0\eta_0\}$, where S_0, $[\eta]$, η_0, $\bar{V}$, and ρ_0 are defined above, and β' is a constant determined by the shape of the biopolymer.] However, the theoretical basis for this method is not particularly sound.

3.3.5. Other techniques

A number of other techniques are available, some specific for one type of polymer only. The more important of these are outlined here.

3.3.5.1. Gel electrophoresis

Gel electrophoresis (in the presence of sodium dodecyl sulphate (SDS)) is used only for proteins, and is useful for the determination of the molecular weight of single protein chains. In cases where the chains of a molecule are held together by disulphide bonds (see Section 4.2.3, p. 83), these bonds must first be reduced and the original three-dimensional structure of the molecules destroyed. The SDS molecules then associate with each protein chain to give a fairly rigid rod. It is believed that the ratio of SDS to protein in the rod is fairly constant for different proteins, i.e. the larger the protein molecule, the greater the number of SDS molecules which become attached to it. The sulphate groups of the SDS are believed to be on the surface of the complex, giving it an overall negative charge.

The protein–SDS complex is then subjected to electrophoresis on a polyacrylamide gel (see Section 2.3.2.2, p. 22), and the distance a particular protein moves relative to a marker substance depends on the size of the SDS–protein rod, which in turn is a function of the molecular weight of the protein chain. Here the SDS–protein complexes must move through the gel pores, for the gel is *not* in the bead form used for gel-permeation techniques. Thus the smaller macromolecules move more freely than the larger ones, i.e. in this type of electrophoresis, the smaller proteins move the greater distance.

Calibration with protein chains of known molecular weight shows that the distance moved during electrophoresis varies linearly with the logarithm of the molecular weight. Thus, from a graph of distance moved *versus* log M, the molecular weight of an unknown chain can be found, once the distance travelled by it on electrophoresis has been measured.

Glycoproteins tend to move shorter distances than proteins of the same molecular weight, because of lower SDS-binding. Thus a separate calibration graph must be constructed for glycoproteins.

If the protein molecule consists of a single chain, this method gives the weight of the molecule directly. Where a molecule consists of several chains, the number of chains must be known before the weight of the complete molecule can be calculated, as only the weight of single chains can be found from the electrophoresis.

The method is easily performed, the equipment required is relatively simple, and only small quantities (micrograms) of protein are needed. By varying the porosity of the polyacrylamide gels, different molecular weight ranges can be studied, and the technique is useful overall for the range 12,000–200,000. The method is, therefore, one of the commonest used at present for the measurement of protein molecular weights.

3.3.5.2. Gel-permeation chromatography

This method can be used for different types of polymers, and, indeed proteins, polysaccharides, nucleic acids, and lignin have been investigated in this way.

The technique involves the use of either a column of gel beads or a thin layer of gel beads on an inert support (compare Section 2.3.1.3, p. 15).

But the method is not absolute, and it requires calibration with polymers of known molecular weight, similar in nature to the unknown.

If a polymer solution is applied to a column of gel, the volume of liquid required to elute the macromolecules from the column depends on the size and shape of the polymer molecules. The elution volume for the unknown (or K_{av}, see Section 2.3.1.3, p. 17), is compared to the elution volumes (or K_{av}) for a series of known macromolecules. For good results, the unknown macromolecules must be the same *shape* as the known molecules. This requirement is not always met, and so introduces some uncertainty into the result—in general, a molecular weight determined in this way has an associated uncertainty of approximately 10%.

It has been found for any polymer type e.g., globular proteins, that the elution volume (or K_{av}) varies linearly with the logarithm of molecular weight (except at the high and low molecular weight limits-of-usefulness of one particular gel). Thus the molecular weight of an unknown sample may be found from its elution volume and this type of calibration graph (see Figure 3.8). A separate calibration graph must be obtained for each gel,

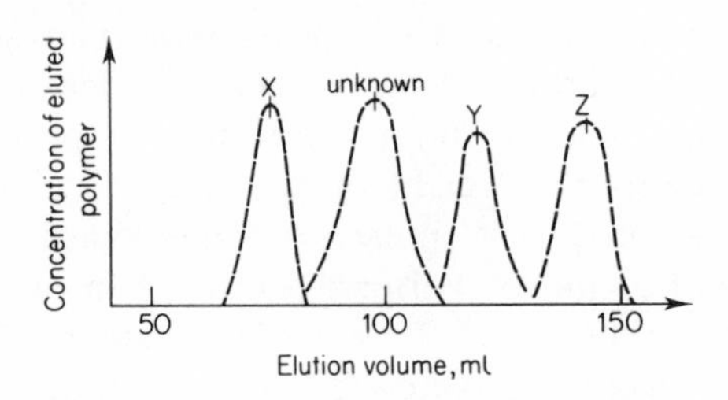

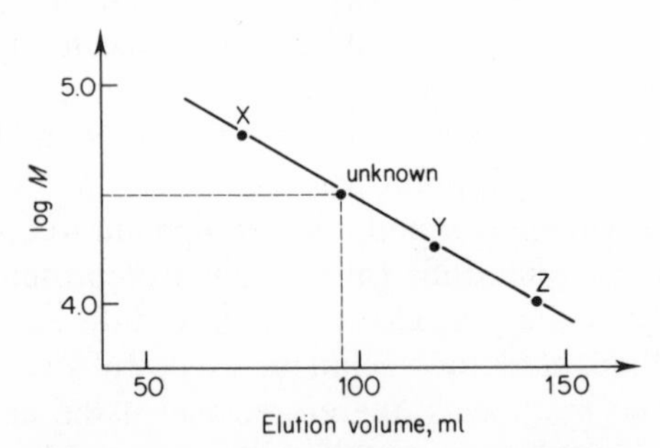

FIGURE 3.8. Determination of molecular weight by gel-permeation chromatography. (a) Chart obtained from gel-permeation equipment fitted with detector and recorder; X, Y, and Z are standards of same polymer type and of known molecular weight: 60,000, 18,000, and 10,000, respectively. (b) Calibration graph: $\log M_{unknown} = 4.5$, and hence $M_{unknown} \sim 31{,}600$

and for each kind of polymer, e.g. globular proteins, glycoproteins, nucleic acids, etc.

Because proteins of many shapes exist, a better value for protein molecular weight may be obtained by reducing disulphide bonds in the protein molecule, and completely destroying the native three-dimensional structure (in, for example, 6M guanidinium chloride solution), prior to introduction of the protein onto the column. If the standard proteins are treated in the same way, it is reasonable to assume that the macromolecules then all have the same shape, a random coil. However, if a protein molecule consists of several chains, this procedure gives the single-chain weight only.

If a sample is polydisperse, an average molecular weight may be obtained from the elution volume in which the maximum amount of polymer appears. This average is often between a number- and weight-average value.

Thin-layer gel-permeation has the added advantage over the column procedure that it is extremely fast and only very small quantities of polymer are required. Again the distance moved by a polymer through the gel depends on macromolecular size and shape. In this method, the distances moved by polymers on the gel plate vary linearly with the logarithm of the molecular weight for a series of related macromolecules of the same shape. Thus the weight of an unknown may be found from the distance it moves on the gel and an appropriate calibration graph.

Gel-permeation techniques in general are extremely useful for molecular weight determination, because they are rapid and simple. Provided that standards are available for calibration, the method may be used for weights in the range 1,000 to several million. An additional advantage is that the polymer to be studied need not be completely pure; if it does not interact with contaminants present, its molecular weight can be obtained, as long as a method of detection is available which is specific for the required polymer. The more sensitive the detection method, the less polymer is needed for a molecular weight determination. In consequence, this technique is very important when limited amounts of a natural polymer are available.

Association and dissociation of biopolymers may also be studied by gel-permeation.

3.3.5.3. Chemical methods

In this technique, a quantitative assay is made either of all the constituent monomers, or of a specific grouping in a macromolecule.

If the fraction of each monomer in a polymer is known, and it is assumed that the macromolecule must contain at least one residue of the monomer present in the smallest amount, then a minimum molecular weight can be found simply by adding together the contribution of each

residue to the molecular weight (i.e. Σ molecular weight of monomer residue, y, multiplied by number of molecules of monomer residue, y, assumed in the macromolecule). Where the complete structure is known, as would be the case of a protein after X-ray analysis (see Section 3.6), the molecular weight is calculated by adding together the molecular weight of all the monomer residues known to be present.

In cases where a polymer molecule contains a particular functional group, the molecular weight may be determined by assay of this group, provided the number of such groups per molecule is known. For example, it was found that the protein, myoglobin, contained one atom of iron per molecule, and that this iron constituted 0.335% by weight of the total macromolecule. Thus we have 0.335 g of iron or 0.335/55.8 g atoms of iron in 100 g of protein. Because there is one atom of iron per protein molecule, there must therefore be 0.335/55.8 moles of protein in 100 g, i.e. the molecular weight of myoglobin must be $(100 \times 55.8)/0.335 =$ 16,660 g. In general, if there are n functional groups per macromolecule and it requires R moles of reagent to react with weight, w, of polymer, the molecular weight, M, of the polymer is given by

$$M = nw/R. \tag{3.26}$$

Often a macromolecule has a group at the end of the polymer chain which may be assayed fairly readily. Then molecular weights are said to be determined by *end-group assay*. For example, many polysaccharides contain one reducing group at a chain-end, and this may be measured quantitatively. Nucleic acids often have a phosphate group at one end of the polymer chain only, and so molecular weights can be determined from assays of this phosphate. The method of assay must ensure complete reaction with the functional group, no side-reactions, and no degradation of the polymer. The larger the polymer, the more difficult it is to determine end-groups accurately, and so these methods cannot be used for large polymers of molecular weight above 10^5.

Because these methods effectively count the number of molecules in solution, a number-average molecular weight is obtained for polydisperse samples.

3.4. DETERMINATION OF MOLECULAR WEIGHT DISTRIBUTION

Polymers such as polysaccharides, lignin, rubber, and gutta percha are usually polydisperse, and so, to characterize the sample completely, the molecular weight distribution should be investigated.

One of the earliest methods of studying molecular weight distributions involves splitting the polydisperse sample up into a number of fractions of narrow molecular weight distribution, and studying these fractions. This fractionation of a polymer is very often based on solubility differences: a non-solvent is added to a polymer solution, and the precipitated polymer removed at various concentrations of non-solvent. Large polymer molecules tend to be less soluble than small ones, so that the first precipitates contain the highest molecular weight molecules, while later precipitates contain the smaller macromolecules. Each fraction is then weighed, and its molecular weight determined to enable a graph to be constructed showing the amount of polymer in each molecular weight interval as a function of molecular weight (see Figure 3.9). This method is, however, tedious and time-consuming.

A more direct method of determining the molecular weight distribution is now available, i.e. gel-permeation chromatography (see previous section). If a calibrated column is used, an estimate of the molecular weight distribution can be obtained from the weight of polymer appearing in each tube of eluate collected from the column. Errors are of course introduced by diffusion; thus each tube of eluate contains macromolecules of a certain size plus some smaller and larger molecules which leave the column at the same time because of diffusion. In many cases, the error in the overall distribution is not large because the error caused by diffusion is approximately the same for all sample tubes. Commercial equipment is currently available which will record this distribution directly.

Problems may be encountered if the molecular weight distribution is very wide; thus one gel may not be capable of separating out both the smallest and largest molecules of a very polydisperse polymer.

Molecular weight distributions may also be investigated using the ultracentrifuge, but the calculations can be complex and time-consuming without the use of a computer.

The photographs obtained from a sedimentation velocity run (see Figure 3.5c, d and e) may be analysed for the distribution of apparent sedimentation coefficients, using the following equation:

$$g(S) = \frac{\omega^2 t}{x_m^2 c_0} \cdot x^3 \cdot \frac{dc}{dx}, \tag{3.27}$$

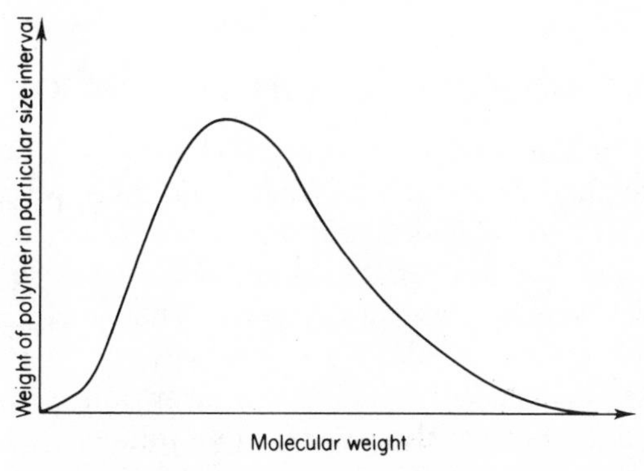

FIGURE 3.9. Typical molecular weight distribution curve for a natural polymer showing a 'skew' distribution towards higher values

where g(S) is the amount of polymer having a sedimentation coefficient within a certain interval of size, ω is the rotor speed, t is the time after the start of the run, x_m is the distance of the meniscus of the solution from the axis of rotation, c_0 is the original solution concentration, x is the distance in the cell from the axis of rotation, and dc/dx is the concentration gradient at that distance x in the cell.

Equation (3.27) gives an apparent distribution which must then be corrected for diffusion and the effects of concentration and pressure (if any). The effect of pressure may be small, but if not, the distribution should be investigated at different rotor speeds, and extrapolated to zero speed. At infinite time, the effect of diffusion on the spreading of a solvent–solution boundary is negligible compared to the spreading due to polydispersity. Thus errors in distribution caused by diffusion can be minimized by extrapolating the distribution to infinite time; this correction is normally made by graphing g(S), for each value of S, against $1/t$, and extrapolating to $1/t = 0$. For very large molecules such as large nucleic acids the rate of diffusion is very small, and this correction may be ignored.

Lastly the distribution should be examined at a number of polymer concentrations, and each g(S) extrapolated to zero polymer concentration. If very low polymer concentrations are used with absorption optics, this correction may also be ignored.

The corrected sedimentation coefficient distribution can then be related to the molecular weight distribution by use of equation (3.16). This transformation can be difficult if the influence of size on D or $\bar{V}$ is not known.

Sedimentation equilibrium runs in the ultracentrifuge may also give information on molecular weight distributions. The concentration distribution throughout the cell is related to the size distribution, but the calculations required to correct for diffusion, pressure, and concentration effects may be exceedingly complex, and are best accomplished with the aid of a computer.

3.5. MOLECULAR SHAPE IN SOLUTION

The process of dissolving a polymer into solution is not as straightforward as that for a small molecule, and takes place in two stages. First the solid polymer imbibes liquid and swells to form a *gel*, in which parts of the macromolecules remain in contact, to give a loose structure through which the solvent can move freely. Then further addition of a good solvent breaks the residual polymer/polymer contacts, and each polymer molecule becomes surrounded by a layer of solvent molecules, i.e. is solvated, as it changes from the gel state to a solution.

For many natural polymers, i.e. proteins, polysaccharides, and nucleic acids, aqueous solutions are extremely important, and it is, therefore, appropriate to comment on water as a solvent.

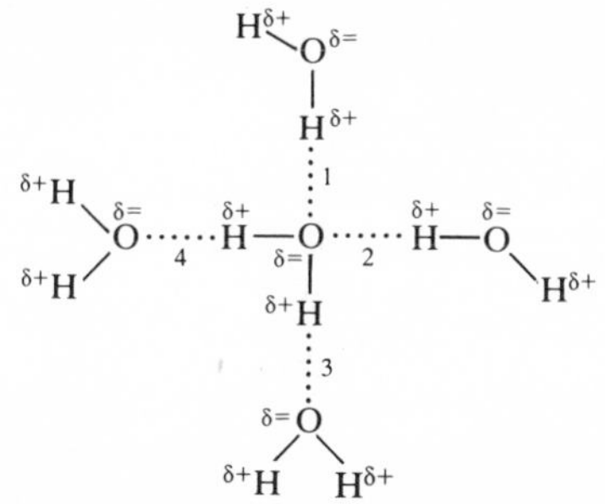

$$
\begin{array}{c}
H^{\delta+}\!\searrow O^{\delta=} \\
| \\
H^{\delta+} \\
\vdots\,1 \\
{}^{\delta+}H\searrow O^{\delta=}\cdots\cdots_{4}\ H^{\delta+}\!-\!O^{\delta=}\cdots\cdots_{2}\ H^{\delta+}\!-\!O^{\delta=}\searrow H^{\delta+} \\
{}^{\delta+}H\nearrow \qquad\qquad | \\
{}^{\delta+}H \\
\vdots\,3 \\
{}^{\delta=}O \\
{}^{\delta+}H \quad H^{\delta+}
\end{array}
$$

FIGURE 3.10. Hydrogen bonding in water. This is a schematic planar representation of a tetrahedral arrangement. If bonds 3 and 4 lie in the plane of the paper, one of bonds 1 and 2 would be above and the other below the plane of the paper

Because of the electron distribution, the water molecule has the ability to form up to four hydrogen bonds; one involving each hydrogen atom and two involving the lone electron pair on the oxygen (see Figure 3.10). Indeed, such an arrangement forms the basis of the three-dimensional structure of ice. Much of this order is thought to be retained in liquid water. The exact structure of liquid water is unknown, but it is believed that there are highly-ordered regions, where many molecules are hydrogen-bonded together, as well as regions of single, non-bonded molecules. The ordered regions are short-lived (the average lifetime is 10^{-11} sec), but are continually being formed and reformed. When a solute is added to water, the arrangement of ordered and disordered regions is altered.

In general, ionic and polar polymers dissolve readily in water, the former because the partial charges on water molecules interact readily with charges on ionic polymers; the latter because they can easily hydrogen bond to water molecules. Non-polar macromolecules tend to interact with each other rather than water, and so can be dissolved in aqueous solution only with difficulty, if at all. Non-polar groups of soluble polymers also tend to interact with one another, rather than with water, for their exposure to solvent water would cause an unfavourable decrease in entropy.

A difficulty arises when discussing polymer shape in solution, for most polymer molecules are able to adopt different conformations when dissolved in a liquid, rotation being possible about many of the single bonds of a polymer chain. Thus a dissolved polymer molecule does not have a single static shape, but can continually undergo changes in conformation. What we measure, therefore, is only an *average shape*. At one extreme, a molecule may be fully extended, and a suitable model for this would be a rigid rod; short stretches of double-stranded nucleic acid

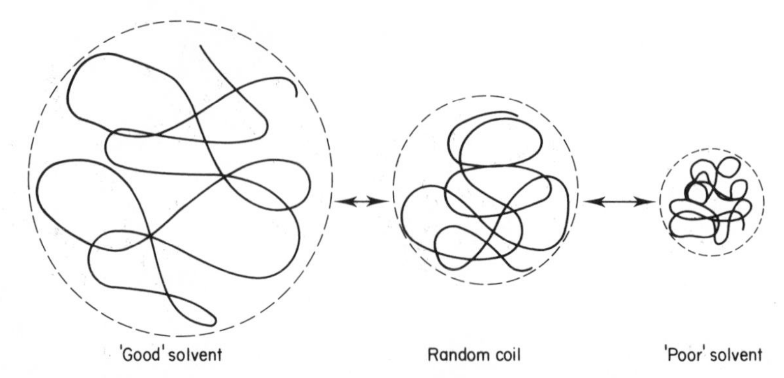

FIGURE 3.11. Schematic representation of the random polymer coil in various solvents

can behave in this way. It is more common, however, for a linear polymer chain to behave like a tangled piece of thread, i.e. as a random coil. On average, the shape of this random coil will be a sphere (see Figure 3.11); in 'good' solvents where polymer–solvent contacts are preferred, the molecule expands, whilst in 'bad' solvents where polymer–polymer contacts are favoured, the molecule contracts (see Figure 3.11) (see also Section 3.8). Branched polymers are less extended in space than linear polymers, and highly-branched macromolecules probably have a fairly dense, spherical structure in solution.

For some biopolymers, e.g. many globular proteins, the forces determining the three-dimensional structure of the molecules are believed to be so strong that few rotations about single bonds are permitted. In such polymers, the molecular shape is believed to be more static, with only minor conformational changes able to take place. This situation may also be true for the smaller ribonucleic acids.

Information on the shapes of biopolymer molecules in solution can be obtained by a variety of methods, including some of those described above for molecular weight determinations.

Light-scattering measurements (see Section 3.3.2, p. 43) yield the root-mean-square radius of gyration, the square root of the weight average of r^2 for all mass elements of a polymer, where r is the distance of a mass element from the centre of mass of the molecule. Using this value, the intensity of scattered light as a function of scattering angle can be calculated theoretically for certain model molecular shapes, i.e. rigid rods, random coils, and dense spheres. These intensities can then be compared to the values obtained experimentally on a Zimm graph (see Figure 3.4) for the zero concentration line, and the shape of the molecule inferred from whichever theoretical model predicts intensities closest to the experimental values. This type of calculation has been carried out for proteins, polysaccharides, and nucleic acids.

Studies of *viscosity* may also give information on macromolecular shape.

The value of α (see equation 3.23) depends on the shape of the polymer in a particular solvent. Thus in general, $\alpha \sim 0.5$ in a poor solvent where the polymer molecules behave as dense spheres, $\alpha \sim 0.75$ for random coils, and α is greater than one for an elongated rod.

Viscosity measurements are especially useful for indicating *changes* in shape; thus the viscosity of a protein, polysaccharide, or nucleic acid solution may change dramatically if conditions, e.g. pH or temperature, are altered so that the native conformation of the macromolecule is destroyed.

Measurements of *sedimentation coefficient* (see Section 3.3.3, p. 48) can also give indications of changes in shape. Again changes in sedimentation coefficients of some proteins and large nucleic acids can be observed when the original structures of these polymers are altered.

Optical rotatory dispersion (ORD) (the variation of optical rotation of a solution with the wavelength of polarized light) as well as *circular dichroism* (CD) (the polarization of the transmitted fraction of circularly polarized light incident on a solution) have both been studied for proteins in an attempt to gain information about the types and extent of secondary structure (see Section 4.4.2, p. 97). It was found, for regular synthetic polymers, that certain ORD and CD spectra can be related to definite types of molecular folding. For some time it was assumed that the molecules of natural polymers (particularly proteins) giving similar spectra had these same types of folding. Indeed, calculations of the extent of such folding were made, based on similarities between the spectra of natural and synthetic macromolecules. But it is now believed that these results are of limited value, having been based on too simple a model, i.e. the folding of a protein molecule is much more complex than that of a regular synthetic macromolecule. However, *changes* in optical rotatory dispersion or circular dichroism are useful for indicating changes in polymer conformation, i.e. an alteration from a helix to, say, a random coil.

Nuclear magnetic resonance studies, investigating both protons and ^{13}C, are being carried out increasingly on macromolecules. However, currently, the complex spectra obtained are exceedingly difficult to interpret. Nevertheless, changes in local environment within a macromolecule can be studied during, for example, the binding of a small molecule, or during conformational changes brought about by alterations in the pH of the solution. Also information concerning intra- and inter-molecular hydrogen bonding can sometimes be obtained.

3.6. MOLECULAR SHAPE IN THE SOLID STATE

The constraints on molecular movement are naturally very much greater in a solid than in a liquid and only some minor vibration is possible. Hence the shape of a polymer molecule in a solid may be considered to be close to the actual shape of the macromolecule, rather than a time-average shape as obtained for molecules in solution.

Hypothetically, the easiest way to determine the shape of polymer molecules in the solid is to 'look' at individual molecules in a microscope of suitably high magnification. In practice, such a microscope would have to operate with electromagnetic radiation of wavelength smaller than the molecules to be studied, i.e. wavelengths of 1 Ångstrom (10^{-10} m) or less: this effect can be achieved by the *electron microscope*. In this equipment (see Figure 3.12) electrons can behave like radiation of wavelength 0.5–1 Å. The electron beam is focused using magnetic and electric fields and allowed to impinge on the sample—usually obtained by spraying a solution of the substance under study onto a grid and drying in a vacuum. The molecules of the sample scatter the incident electrons, which, after refocusing, strike a fluorescent screen where an image of the sample may be seen. The resolution of such instruments is 5–10 Å, which enables large macromolecules and subcellular structures to be visualized, but fine details of molecular structure cannot be distinguished.

In practice, the scattering of electrons depends on the type of atoms within the polymer—atoms of high atomic number scatter more effectively. Because biological polymers contain mostly atoms of low atomic number, these molecules have low scattering power and, by themselves, give poor images in electron microscopes. These molecules may be more readily

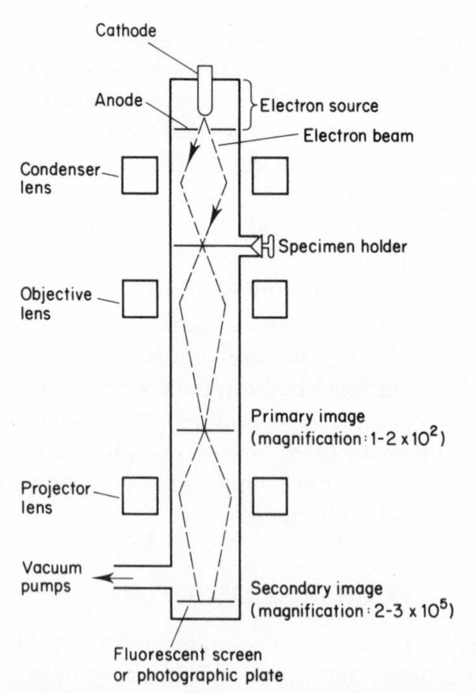

FIGURE 3.12. Schematic representation of an electron microscope. Note: modern microscopes have more lenses

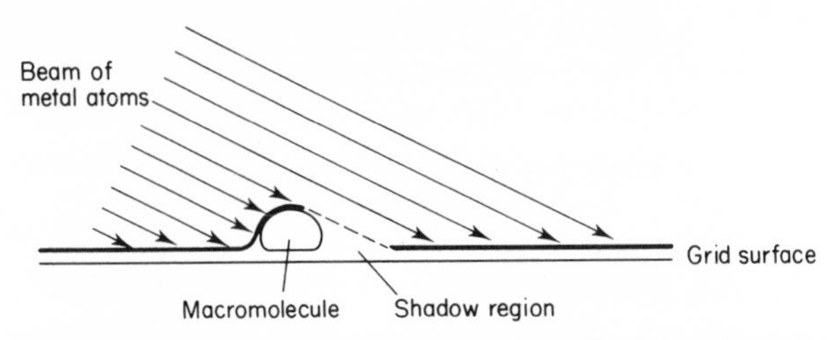

FIGURE 3.13. Heavy-metal 'shadowing' for electron microscopy

visualized, however, after being subjected to a process known as 'shadowing' with a heavy metal. The sample is first applied to a grid, as usual, then a beam of heavy metal atoms such as platinum or palladium, evaporated from a heated filament, is directed obliquely at the sample surface. The metal atoms coat the surface, but leave an uncoated 'shadow' where a large molecule rises above the grid surface (see Figure 3.13). On the subsequent photograph of the image on the fluorescent screen (electron micrograph), the macromolecule can be seen because of its shadow contrasting with the background of electron scattering by the heavy metal ions. For very large molecules, both shape and size can be estimated by examination of appropriate electron micrographs. This technique has been used to study large biopolymers such as structural proteins, polysaccharides, and deoxyribonucleic acids, as well as inorganic polymers such as graphite. A disadvantage of the procedure, however, is that very large molecules may be fragmented during the drying of the sample on the grid. The lengths of large deoxyribonucleic acid molecules can be measured and their molecular weights determined, since the weight per unit length of these molecules is almost constant.

Polymer shapes may also be obtained indirectly using *X-ray diffraction*. Indeed, this is the only method at present for determining the detailed three-dimensional structures of macromolecules. In the crystalline state, molecules of a substance are arranged in a regular way on a three-dimensional lattice. The lattice is built-up from a simple repeating unit, the unit cell of the crystal. Because atomic diameters are comparable in size to the wavelengths of X-rays, crystals with their regular arrangements of atoms behave towards X-rays as three-dimensional diffraction gratings. X-rays directed on a crystal are scattered by the atoms in the crystal. In certain directions the scattered rays interfere destructively and no scattered radiation would be detected, whilst in other directions the waves reinforce each other. Consider an array of atoms as in Figure 3.14. Constructive interference of scattered X-rays occurs whenever the pathlength of rays scattered from successive layers differs by an integral multiple of wavelengths, and this radiation may be detected on, say, a photographic plate. (Constructive interference is obtained when scattered waves are exactly in phase with one another; complete destructive

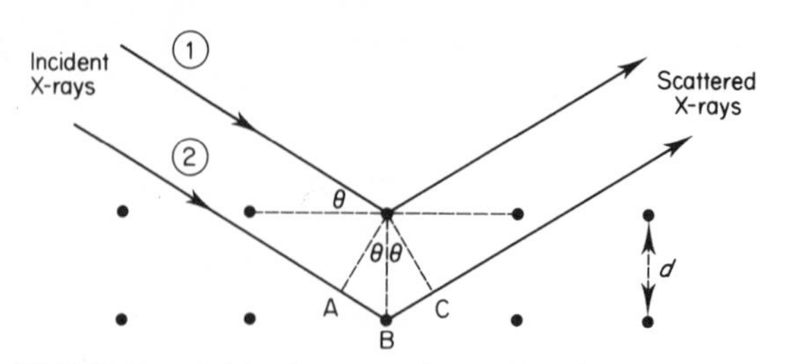

FIGURE 3.14. Constructive interference of X-rays

interference results when waves are exactly out of phase.) For beams striking planes of atoms at angle θ, the path-length of beam 2 exceeds that of beam 1 by the distance $AB + BC$ (Figure 3.14). If the interplanar distance is d, then $AB + BC = 2d \sin\theta$. This function must be a whole number of wavelengths for constructive interference, i.e.

$$2d \sin\theta = n\lambda, \tag{3.28}$$

where n is an integer and λ is the wavelength of the X-rays. This equation is known as the Bragg condition for diffraction of X-rays. [Equation (3.28) can be rearranged to give $\lambda = (2d/n)\sin\theta$. Thus higher order ($n > 1$) reflections from planes of spacing d can be regarded as first-order reflections from planes of spacing d/n. Then the fundamental equation can be considered as $\lambda = 2d \sin\theta$.]

If a single crystal of a substance can be grown to sufficient size and then rotated (about, say, an axis of symmetry of the crystal) in a beam of X-rays, scattered X-rays may be detected as a pattern of spots on a photographic film placed behind the sample. The positions of the spots are related to the planes of molecules in the crystal and the interplanar spacings by the Bragg equation. From the positions of the spots and a knowledge of crystal symmetry, the size and shape of the unit cell of the crystal can be obtained. When the molecular weight of a substance is known, the number of molecules in the unit cell can be calculated from the dimensions of the unit cell and the crystal density.

Where only numbers of small crystals are available, a diffraction pattern can still be obtained. The pattern takes the form of lines, rather than spots, on the photographic plate, but from these lines, knowledge of the unit cell can be gained.

Many biopolymers, in fact, are not completely crystalline in the solid state. X-ray diffraction indicates some crystallinity, i.e. some regular arrangements within the solid, but these polymers are believed to be made up of small regions of ordered structure in the midst of non-crystalline, or amorphous, material. Stretched rubber, for example, is much more crystalline than unstretched rubber, because of the increase in the ordered arrangements of parts of the molecules on stretching (see Section 7.2.2, p. 331).

Reasonably good diffraction patterns can be obtained from many fibrous biopolymers, and this has been interpreted as indicating a regular arrangement along the fibres, whilst structures at right-angles to the fibre axis may be more disordered. Some polysaccharides, structural proteins, and long deoxyribonucleic acids have been studied in this way, and it has often been possible to determine the size of the repeating unit of the polymer chain backbone and the interchain distances. For some types of biopolymers, these studies have indicated the presence of organized structures such as helices or pleated sheets (see Section 4.4.2, p. 97).

Certain natural polymers, notably some proteins, ribonucleic acids, and the inorganic polymers, can exist in a true crystalline state, and single crystals of these substances can be studied using X-rays. The diffraction patterns consist of sharp spots, and by laborious analysis of both the positions and intensities of these spots, the details of three-dimensional structure of polymer molecules can be obtained. Hydrogen atoms have very low scattering power for X-rays, and so the positions of hydrogen atoms within macromolecules cannot usually be determined in this way; hydrogen atom placement is usually decided from knowledge of the geometry of the monomers which constitute a polymer.

The intensity of a spot in a diffraction pattern depends on the angle of diffraction (Figure 3.14 and equation 3.28), on the number of planes giving the spot [several sets of planes with the same value of d (see equation 3.28) may exist in a crystal], and on the distribution of atoms within the polymer molecule and their relative scattering powers—the scattering power of an atom is proportional to the number of electrons it contains.

If the positions of atoms within a molecule and arrangement of molecules within a crystal is known, it is possible to predict the X-ray diffraction pattern. Unfortunately, we are faced with the problem in reverse. The diffraction pattern is known; how can the molecular structure be deduced? Initially, when X-ray crystallography was confined to small, relatively simple, molecules, the structure was obtained by trial and error. A structure was assumed and a diffraction pattern predicted. This was then compared to the experimental pattern, and the postulated structure altered to minimize the differences between the predicted and observed diffraction patterns. Such a method was tedious and time-consuming, and could not be carried out for complicated molecules such as proteins.

As stated above, the intensity of a spot in a diffraction pattern depends on the scattering angle and the number of sets of planes giving that spot. Both of these factors can be accounted for, so that an intensity may be found which is related only to a so-called structure factor, itself a function of atomic position and scattering power. The intensity of X-rays diffracted from one set of planes (to give one spot) is influenced by the positions of atoms within the molecules, because an atom slightly out of the plane gives scattered X-rays slightly out of phase with the waves diffracted from the plane as a whole; these out-of-phase rays cause some destructive interference, and so reduce the intensity of the diffraction spot. [Exact positions of atoms in a crystal are not obtained directly from X-ray diffraction patterns. Instead, a continuous function, $\rho(x,y,z)$ can be calculated. This function is a measure of the electron density at any point (x,y,z) within the crystal, and from it, the positions of atoms can be inferred, because electron-dense regions correspond to atoms. The function, $\rho(x,y,z)$, can be calculated from

$$\rho(x, y, z) = \sum_{h,k,l=-\infty}^{h,k,l=+\infty}\sum\sum | F(h, k, l) | \exp\{-2\pi i(hx + ky + lz - \alpha)\},$$

where F is the amplitude of a scattered X-ray wave and can be found from the square root of the intensity of the corresponding diffraction spot; h, k, and l are whole numbers which define planes in the unit cell—each diffraction spot is associated with unique values of h, k, and l; and α is the phase, which at present cannot be determined directly. Thus even with a knowledge of the h, k, and l values and diffraction spot intensities, $\rho(x,y,z)$ cannot be calculated unless some estimate of α can be made.] It is these phase differences which must be known before a complete molecular structure can be determined by X-ray crystallography, and it is precisely these phase differences which are difficult to obtain and make this whole process of investigation of macromolecular structure both complex and tedious.

Information about phase differences is obtained by the method of *isomorphous replacement*. First, an X-ray diffraction pattern is obtained from a crystal of, say, a globular protein, then a derivative of this protein is prepared, containing a heavy metal atom in such a position that very little perturbation of molecular structure occurs; if a large single crystal of this derivative can be grown, it normally has the same size and symmetry of unit cell as the original protein crystal. An X-ray diffraction pattern of the derivative is compared to the pattern from the original; the differences between the two are of course due to the heavy atoms. From these differences, knowledge of some phase differences can be gained. To evaluate all the phase differences, several different derivatives of a polymer may be required, each containing a different heavy metal ion. Some postulates about the macromolecular structure must also be made, and a final structure reached by making refinements using the trial and error method to minimize differences between predicted and experimental diffraction patterns.

Even with the use of modern computers, the evaluation of the complete three-dimensional structure of a macromolecule is a major undertaking, which may last months or years.

3.7. ADDITIONAL READING

Physical Chemistry of Macromolecules C. Tanford, J. Wiley & Sons, New York and London, 1961.

The determination of molecular weight C. T. Greenwood in *Physical Methods in Organic Chemistry* (Ed. P. Schwarz), Oliver & Boyd, Edinburgh, 1964.

Methods of Enzymology (Eds. S. P. Colowick and N. O. Kaplan), Academic Press, New York and London; particularly Vol. 4 (1957), Vol. 12 (1968), Vol. 26 (1972), and Vol. 27 (1973).

Starch and its Components W. Banks and C. T. Greenwood, Edinburgh University Press, Edinburgh, 1975. (This monograph contains a complete account of the application of polymer theory to the biopolymer situation.)

Structure and Stability of Biological Macromolecules, Vol. 2 of *Biological Macromolecules* (Eds. N. Timasheff and G. D. Fasman), Marcel Dekker, New York, 1969.

Light Scattering from Polymer Solutions M. B. Huglin, Academic Press, New York and London, 1972.

Ultracentrifugation of Macromolecules J. W. Williams, Academic Press, New York and London, 1972.

Molecular Weight Distributions in Polymers L. H. Peebles, Interscience, New York and London, 1971.

Physical Methods in Macromolecular Chemistry, Vol. 2 (Ed. B. Carroll), Marcel Dekker, New York, 1972.

Macromolecules in Solution H. Morawetz, Wiley–Interscience, New York and London, 1975.

X-ray Diffraction Methods in Polymer Science L. E. Alexander, J. Wiley & Sons, New York and London, 1969.

3.8. APPENDIX: THE POLYMER MOLECULE IN SOLUTION

The definition of polymer size: One possible way of characterizing the size of a polymer is to calculate the distance (r_0) between the two ends of the chain. But as the conformation of a polymer in solution may be constantly changing, this distance can change continually. Fortunately, it changes in a completely random way, which enables us to calculate the *average*, or *mean*, distance between the chain ends, i.e. $\bar{r}_0$. In actual fact (as described below) the theoretical calculation does not give the mean distance, but instead yields the mean-square distance between the ends ($\bar{r}_0^2$). We then use the square root of this quantity, i.e. $(\bar{r}_0^2)^{1/2}$, as a measure of the size of the polymer. This is known as the *root-mean-square end-to-end* distance. The difference between the mean value and the root-mean-square value is best illustrated by a simple example. The mean of 3, 5, and 7 is 5; the mean-square is $(3^2 + 5^2 + 7^2)/3$, and the root-mean-square is $[(3^2 + 5^2 + 7^2)/3]^{1/2}$, i.e. 5.26.

A theoretical model for the polymer molecule: The picture of a thread-like polymer molecule must be modified to take into account the fact that, whereas the thread is flexible along its entire length, the individual bonds in the polymer backbone cannot bend. However, rotation about bonds many still occur. The polymer molecule may therefore be regarded as being made up of a number of links, n, each of length l.

Assuming that the direction in space of each link is completely independent of that of its neighbours, we can then calculate the dimensions of this, the random, or statistical coil.

The dimensions of the random, or statistical, coil: The statistical coil is based on the mathematical model of 'random flight' in three dimensions. The problem can be simplified by first considering a similar case in two dimensions—the 'random' or 'drunkard's' walk. Suppose we start at a given point and take n steps of equal length l in a *completely random* fashion, how far shall we be from the starting point? The simplest case is when $n = 2$ shown in Figure 3.15, where OA is the first step, AB is the second, and θ is the angle OAB. The distance from the starting point (OB) can then be obtained from

$$
\begin{aligned}
OB^2 &= OA^2 + AB^2 - 2OA.\,AB\cos\theta \\
&= 2l^2 - 2l^2\cos\theta.
\end{aligned}
$$

Without knowing the value of θ, it is impossible to calculate the length OB. Suppose, however, that the experiment is repeated a very large number of times, the distance OB being measured after each pair of steps. All values of θ will be equally likely to occur, and since $\cos(\theta + \pi) = -\cos\theta$, the sum of all the $\cos\theta$ terms will be zero. Denoting OB by $\bar{r}_0$, we then have

$$\bar{r}_0^2 = l^2.$$

This of course means that the average direction of the second step is at right angles to the first—an answer which is intuitively obvious.

For the more general case of n steps, the relation is

$$\bar{r}_0^2 = nl^2 + 2l^2\Sigma\cos\omega.$$

In this case, the summation covers *all* the interstep angles rather than the

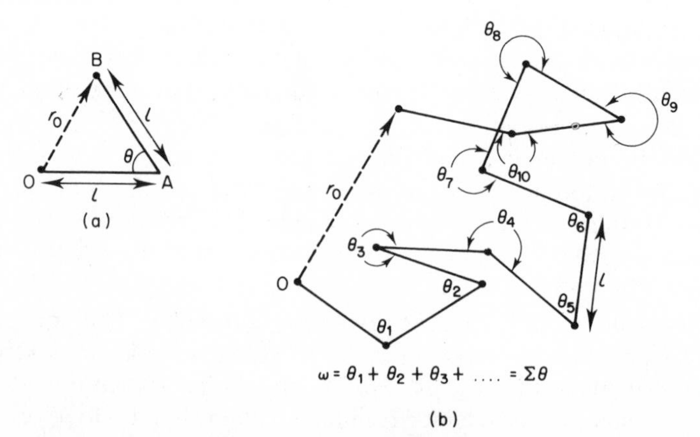

FIGURE 3.15. Random walk experiment for (a) two steps, and (b) eleven steps

angle between successive steps (cf. Figure 3.15b). By the same argument as above $\Sigma \cos \omega = 0$, and

$$\overline{r_0^2} = nl^2.$$

This shows that the mean-square end-to-end distance of the polymer is directly proportional to the number of links, or structural units, in the chain. If this type of calculation is extended to *three dimensions*, i.e. the random flight approach, the relation $\overline{r_0^2} = nl^2$ is again obtained.

How realistic is this as a model for the actual polymer molecule in solution? It is obvious that n represents the number of bonds and l is the length of the structural unit. But is the averaging procedure valid in the case of a real polymer? It is—for the constituent units of the molecules of many polymers are in constant motion as a result of rotation about bonds.

It is perhaps not immediately obvious why it is necessary to use the *mean-square* end-to-end distance, rather than the *mean* separation of the ends. In fact, the latter quantity is zero. Consider the polymer as a loosely coiled mass having spherical symmetry. If we take the centre of this mass as the origin of a three-dimensional coordinate system, then, as in the more familiar two-dimensional graphs, there will be both positive and negative directions. The probability of finding a chain-end at a given distance from the origin in a positive direction will be exactly the same as finding it in the diametrically opposed (negative) direction. The sum of all these probabilities, i.e. the mean separation of the chain-ends, is thus zero. By using squares, the negative signs vanish, and a finite answer results.

Effect of fixed bond angle and hindered rotation: The above model for a polymer molecule is obviously too simple, and we must consider other important modifications to make it more realistic. In the real polymer molecule the angles between successive interunit bonds are fixed in some way, and the direction in space of any bond is not independent of its neighbour. For example, in polyethylene the backbone is a chain of carbon atoms joined by single covalent bonds having the tetrahedral angle of 109°. In the model of the real polymer, there is thus a chain in which there is a fixed angle (θ) between each unit—although free rotation is still allowed about each bond. The mean-square end-to-end distance is then given by

$$\overline{r_0^2} = [(1 + \cos\theta)/(1 - \cos\theta)]nl^2,$$

for large n, and θ not too near zero. However, $\overline{r_0^2}$ is still proportional to n, as it was for the case of the completely random chain.

Another failure of this simple model is that the hypothetical chain can intersect with itself, i.e. two units may well occupy the same point in space. This is clearly impossible in the case of the real polymer. The real polymer is not able to, or is 'excluded' from, taking up all the theoretical conformations, because of the finite volume of a structural unit. We therefore have the concept of the *excluded volume*, and the dimensions of real polymer molecules $(\bar{r}^2)^{1/2}$ are greater than those of the equivalent random model $(\bar{r}_0^2)^{1/2}$ by an *expansion factor*, α, where

$$\alpha = (\bar{r}/\bar{r}_0^2)^{1/2}.$$

The effect of solvent: We must now consider the effect of the solvent on the dimensions of the polymer. A macromolecule dissolves because polymer–solvent contacts are preferred to either polymer–polymer or solvent–solvent contacts. The above theory included the postulate that the various conformations would occur with equal probability; this actually implies that there is no change in *energy* on dissolving the macromolecule, i.e. that the solvent is *athermal* for that particular polymer, which is rather uncommon. For most polymers, energy is involved when they dissolve. When polymer–solvent contacts are preferred, as is the case with a 'good' solvent, conformations which allow more of these contacts will be preferred, i.e. the molecule will expand. Conversely, of course, if polymer–polymer contacts are preferred, as is the case with a 'bad' solvent, the molecule will contract.

'Unperturbed dimensions' of a polymer: The effects of excluded volume and solvent are called 'long-range' forces, because they arise from interactions between parts of the macromolecule which are far removed from each other along the length of the chain, but which are brought close together in space by the coiling of the polymer. They are related to each other, and a theoretical relation has been derived to account for them both, i.e.

$$\alpha^5 - \alpha^3 = A[1 - (\theta/T)]M^{1/2},$$

where A is a constant for a given polymer–solvent system, θ has the dimensions of temperature, T is the temperature, and M is the molecular weight of the polymer. The most important consequence of this relation is that for a polymer in a poor solvent, there is a temperature, the *theta temperature*, at which $\alpha = 1$. At this temperature, therefore, the expansion of the macromolecule as a result of the excluded volume effect is cancelled by the contraction resulting from the use of a solvent which favours polymer–polymer contacts. Under these conditions, the polymer is said to have adopted its *unperturbed* conformation, with characteristic *unperturbed dimensions*. When measurements are made under these θ-conditions, values of $(\bar{r}_0^2/n)$ for many different polymers are found to be constant, as demanded by the theory, but are somewhat greater than expected. This is due to the effect of additional 'short-range' forces. These forces arise as a result of the rotation about interunit bonds being partially hindered. This makes the macromolecule less flexible, and therefore favours more extended conformations.

Branched polymers in solution: The above model is for a linear polymer molecule. If the polymer is branched, there are many chain-ends and so the term 'end-to-end distance' has no real meaning. Furthermore, a branched polymer will be less extended in space than a linear one of the same molecular weight. As the branching becomes more extensive, the number of possible conformations becomes less, and the picture of the random coil is replaced by one in which the polymer molecule is regarded as a fairly dense, spherical structure. Deviations from ideal behaviour are usually very small for highly branched polymers.

Chapter 4

Proteins

4.1. INTRODUCTION

Proteins play an extremely important rôle in all living cells for they perform a large variety of functions, and govern most of the physical and chemical activities necessary for life.

In the human body, for example, proteins act as carriers for the transportation of small, but vital, molecules round the body. The chemical reactions of the body, e.g. digestion and utilization of food, are directed by protein catalysts, the *enzymes*, whilst the *hormones*, another group of proteins, control the overall rate of metabolism. At the same time, immunological mechanisms involve the protein *antibodies*, and, further, protein is one of the chief constituents of structures such as bone, cartilage, and skin. Muscles owe both their structural features and contractile properties to specially designed proteins, whilst proteins associated with genes, the basis of heredity, are essential to the process of cell division.

Protein molecules are extremely complex, but methods now exist for determining their detailed structure. Such information enables the relation between biological function and molecular design to be understood, and indeed much important progress has already been made in this direction.

4.2. THE MONOMERS

Proteins are condensation polymers of the α-amino acids with the general formula;

$$H_2N-\underset{\displaystyle H}{\overset{\displaystyle COOH}{\underset{|}{\overset{|}{C}}}}-R$$

The 23 different amino acids commonly found in proteins differ only in the nature of the side-chain, —R, and a list of these acids is given in Table 4.1, together with their common abbreviations. (It should be noted that two of these acids, proline and hydroxyproline, do not conform to the general formula, and are actually *imino* acids.)

TABLE 4.1. The common amino acids

Name	Abbreviation	Side-chain, R
Glycine	Gly	H—
Alanine	Ala	CH_3—
Valine	Val	$CH_3—CH(CH_3)—$
Leucine	Leu	$CH_3CH(CH_3)CH_2—$
Isoleucine	Ile	$CH_3CH_2CH(CH_3)—$
Serine	Ser	$HO—CH_2—$
Threonine	Thr	$CH_3CH(OH)—$
Aspartic acid	Asp	$HOOC—CH_2—$
Asparagine	Asn	$H_2NOC—CH_2—$
Glutamic acid	Glu	$HOOC—CH_2CH_2—$
Glutamine	Gln	$H_2NOC—CH_2CH_2—$

TABLE 4.1 *continued*

Name	Abbreviation	Side-chain,R
Lysine	Lys	NH_2 on the terminal carbon: $H_2N{-}CH_2CH_2CH_2CH_2{-}$
Hydroxylysine	Hyl	NH_2, OH: $H_2N{-}CH_2{-}CH(OH){-}CH_2.CH_2{-}$
Arginine	Arg	$H_2N{-}C(=NH){-}NHCH_2CH_2CH_2{-}$
Cysteine	Cys	$HS{-}CH_2{-}$
Cystine	$(Cys)_2$	$S{-}CH_2{-}$ / $S{-}CH_2{-}$ (the two S atoms bonded to each other)
Methionine	Met	$CH_3{-}S{-}CH_2CH_2{-}$
Phenylalanine	Phe	Benzene ring: ring carbons HC, CH=CH, CH–CH and C; $C{-}CH_2{-}$ attached to the ring
Tyrosine	Tyr	$HO{-}C$ ring: CH=CH, CH–CH; $C{-}CH_2{-}$ attached to the ring
Tryptophan	Trp	Indole ring: HC, CH, HC, HC, CH, C, C, N–H; $C{-}CH_2{-}$ attached to the CH of the five-membered ring
Histidine	His	Imidazole ring: N—CH, HC, N–H; $C{-}CH_2{-}$ attached to the ring

TABLE 4.1 *continued*

Name	Abbreviation	Side-chain, R Complete formulae
Proline[a]	Pro	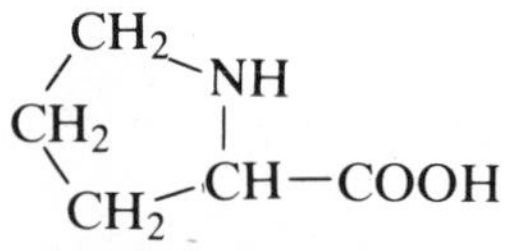 $\begin{array}{l}\ \ CH_2{-}NH \\ /\qquad\quad \vert \\ CH_2 \qquad \vert \\ \backslash \qquad\quad \vert \\ \ \ CH_2{-}CH{-}COOH\end{array}$
4-Hydroxyproline[a]	Hyp	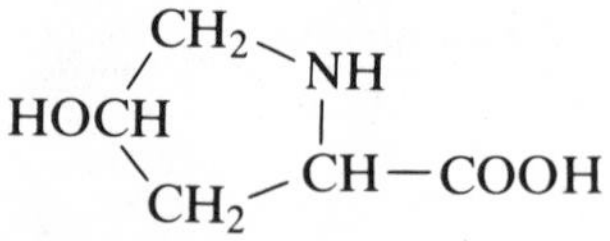 $\begin{array}{l}\ \ CH_2{-}NH \\ /\qquad\quad \vert \\ HOCH \qquad \vert \\ \backslash \qquad\quad \vert \\ \ \ CH_2{-}CH{-}COOH\end{array}$

[a]The cyclic amino acids containing a secondary amino group are known as *imino acids*.

All except the simplest amino acid, glycine, are optically active, but only one optical isomer occurs commonly in proteins, i.e. the L-isomer. D-amino acids are found in bacterial polymers (see, for example, Section 6.5.2.6, p. 310). The two forms are written (in the Fischer convention) thus:

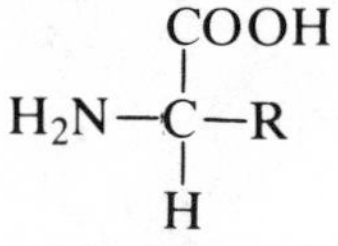

$$\begin{array}{c} COOH \\ \vert \\ H_2N{-}C{-}R \\ \vert \\ H \end{array} \qquad\qquad \begin{array}{c} COOH \\ \vert \\ R{-}C{-}NH_2 \\ \vert \\ H \end{array}$$

L-form 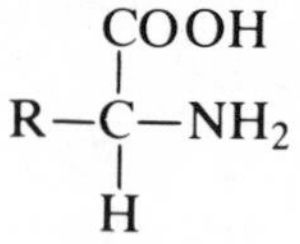 D-form

The unique properties of each amino acid are determined by the nature of the side-chain, but the chemical and physical properties, which are common to all the acids, are due to the spatial arrangement of the amino and carboxyl groups.

4.2.1. Reactions of the monomers

Free amino acids can behave as both acids and bases in aqueous solution, as each contains at least one carboxyl and one amino group. A simple amino acid like glycine can exist as a dipolar ion, or *zwitterion*, in aqueous solution:

$$^+NH_3{-}CH_2{-}COO^-$$

This molecule is electrically neutral, and is termed *isoelectric*, but a change in pH of the solution produces a charged molecule:

$$^+NH_3{-}CH_2{-}COO^- + H_3O^+ \rightleftharpoons {}^+NH_3{-}CH_2{-}COOH + H_2O$$

$$^{+}NH_3{-}CH_2{-}COO^- + OH^- \rightleftharpoons NH_2{-}CH_2{-}COO^- + H_2O$$

The pH at which the molecule has a net change of zero is called the *isoelectric point*.

Some of the condensation reactions of amino acids are important in the determination of protein structure. With ninhydrin, at pH 4, a purple compound, *Ruhemann's Purple*, is formed:

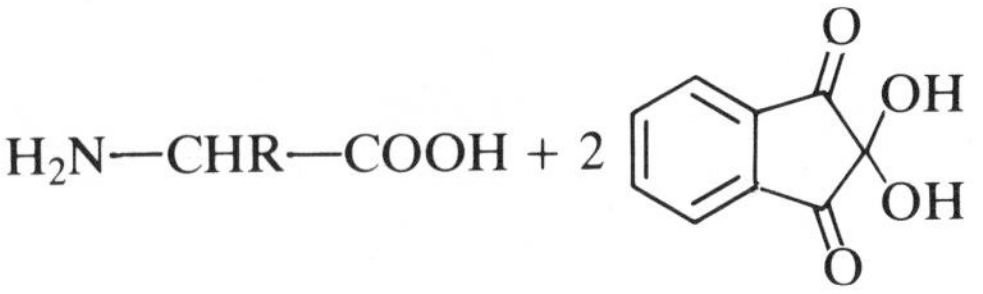

$H_2N{-}CHR{-}COOH + 2$ [ninhydrin structure]

Ninhydrin (2,2 dihydroxy-1,3-indandione)

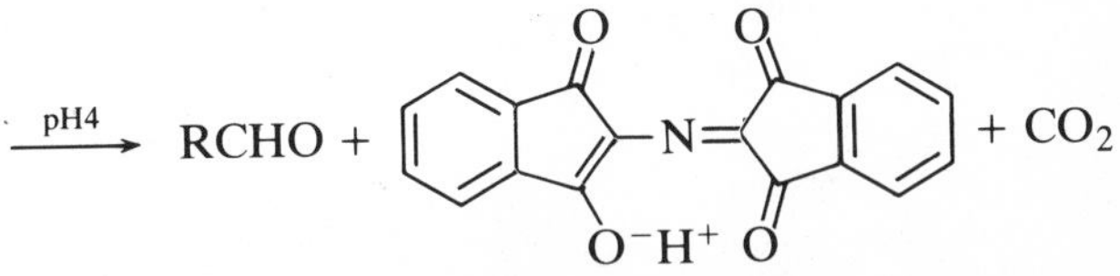

$\xrightarrow{pH4}$ RCHO + [Ruhemann's Purple structure] + CO_2

Ruhemann's Purple
(Indandion-2-*N*-2′-indanone enolate)

The reaction with amino acids is not stoichiometric and one mole of different amino acids produces different amounts of coloured compound. However, the reaction and amount of colour produced by any one particular acid are reproducible, and so, after standardization, this method can be used for the estimation of amino acids. It should be noted that the amino acids, proline and hydroxyproline (a less commonly occurring acid, important in collagen), do not form Ruhemann's Purple, but give a yellow-brown product, believed to have the structure

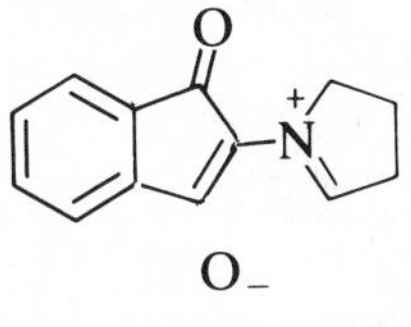

[structure]

Amino acids also react with fluorodinitrobenzene in alkaline solution to give a stable, yellow *dinitrophenyl derivative*, which can easily be detected and identified:

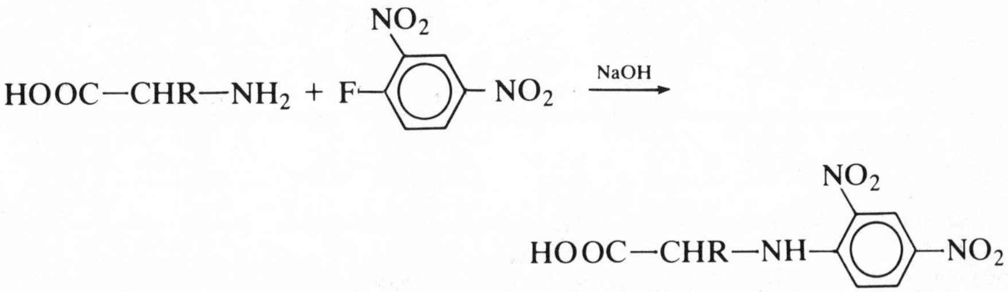

$$HOOC{-}CHR{-}NH_2 + F{-}C_6H_3(NO_2)_2 \xrightarrow{NaOH} HOOC{-}CHR{-}NH{-}C_6H_3(NO_2)_2$$

A similar reaction can be obtained with dansyl chloride in alkaline solution:

$HOOC{-}CHR{-}NH_2 + ClSO_2{-}C_{10}H_6{-}N(CH_3)_2 \longrightarrow$

$HOOC{-}CHR{-}NH{-}SO_2{-}C_{10}H_6{-}N(CH_3)_2$

This derivative is fluorescent.

A further fluorescent product can be obtained by the reaction of an amino acid with fluorescamine:

$+ HOOC{-}CHR{-}NH_2 \longrightarrow$ HOOC—CHR—N ... OH ... C—OH, ‖ O

Fluorescamine

Fluorescent derivative

(Proline and hydroxyproline do not react, because they contain no primary amino group.)

Another important reaction of the amino group of amino acids involves the formation of *thiazolinones* with phenylisothiocyanate. These derivatives, which rearrange in aqueous acid to form *thiohydantoins*, can be readily identified:

$HOOC{-}CHR{-}NH_2 + C_6H_5{-}NCS \longrightarrow$

$C_6H_5{-}NH{-}C({=}S){-}HN{-}CHR{-}CO.OH \longrightarrow$

$C_6H_5{-}NH{-}C{-}S$, N CO, CH, R $\longrightarrow$ $C_6H_5{-}N{-}C{=}S$, C NH, O CH, R

R-Phenylthiazolinone

R-Phenylthiohydantoin

During chemical synthesis of proteins, amino and carboxyl groups of amino acids must often be protected to prevent side-reactions. A typical reaction for protecting amino groups involves the formation of a t-*butyl oxycarbonyl* (*Boc*) derivative of an amino acid:

$$(CH_3)_3CO{-}\overset{\overset{\displaystyle O}{\|}}{C}{-}Cl + H_2N{-}CHR{-}COOH \longrightarrow$$

t-Butyloxycarbonyl chloride

$$(CH_3)_3{-}CO{-}\overset{\overset{\displaystyle O}{\|}}{C}{-}NH{-}CHR{-}COOH$$

Boc amino acid

On the other hand, ester groupings are frequently used for carboxyl protection.

$$C_6H_5{-}CH_2OH + H_2N{-}CHR{-}COOH \xrightarrow{H^+}$$

Benzyl alcohol

$$H_3\overset{+}{N}{-}CHR{-}COOCH_2{-}C_6H_5$$

Benzyl ester of amino acid

4.2.2. The peptide bond

In proteins, the amino acids are linked together by similar covalent bonds, the peptide bonds. Consider the hypothetical condensation of two amino acids with the elimination of water to form a *dipeptide* containing one peptide bond:

$$H_2N{-}CHR^1{-}COOH + H_2N{-}CHR^2{-}COOH$$

$$\downarrow$$

$$H_2N{-}CHR^1 \vdots {-}CO{-}HN{-} \vdots CHR^2{-}COOH + H_2O$$

Peptide bond

In Nature, the formation of peptide bonds is much more complex, but the end result is that shown here. (Protein biosynthesis is discussed in detail in Chapter 5.) If further amino acids are added, *tripeptides*, *tetrapeptides*, etc. are formed. A chain of many amino acids is a *polypeptide*, but the distinction between a protein and a polypeptide is arbitrary, for a protein is simply a large polypeptide:

$$NH_2{-}CHR^1{-}CO{-}NH{-}CHR^2{-}CO{-}NH{-}CHR^3{-}CO \ldots . NH{-}CHR^n{-}COOH$$

Although written formally as a single bond, the peptide bond has some double-bond characteristics, and may be considered being intermediate between the two extreme resonance forms:

$$\cdots{-}C{-}\overset{\displaystyle O}{\overset{\|}{C}}{-}N(H){-}C{-}\cdots \quad \text{and} \quad \cdots{-}C{-}\overset{\displaystyle {}^{-}O}{\overset{|}{C}}{=}\overset{+}{N}(H){-}C{-}\cdots$$

That is, the C=O bond is less than double in character, and the C—N bond is a partial double bond, being shorter in length than a 'normal' C—N single bond. Thermodynamic measurements on simple amide models indicate that more energy is required to twist the amide link around the C—N axis than around a 'normal' single bond. Hence, it is not surprising that X-ray studies of solid peptides have shown that the atoms involved in the peptide link are almost coplanar (see Figure 4.1). These studies have also demonstrated that the side-chains of the amino acid residues project on either side of the main polypeptide chain.

All proteins have the common feature of at least one long chain of amino acid residues linked by peptide bonds. This chain, like the constituent amino acids, can have a free amino group at one end (*N-terminal end*) and a free carboxyl at the other (*C-terminal end*). By convention, the polypeptide chain formula is always written with the *N*-terminal end at the left and the *C*-terminal end at the right:

$$H_2N{-}CHR^1{-}CONH{-}CHR^2{-}CONH{-}CHR^3{-}CONH{-}CHR^{n-1}{-}CONH{-}CHR^n{-}COOH$$

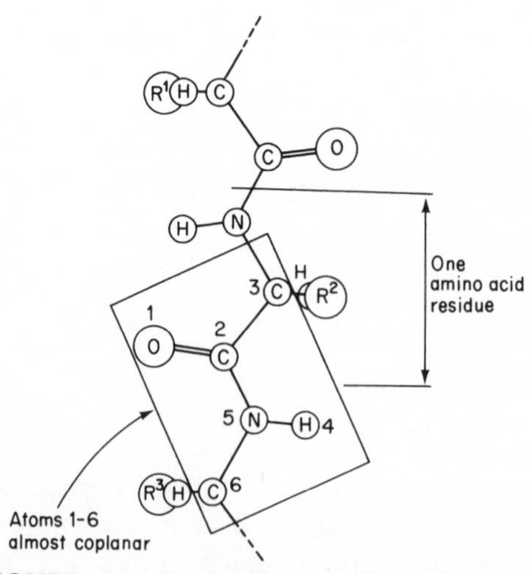

FIGURE 4.1. Spatial arrangement of atoms in an extended polypeptide chain. Atoms 1–6 are almost coplanar

Proteins contain many ionizable side-groups of amino acid residues (e.g. Asp, Glu, Lys, Arg, etc.) as well as free amino and carboxyl groups at the polypeptide chain-ends, and thus behave as polyelectrolytes. For each protein there is a pH at which the net charge on the macromolecule is zero, and this is known as the *isoelectric point* of the protein. The protein is usually least soluble at its isoelectric point—a simple explanation is that electrostatic repulsions between the molecules are at a minimum, so that the polypeptide chains can aggregate and precipitate from solution.

At other values of pH, the protein molecules carry a net charge, and hence will move in an electric field. This capability to undergo electrophoresis is now widely used in the purification of proteins (see Section 2.3.2, p. 21).

Ionic strength also affects the solubility of proteins. At low ionic strengths, increasing salt concentration lowers the effective activity of a polyelectrolyte, such as a protein, and so increases its solubility. At high ionic strengths, protein solubility decreases, possibly because of 'competition' for water of solvation between the salt and the protein. These phenomena are utilized in the purification of proteins by fractional precipitation (see Section 2.3.4.2, p. 30).

Proteins differ from one another in the actual arrangement, or sequence, of amino acids along the chain, and also in the way in which the amino acid side-chains interact. In fact the biological function of each protein is determined by the particular order of the amino acid sequence.

Some proteins, the conjugated proteins, contain atoms and molecules in addition to amino acids. The non-protein portion of the molecule is known as the *prosthetic group*, and may be a simple, single metal atom, or a complex, large organic pigment. Thus, *glycoproteins* are conjugated proteins containing carbohydrate, whilst haemoglobins contain a haem group as the non-protein part of the molecule.

4.2.3. Other bonds

A protein molecule may consist of a single polypeptide chain, or of several chains held together by various types of bonds. Such bonds may not only link together two separate chains, but also two parts of the *same* chain.

The most common covalent cross-chain link in proteins is the *disulphide bridge* of cystine (see Figure 4.2a).

The side-groups of many amino acids can ionize, thus giving a protein an electric charge. When positive and negative charges from different amino acid residues come close together, they can interact to give a strong attraction between chains. This type of *ionic bond* can occur between the amino group of the side-chain of lysine, or arginine, and the side-chain carboxyl group of glutamic, or aspartic acids, as both the basic and acidic groups would be ionized at physiological pH (see Figure 4.2b). However,

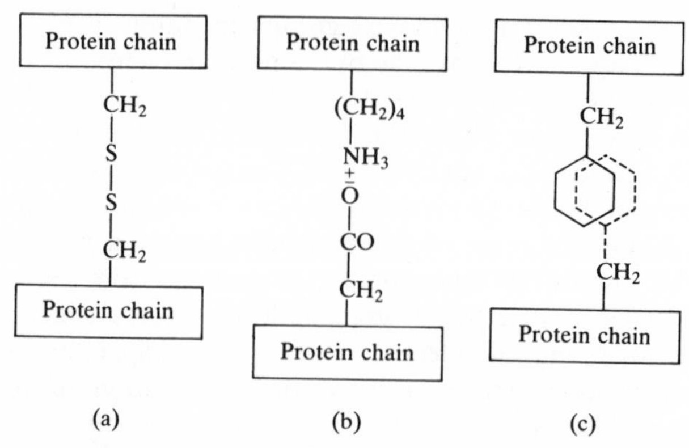

FIGURE 4.2. Schematic representation of bonds stabilizing the tertiary structure of proteins: (a) disulphide bridge; (b) ionic bonding; (c) hydrophobic interaction

this attraction between NH_3^+ and COO^- may be strong only in the absence of water, i.e. away from the surface of a water-soluble protein, for both ions can be strongly hydrated. Changes in pH may neutralize the charge on one ion involved in an ionic bond, effectively destroying the bond. Where these bonds are responsible for maintaining the structure of the protein molecule, a change in pH can alter the three-dimensional structure, and often destroy the biological activity. This alteration of structure and concomitant loss of activity is known as denaturation (see Section 4.4.3, p. 101).

Hydrogen bonding is also extremely important for maintaining the correct folding of polypeptide chains. Although single hydrogen bonds are weak, they can reinforce each other when many are present in one polymer molecule. Hydrogen bonding is possible between side-chains of amino acid residues, but in many proteins important hydrogen bonds are formed between the hydrogen adjacent to one peptide bond and the oxygen adjacent to another:

```
        R¹          H
        |           |
   —N—C—C—N—C—
     |    |    ||   |    |
     H   H   O   H   R²
     :               :
     :               :
R³  O          H   O
|     ||          |    ||
—C—C—N—C—C—N—
|           |    |          |
H          H   R⁴        H
```

Heating disrupts hydrogen bonds, and in cases where these are important for the three-dimensional structure, the protein may be denatured. High concentrations of hydrogen-bonding compounds, such as urea, which may compete with the hydrogen bonds of the protein, can also cause denaturation.

Association of hydrocarbon side-chains of amino acid residues from different parts of a polypeptide chain can also occur (see Figure 4.2c), and contribute greatly to the stability of the chain-folding. This *hydrophobic interaction* (also often called hydrophobic bonding) effectively excludes water from the vicinity of the hydrocarbon groups, and it is this exclusion of water which is believed to be the 'driving force' in these interactions. Water surrounding a hydrocarbon group has a more ordered structure than in the absence of such a grouping. Thus transfer of a hydrophobic group to an aqueous environment is accompanied by a negative entropy change, while a positive entropy change is usually found if the reverse process is carried out, i.e. entropy increases when water is excluded from close contact with hydrophobic groups. This tendency to increased entropy is one of the principal effects contributing to hydrophobic interactions; such interactions are believed to be very important in determining and stabilizing the 'native' conformations of polypeptide chains.

In water-soluble proteins, hydrophobic interactions take place in the interior of the molecule, leaving polar side-chains on the exterior. It is now thought that the lessening of hydrophobic interactions is more important in the denaturing action of urea than effects on hydrogen bonds.

In addition to the bonding discussed above, all atoms in close contact have a weak tendency to attract each other. These attractive forces, called *van der Waals' forces*, can be important in the interior of a protein molecule and can help to stabilize chain-folding.

4.3. SIZE AND COMPOSITION

4.3.1. Molecular size

The methods described in Chapter 3 can be used to measure the molecular weight of proteins. These can vary from several thousands to many millions, although the lower limit, i.e. the distinction between a protein and a polypeptide, is arbitrary. Many of the larger protein molecules are composed of subunits, which may separate under certain experimental conditions, and thus complicate the molecular weight determination. In general, proteins have a unique structure, and hence a uniform molecular weight and no molecular weight distribution (unless dissociation, or association, of subunits is actually taking place during measurements).

4.3.2. Amino acid composition

To study the amino acid composition, the biopolymer must first be hydrolysed to its constituent amino acids. Hydrolysis is commonly brought about with hydrochloric acid, but this reagent concurrently destroys the tryptophan residues present. Tryptophan is not destroyed, however, if a sulphonic acid, such as mercaptoethane sulphonic acid, is used instead of hydrochloric acid. Tryptophan may also be determined after a separate alkaline hydrolysis.

Cystine residues may be destroyed by acid hydrolysis, and so are often oxidized to cysteic acid prior to hydrolysis:

$$\begin{matrix} \vdots & & \vdots \\ CO & & CO \\ | & & | \\ CH-CH_2-S-S-CH_2- & & CH \\ | & & | \\ NH & & NH \\ \vdots & & \vdots \end{matrix} \xrightarrow[\text{acid}]{\text{performic}} 2\ \begin{matrix} \vdots \\ CO \\ | \\ CH-CH_2-SO_3H \\ | \\ NH \\ \vdots \end{matrix}$$

Cysteine residues may be treated in the same way. The number of cysteine residues in a polypeptide chain can be estimated independently, however, using Ellman's reagent [5,5′-dithio(bis)-2-nitrobenzoic acid]:

$$O_2N-C_6H_3(COO^-)-S-S-C_6H_3(COO^-)-NO_2 + HS-CH_2-\begin{matrix} \vdots \\ NH \\ | \\ CH \\ | \\ C=O \\ \vdots \end{matrix} \longrightarrow$$

Ellman's reagent (DTNB)

$$O_2N-C_6H_3(COO^-)-S-S-CH_2-\begin{matrix} \vdots \\ NH \\ | \\ CH \\ | \\ C=O \\ \vdots \end{matrix} + {}^-S-C_6H_3(COO^-)-NO_2 + H^+$$

Thiol anion

The thiol anion concentration can be estimated by light absorption, and hence the number of —SH groups in the protein can be deduced.

The amino acids obtained by protein hydrolysis must be separated, and are then quantitatively estimated using the ninhydrin reaction, or the fluorescamine reaction (see Section 4.2.1, p. 79). It is now usual for these procedures to be carried out automatically on an amino acid analyser, where the acids are separated on sulphonated polystyrene cation-exchange resins. Each acid is identified by the time taken to pass through the resins, and its amount estimated spectrophotometrically by the depth of colour given when the eluate reacts with ninhydrin, or by the fluorescence of the fluorescamine–amino-acid derivative. A second fluorescent derivative is now often used for amino acid determination, the product resulting from reaction of an amino acid with *o*-phthalaldehyde

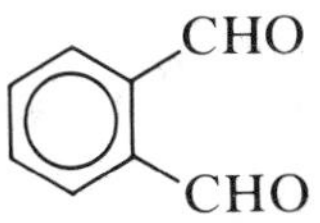

$$C_6H_4(CHO)_2$$

in a reducing medium. Using this method, amounts of amino acid less than 10^{-10} moles can be measured accurately. Unfortunately, procedures involving fluorescamine and *o*-phthalaldehyde both suffer from the drawback that imino acids such as proline cannot be determined directly.

Amino acid composition alone cannot give much information about the molecular structure of a protein. For this purpose, the *amino-acid sequence* of the polypeptide chain is much more important.

4.4. STRUCTURE

We shall discuss first the *primary structure*, i.e. the amino acid sequence in the polypeptide chain; then the *secondary structure*, i.e. the α-helix and β-sheet content of the molecules; then the *tertiary structure*, i.e. the overall folding of the polypeptide chains; and finally the *quaternary structure*, i.e. the arrangement of subunits in the molecules.

4.4.1. Primary structure—the amino acid sequence

Sequence determination is a complex operation involving many stages.

4.4.1.1. Determination of the number of chains

First, the number of polypeptide chains in the protein molecule must be determined. One way of achieving this is to use fluorodinitrobenzene, or dansyl chloride, to form a modified protein containing dinitrophenyl, or dansyl, derivatives of the amino acid at the *N*-terminal end of each chain of the molecule, and also of the lysine side-chains. The modified protein is hydrolysed by acid, and the liberated dinitrophenyl, or dansyl, amino acids are separated, and identified by paper chromatography or electrophoresis. The number of different amino acids (other than lysine) found as the modified derivative gives a good indication of the number of polypeptide chains in the protein molecule. In some proteins, no derivative of the *N*-terminal acid is formed, indicating that the amino group is already modified in the native protein. One of the most common of such modifications is the presence of an acetyl group attached to the nitrogen of the *N*-terminal acid.

Additional evidence concerning the number of chains in the protein molecule may be obtained by studying the amino acids at the *C*-terminal end of the chain or chains. One method involves the use of hydrazinolysis:

$$H_2N{-}CHR^1{-}CO \ldots . NH{-}CHR^2{-}CO{-}NH{-}CHR^3{-}COOH + H_2N{-}NH_2 \longrightarrow$$

$$H_2N{-}CHR^1{-}CO{-}NH{-}NH_2 + \ldots + H_2N{-}CHR^2{-}CO{-}NH{-}NH$$

$$+ H_2N{-}CHR^3{-}COO$$

The *C*-terminal residue produces a free amino acid, whilst the other residues give the hydrazides. The free amino acid can be separated from the mixture by chromatography on a strong cation-exchange resin, as the basic hydrazides are strongly bound on the resin whilst the free acid is readily eluted. Identification of the free acid(s) produced by hydrazinolysis will again give a measure of the number of different chains in the polymer molecule.

Another method which may be employed to study the *C*-terminal acids of a protein involves the enzyme carboxypeptidase (see Section 4.5.2.5, p. 130), which catalyses the stepwise hydrolysis of proteins from the *C*-terminal end of chains. The reaction must be studied kinetically, for the acid second from the *C*-terminal end will begin to be released just after the terminal acid has been removed. For example, if a polypeptide chain ending in *alanine—leucine—phenylalanine* is considered, before all the phenylalanine has been released by the enzyme, leucine, and then alanine will begin to appear (see Figure 4.3). Consequently, this method not only gives information about the *C*-terminal acid, but can also yield the partial sequence at that end of the polypeptide chain.

If two chains are present with different *C*-terminal acids, the two acids

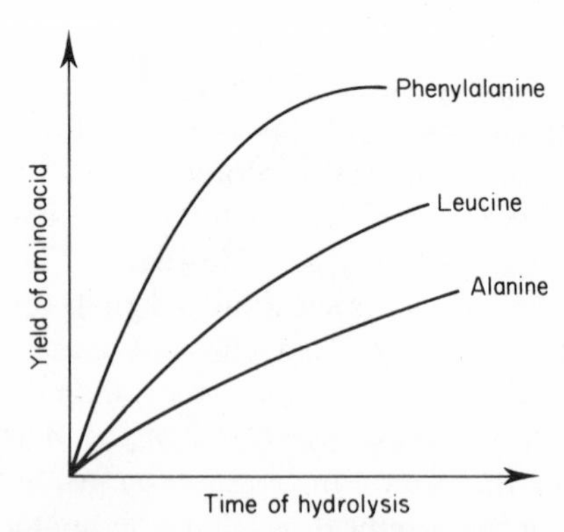

FIGURE 4.3. Time-course of release of amino acids by carboxypeptidase from a protein ending in the sequence —alanine—leucine—phenylalanine

will be produced simultaneously. It has to be noted, however, that the enzyme will not release lysine, arginine, or proline residues, and so the method cannot be used invariably.

Where a protein molecule contains two or more identical chains, the number of chains making up the molecule can be obtained from molecular weight measurements, provided the weights of the whole molecule and single chains can be determined.

4.4.1.2. Separation of chains

When a protein molecule is shown to contain more than one polypeptide chain, these must be separated before the amino acid sequence of each chain can be determined. Chains held together by non-covalent interactions may be dissociated by acid, base, high salt concentration, or denaturing agents such as 8 M urea. But, if covalent disulphide bridges are present, they must be broken by either oxidation (see Section 4.3.2), or reduction and modification:

$$\begin{array}{ccccc}
\cdots\!-\!NH\!-\!CH\!-\!CO\!-\!\cdots & & \cdots\!-\!NH\!-\!CH\!-\!CO\!-\!\cdots & & \cdots\!-\!NH\!-\!CH\!-\!CO\!-\!\cdots \\
| & & | & & | \\
CH_2 & & CH_2 & & CH_2 \\
| & \text{mercaptoethanol} & | & \text{iodoacetate} & | \\
S & HS.(CH_2)_2.OH & SH & ICH_2.COOH & S.CH_2.COOH \\
| & \xrightarrow[\text{reduction}]{} & & \xrightarrow[\text{modification}]{} & \\
S & & SH & & S.CH_2.COOH \\
| & & | & & | \\
CH_2 & & CH_2 & & CH_2 \\
| & & | & & | \\
\cdots\!-\!NH\!-\!CH\!-\!CO\!-\!\cdots & & \cdots\!-\!NH\!-\!CH\!-\!CO\!-\!\cdots & & \cdots\!-\!NH\!-\!CH\!-\!CO\!-\!\cdots
\end{array}$$

The sulphydryl groups produced by reduction must be protected before further work can be carried out, for such groups are extremely reactive, and may be oxidized by atmospheric oxygen to form new disulphide bridges. Reduction can also be brought about by using dithiothreitol or dithioerythritol [$CH_2.SH.(CH_2OH)_2.CH_2.SH$]. Acrylonitrile may be employed to modify the resultant —SH groups, instead of iodoacetic acid:

$$\cdots-CH_2-SH + \underset{\text{Acrylonitrile}}{CH_2{=}CHCN} \longrightarrow \cdots-CH_2-S-CH_2-CH_2-CN$$

'Native' —SH groups in the protein are usually modified as above prior to sequence determination in order to prevent accidental oxidation.

The single polypeptide chains of the protein molecule formed in this way may then be separated by gel-permeation chromatography if they differ in size, or by ion-exchange chromatography if they differ in electrical charge.

Non-protein prosthetic groups have also to be removed before amino acid sequencing is attempted.

4.4.1.3. Amino acid sequence of each chain

The amino acid sequence of each chain in turn is then studied. At the present time, the most commonly used method of sequence determination involves partial hydrolysis of the polypeptide chain to give shorter peptides, which are then separated, and studied individually. Different methods of hydrolysis are used to give scission at different points in the chain, and hence to obtain different peptides. The protein may be partially hydrolysed using dilute acid. A more specific reaction involves the use of cyanogen bromide, which cleaves the chain at methionine residues:

$$\cdots\!-\!CO\!-\!NH\!-\!\underset{\displaystyle|\atop{\displaystyle CH_2 \atop {\displaystyle | \atop {\displaystyle CH_2 \atop {\displaystyle | \atop {\displaystyle S \atop {\displaystyle | \atop \displaystyle CH_3}}}}}}}{CH}\!-\!CO\!-\!NH\!-\!CHR\!-\!CO\!-\!\cdots \xrightarrow{BrCN} \cdots\!-\!CO\!-\!NH\!-\!\underset{\displaystyle|\atop{\displaystyle CH_2 \atop {\displaystyle | \atop {\displaystyle CH_2 \atop {\displaystyle | \atop {\displaystyle {}^{+}S\!-\!C\!\equiv\!N \atop {\displaystyle | \atop \displaystyle CH_3}}}}}}}{CH}\!-\!CO\!-\!NH\!-\!CHR\!-\!CO\!-\!\cdots + Br^-$$

$$\longrightarrow \cdots\!-\!CH\!-\!NH\!-\!CH\!-\!C\!=\!\overset{+}{N}H\!-\!CHR\!-\!CO\!-\!\cdots \text{ (CH–C ring: } H_2C\!-\!CH_2\!-\!O\text{)} + CH_3SCN$$

$$\longrightarrow \cdots\!-\!CO\!-\!NH\!-\!CH\!-\!C\!=\!O \text{ (ring: } H_2C\!-\!CH_2\!-\!O\text{)} + H_3N^+\!-\!CHR\!-\!CO\!-\!\cdots$$

Homoserine lactone

Other chemical modifications are now available for splitting polypeptide chains at specific residues, or sequences of residues.

Another method of partial cleavage involves proteolytic enzymes (see Section 4.5.2.4, p. 126), such as trypsin and chymotrypsin, which hydrolyse proteins at specific points; i.e. trypsin cleaves the chain almost exclusively at the peptide bonds involving the carboxyl groups of the basic amino acids lysine and arginine, whilst chymotrypsin, on the other hand, hydrolyses most rapidly the peptide bonds associated with the carboxyl groups of tryptophan, tyrosine, phenylalanine, and leucine. Trypsin hydrolysis can be confined to bonds adjacent to arginine residues by a prior modification of lysine side-chains, e.g.

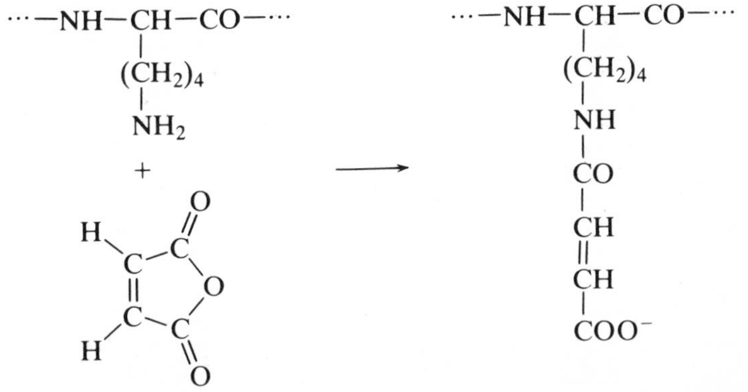

$$\cdots\text{—NH—CH—CO—}\cdots \; (\text{side chain } (CH_2)_4\text{—}NH_2) \; + \; \text{Maleic anhydride} \longrightarrow \cdots\text{—NH—CH—CO—}\cdots \; (\text{side chain } (CH_2)_4\text{—NH—CO—CH=CH—}COO^-)$$

Maleic anhydride

These proteolytic enzymes cannot act directly on the native protein if the molecule is too tightly folded, and, in such cases, the protein must be denatured, and its three-dimensional structure destroyed, before enzymic hydrolysis can take place. This denaturation may be brought about by heat, or 8M urea.

After partial hydrolysis, the resultant peptides may be separated by ion-exchange or paper chromatography, or electrophoresis. Each pure peptide is then examined for amino acid composition and sequence. (If the peptide is too large for a complete sequence determination, it must be broken down further by hydrolysis, and the sequences in the fragments determined).

The position of the amides, glutamine and asparagine, *must* be determined by enzymic hydrolysis, for any procedure involving acid treatment converts them into glutamic and aspartic acids, respectively.

Dinitrofluorobenzene will give information about the amino acid at the *N*-terminal end of the chain (see Section 4.2.1, p. 79), and, indeed, this method was used during the first determination of the complete amino acid sequence of a protein, i.e. that of insulin. In 1958, Sanger received the Nobel Prize for this pioneering work in amino acid sequencing. Currently, dansyl chloride (see Section 4.2.1, p. 80) is the reagent which is used more often as the product is more easily detected.

The Edman procedure using phenylisothiocyanate (see Section 4.2.1, p. 80) has proved to be a powerful tool for sequence determination, and this method is described in detail. Under alkaline conditions, phenylisothiocyanate reacts with the *N*-terminal amino acid of a polypeptide chain. This modified amino acid may be split from the chain by acid, leaving the remainder of the polypeptide chain intact. The resultant thiazolinone may be extracted, and in aqueous acid this compound rearranges to form the phenylthiohydantoin (PTH), which can be identified by high-pressure liquid or thin-layer chromatography:

$$C_6H_5{-}N{=}C{=}S + H_2N{-}CHR^1{-}CO{-}NH{-}CHR^2{-}CO \ldots NH{-}CHR^n{-}COO^-$$

$$\downarrow OH^-$$

$$C_6H_5{-}NH{-}C({=}S){-}HN{-}CHR^1{-}CO{-}NH{-}CHR^2{-}CO{-}\ldots{-}NH{-}CHR^n{-}COO^-$$

$$\downarrow H_3O^+$$

$$\text{(thiazolinone: } C_6H_5{-}NH{-}C{=}N{-}CH(R^1){-}CO{-}S \text{ ring)} + H_3N^+{-}CHR^2{-}CO{-}\ldots{-}NH{-}CHR^n{-}COOH$$

A thiazolinone

Then

$$\text{thiazolinone} \xrightarrow[H_2O]{H_3O^+} \text{(phenylthiohydantoin: } C_6H_5{-}N{-}C({=}S){-}NH{-}CH(R^1){-}C({=}O) \text{ ring)}$$

A phenylthiohydantoin

The remaining polypeptide chain, now shorter by one amino acid, may be subjected again to the same treatment, and so sequential degradation from the *N*-terminal end may be carried out.

This method is widely used to determine the sequences of peptides, and it has been possible by using a '*protein sequenator*'—an instrument which enables the reactions to be carried out automatically—to determine sequences of more than 50 amino acids in a polypeptide chain.

In some cases, the Edman procedure is carried out on a peptide attached to a solid support, which facilitates washing the peptide free of other chemicals between steps of the Edman degradation. A peptide containing a lysine residue can be coupled to a support, such as an aminopolystyrene, using, for example, *p*-phenylene diisothiocyanate:

$$\text{Polystyrene}-C_6H_4-NH_2 + S{=}C{=}N-C_6H_4-N{=}C{=}S + H_2N-X$$

Aminopolystyrene resin — *p*-Phenylene diisothiocyanate — Peptide

$$\downarrow$$

$$\text{Polystyrene}-C_6H_4-NH-\overset{\overset{\displaystyle S}{\|}}{C}-NH-C_6H_4-NH-\overset{\overset{\displaystyle S}{\|}}{C}-NH-X$$

The peptide becomes attached to the resin both at the side-chain of lysine and at the amino group of its *N*-terminal end. In trifluoroacetic acid, however, the amino acid residue at the *N*-terminal end of the peptide is cleaved off the resin and the peptide, leaving the peptide attached to the resin by the side-chain of lysine only. The peptide can then be subjected to several rounds of Edman degradation until the lysine residue is released as the PTH derivative. This method, therefore, is useful only for peptides containing lysine near the *C*-terminal end. (If lysine is at the *N*-terminal end initially, trifluoroacetic acid would release the whole peptide from the resin.)

The formation of dansyl derivatives can also be used in conjunction with the Edman procedure, for methods of detection of dansylated amino acids are more sensitive than methods for phenylthiohydantoins. Consider the peptide Ala—Asp—Lys—Thr—Val. Initial dansylation and hydrolysis gives dansyl-Ala. Thus alanine is the *N*-terminal residue of the peptide. If, however, the peptide is subjected to one round of the Edman procedure, it becomes shortened by one residue to Asp—Lys—Thr—Val. Dansylation and hydrolysis of a sample of the shortened peptide gives dansyl-Asp. Thus it can be deduced that the peptide under study begins with the sequence Ala—Asp—. Subjection of the remaining tetrapeptide to a round of Edman degradation, and dansylation of the resulting tripeptide gives dansyl-Lys. Thus the sequence is Ala—Asp—Lys—. If this process is continued, the complete sequence of the pentapeptide can be obtained from the order of appearances of dansyl derivatives, without the need to identify the PTH derivatives released at each round of the Edman procedure.

For small peptides (up to 12 residues in some cases), the sequence may be studied by mass spectrometry. A peptide derivative is bombarded with electrons, and loses one or more of its own electrons to become positively charged. This positive ion fragments, most often at the peptide bonds, so that positive ions are obtained corresponding to the whole peptide, the whole peptide minus the *C*-terminal residue, the whole peptide minus two residues (the *C*-terminal and adjacent residues), etc. These ions can be separated on the basis of their mass/charge ratio, and the mass of each fragment can be determined. Then the amino acid residue at the *C*-terminal end of the original peptide can be identified from the difference in mass of the ions corresponding to the whole peptide and the whole peptide minus the terminal residue. (The amino acids can be distinguished from each other on the basis of mass, except for leucine and isoleucine.) Similarly the acid penultimate to the *C*-terminal end can be identified from the mass difference between the ions of the peptide minus one residue and the peptide minus two residues. In favourable cases the whole sequence of the peptide can be obtained from one mass spectrum.

A volatile peptide derivative must be used for this procedure, as the initial electron bombardment takes place in the gas phase. The amino group at the *N*-terminal end of the peptide and those of lysine side-chains are usually acetylated, whilst carboxyl groups are esterified (as methyl, ethyl or benzyl esters). Derivatives of all polar side-chains must also be formed to increase volatility. Lastly, permethylation of a peptide further increases volatility. This reaction is carried out using methyl iodide in the presence of silver oxide, and gives methylation at the peptide bond:

$$\cdots\!-\!CHR^1\!-\!\underset{\underset{O}{\|}}{C}\!-\!\underset{\underset{H}{|}}{N}\!-\!CHR^2\!-\!\cdots \xrightarrow[Ag_2O]{CH_3I} \cdots\!-\!CHR^1\!-\!\underset{\underset{O}{\|}}{C}\!-\!\underset{\underset{CH_3}{|}}{N}\!-\!CHR^2\!-\!\cdots$$

Only peptide bonds involving imino acids like proline are unaffected.

When the sequence in each peptide is known, an attempt is made to deduce the sequence of the whole chain by studying overlapping peptides, i.e. sequences which are common to several peptides.

Series of peptides are prepared by different methods of cleavage of the polypeptide chain. After determination of the amino acid sequence of each peptide, sequences from one series of peptides are compared to sequences from another in an attempt to find sequences common to both. Thus in the example shown in Figure 4.4, if peptides 1 and 2 both contain sequences found in peptide 4, it can be concluded that 1 and 2 were close, or adjacent, to each other in the original polypeptide. This method is continued until the positions of all the peptides in the original polypeptide have been located. The procedure can be difficult and time-consuming, and, for a long polypeptide chain, is usually the most difficult task in sequence determination.

We shall see in Chapter 5 that the amino acid sequence for a protein is controlled by a stretch of deoxyribonucleic acid (DNA) in the organism concerned. Methods for determining the sequence of monomers in DNA

Fragments:

Tryptic peptides
(1) Val—Leu—Ser—Pro—Ala—Asp—Lys
(2) Thr—Asn—Val—Lys
(3) Ala—Ala—Trp—Gly—Lys

Chymotryptic peptide (4) Ser—Pro—Ala—Asp—Lys—Thr—Asn—Val—Lys—Ala—Ala—Trp

Peptides arranged to show overlap:

(4) Ser—Pro—Ala—Asp—Lys—Thr—Asn—Val—Lys—Ala—Ala—Trp

Val—Leu—Ser—Pro—Ala—Asp—Lys (1) Thr—Asn—Val—Lys (2) Ala—Ala—Trp—Gly—Lys (3)

Complete sequence:

Val—Leu—Ser—Pro—Ala—Asp—Lys—Thr—Asn—Val—Lys—Ala—Ala—Trp—Gly—Lys

FIGURE 4.4. Fitting peptide fragments together to give the sequence of a larger section of α-chain of human haemoglobin

(Section 5.4.1.1, p. 180) have now advanced to such a stage that, in some cases, it is easier to find an amino acid sequence from examination of the base sequence of the corresponding DNA than to go through the processes described above.

4.4.1.4. *Location of disulphide bonds and prosthetic groups*

After the determination of the amino acid sequence of each separated chain of a protein, the position of disulphide bridges in the original protein must be investigated. The original protein is hydrolysed enzymically to small peptides, leaving the disulphide bridges intact. (Hydrolysis of proteins with strong acid or alkali usually gives disulphide interchange, e.g.:

$$\text{Protein}^1\text{—Cys—S—S—Cys—Protein}^2 + \text{Protein}^3\text{—Cys—S—S—Cys—Protein}^4 \longrightarrow$$

$$\text{Protein}^1\text{—Cys—S—S—Cys—Protein}^3 + \text{Protein}^2\text{—Cys—S—S—Cys—Protein}^4$$

and is therefore avoided.)

The peptides containing the cystine residues may be separated chromatographically, and each fragment is then oxidized to cysteic-acid-containing peptides, which are separated and characterized:

$$\begin{array}{ccccc}
\cdots\text{—R}^1\text{—R}^2\text{—R}^3\text{—Cys—R}^4\text{—R}^5\text{—}\cdots & & \text{R}^2\text{—R}^3\text{—Cys—R}^4 & & \text{R}^2\text{—R}^3\text{—Cys—R}^4 \\
\text{S} & \xrightarrow{\text{hydrolysis}} & \text{S} & \xrightarrow{\text{oxidation}} & SO_3H \\
 & & & & + \\
\text{S} & & \text{S} & & SO_3H \\
\cdots\text{—R}^6\text{—R}^7\text{—Cys—R}^8\text{—R}^9\text{—}\cdots & & \text{Cys—R}^8\text{—R}^9 & & \text{Cys—R}^8\text{—R}^9
\end{array}$$

+ other peptides

As the amino acid sequence in the immediate vicinity of each half of a disulphide bridge is already known, it is then possible to identify the two half-cystine residues which are associated with each other.

An elegant procedure, the *diagonal method*, can be used to obtain the cysteic-acid-containing peptides. Peptides, obtained by enzymic hydrolysis of a polypeptide chain, are subjected to paper electrophoresis. The paper is exposed to performic acid vapour, which oxidizes and cleaves the disulphide bonds, forming cysteic-acid-containing peptides on the paper. Electrophoresis is then carried out in a direction at right-angles to the first electrophoretic run. Peptides unmodified by the performic acid appear on a diagonal line across the paper, while the cysteic-acid-containing peptides are found at other positions (see Figure 4.5). These peptides may be eluted from the paper and their amino acid composition determined.

Figure 4.6 shows the complete amino acid sequence of *bovine insulin*, the first protein of which the primary structure was determined.

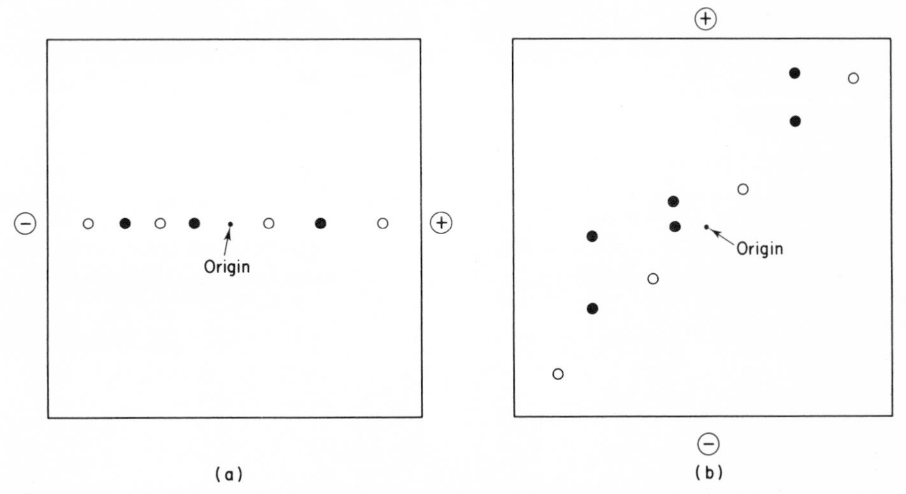

FIGURE 4.5. Diagonal electrophoresis of peptides to locate disulphide bonds. (a) After first electrophoresis: ● disulphide-containing peptides, ○ other peptides. (b) After second electrophoresis: ● cysteic-acid-containing peptides, ○ other peptides. Note that after performic acid treatment each disulphide-containing peptide gives *two* cysteic-acid-containing peptides

In conjugated proteins, the mode of linkage of the protein to the prosthetic group must also be determined. If the prosthetic group is not covalently bound, X-ray analysis of the solid protein is one means of studying the interaction between protein and such a prosthetic group. Residues close to the prosthetic group can be studied by using reagents which cross-link the prosthetic group to neighbouring amino acid residues. Identification of the modified amino acids and their positions in the primary sequence gives an indication of which sections of the polypeptide chain come together to form the environment of the prosthetic group.

Where a covalent linkage exists between the protein and non-protein parts of the molecule, partial hydrolysis of the protein will give fragments in which the prosthetic group is attached to a small number of amino acids. Studies of these fragments will then give information about the linkage of the prosthetic group to polypeptide in the original protein.

4.4.2. Secondary structure—hydrogen bonds and chain-folding

Polypeptide chains of proteins are not found in living organisms as long, linear structures, but are folded or interconnected in the manner required for the correct functioning of that protein. Folding of the chain is stabilized by the disulphide bridges, hydrophobic interactions, van der Waals' forces, and ionic and hydrogen bonds (see Section 4.2.3, p. 83).

Several types of ordered structure of the polypeptide chains are stabilized by hydrogen bonding between the C=O and N—H groups of

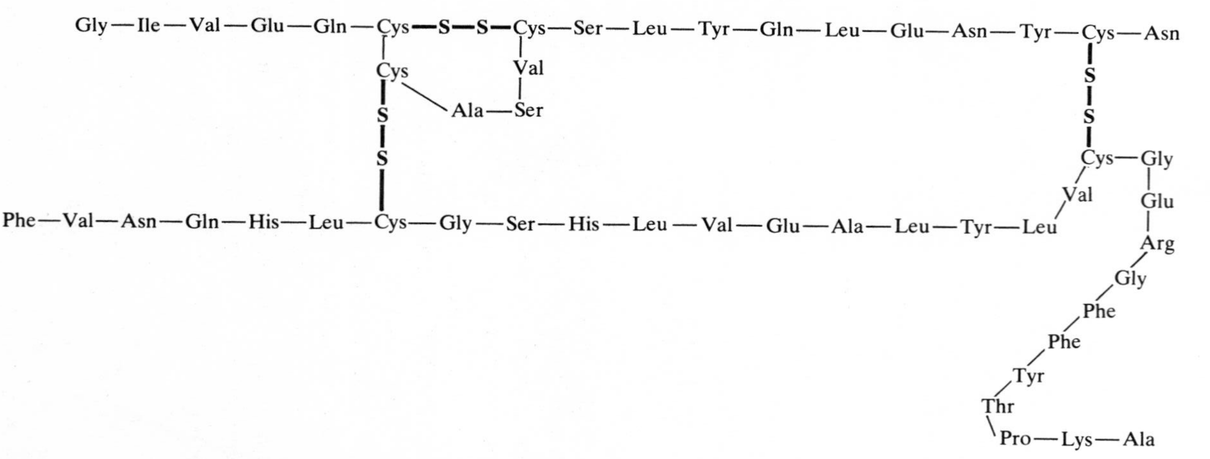

FIGURE 4.6. Amino acid sequence of bovine insulin

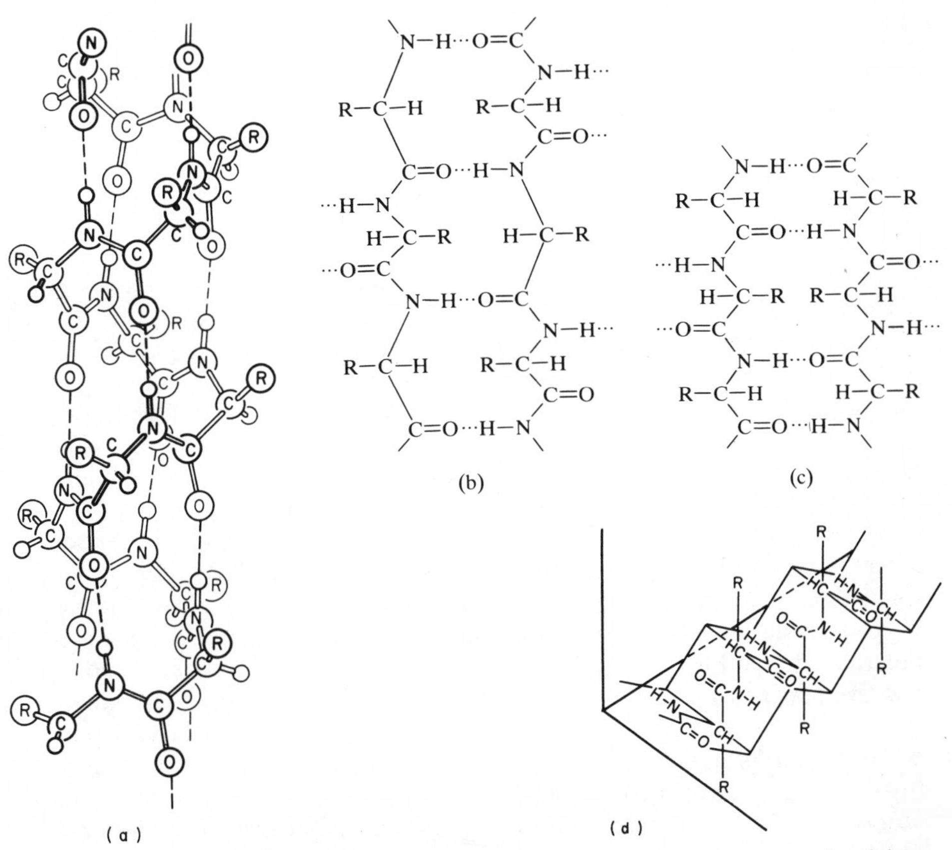

FIGURE 4.7. Representation of secondary protein structure involving chain-folding and hydrogen bonds: (a) the α-helix of Corey and Pauling (from B. W. Low and J. T. Edsall in D. E. Green (Ed.) *Currents in Biochemical Research*, Wiley–Interscience, New York, 1956), (b) a parallel pleated sheet with polypeptide chains running in the same direction; (c) an antiparallel pleated sheet with polypeptide chains running in opposite directions; (d) the polypeptide chain showing the 'pleating'

the peptide link. Such structures constitute the secondary structure of a protein.

Two types of secondary structure have received much attention—the α-helix and the β-sheet. Other types of helix, and structures such as 'hairpin loops', are now considered important in polypeptide chain-folding.

The *α-helix* was proposed by Pauling and Corey (Figure 4.7) and has been confirmed by X-ray diffraction studies. (Pauling received the Nobel Prize in 1954 for his work on the forces stabilizing protein structure.) The right-handed helix is favoured for L-amino acids, and is shown in Figure 4.7a; in a left-handed helix the chain would spiral in the opposite direction. However, fewer H-bonds can be formed in a left-handed helix, and the structure has been shown by theoretical calculations of interaction energies

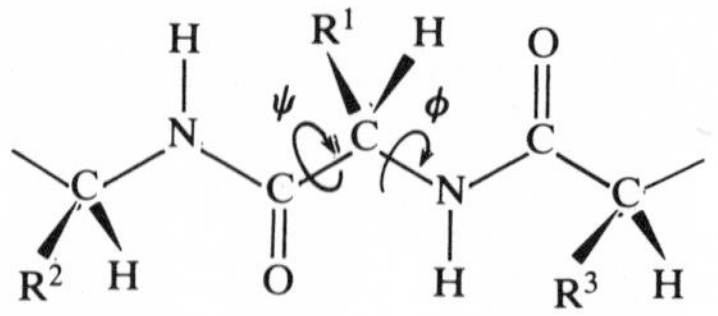

FIGURE 4.8. Two amino acid residues in the extended conformation, where $\psi = \phi = 180°$

to be less stable. In the right-handed helix, there are 3.7 amino acids in each complete turn.

A second important type of ordered arrangement is the *pleated*, or *β-sheet*, where the polypeptide chains take a more extended form than that in the α-helix (Figure 4.7b, c, and d). Separate chains lie parallel to one another, running in the same or opposite directions; parts of the same chain may also fold to give a pleated sheet. The presence of such structures in proteins has been confirmed by X-ray crystallography.

In general, we can think of each amino acid residue in a polypeptide chain as having two single bonds about which rotation can occur. If the torsion angle about both of these bonds is known for every residue in the protein molecule, then the conformation of the complete chain is determined (see Figure 4.8). These angles are given the symbols ϕ and ψ, and are assigned the value 180° for the fully extended chain. (The angles may, in theory, vary from $-180°$ to $+180°$.) Both ϕ and ψ can vary, and so many conformations are possible. By steric considerations, however, many of these conformations can be eliminated because they would bring atoms or bulky side-chains too close together. Right-handed helices are obtained if both angles are approximately $-60°$ (for the α-helix $\phi = -57°$, $\psi = -47°$), whilst pleated sheets result when $\phi = -120°$ to $-140°$, $\psi = 110$ to $135°$.

4.4.3. Tertiary structure—the overall folding of the polypeptide chain

Proteins do not have structures which are entirely α-helical or pleated, but the molecules contain short helical regions, or sheets, interspersed with apparently randomly coiled areas. The resultant structure is often highly folded and compact.

At present, X-ray diffraction is the only method available which can give complete information on the tertiary structure of proteins, and many solid proteins have been investigated. Figure 4.9 shows the complex folding of *myoglobin*, a conjugated protein molecule; the prosthetic group can be seen in the upper part of the molecule. The straight 'tubes' in the diagram coincide with α-helical sections of the polypeptide chain. Between the stretches of α-helix are regions of essentially random folding.

The first protein structure to be determined by X-ray diffraction methods was that of myoglobin, and J. C. Kendrew received the Nobel

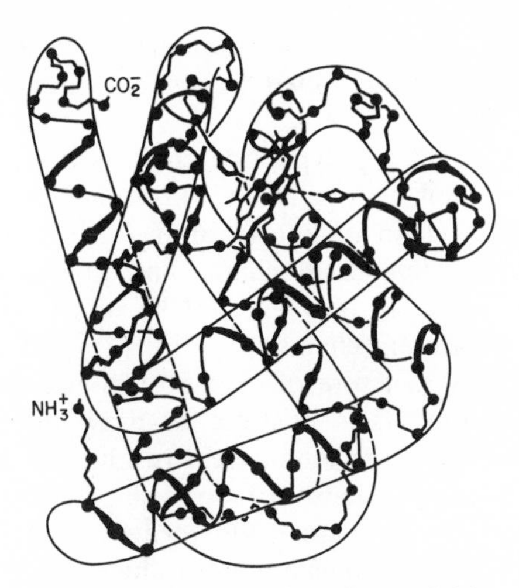

FIGURE 4.9. A model of myoglobin showing the complex folding of the polypeptide chain. Reproduced by permission of M. F. Perutz

Prize in 1962 for this work. Myoglobin is now considered 'unusual', for most protein molecules contain much less α-helix and more apparently random folding. Other tertiary structures are discussed later in the chapter (see, for example, Sections 4.5.2 and 4.5.4, pp. 114 and 140).

The intact, tertiary structure controls the biological function of any protein, for usually all biological activity is lost when the bonds maintaining the tertiary structure are broken, and the protein is *denatured*. Sometimes, removal of the denaturing agent results in the original tertiary structure of a protein being recovered and the return of biological activity, and the protein is then said to be *renatured*. Such studies have shown that the primary structure, i.e. the amino acid sequence of a protein, contains all the information necessary to specify the correct folding of the chain for biological activity.

Difficulties arise when sulphydryl groups are involved in the reformation of disulphide bridges, for these bridges may be formed at random, e.g. if a protein contains 4 sulphydryl groups, three possible pairs of bridges can result. In Nature, proteins are often synthesized as longer precursors which direct formation of the correct disulphide bonds (cf. Section 4.5.3.1, p. 135).

In general, the polypeptide chains of the water-soluble proteins fold in such a way that the molecules are compact with the hydrophobic side-chains in the interior of the molecule, and polar side-chains on, or near, the molecular surface. Furthermore, an α-helix is interrupted whenever a proline residue appears in a chain, for in this residue the δ-carbon of the pyrrolidine ring takes the place of the hydrogen atom required for bonding in the helix. Many proteins have 'domains'—substructures which are regions of polypeptide chain folded to bring hydrophobic groups to the interior.

Although the folding of a polypeptide chain is determined by its amino acid sequence, it is not yet possible to predict the three-dimensional structure of a peptide chain of known amino acid sequence. Indeed, the tertiary structure of haemoglobin chains and myoglobins from different animals are similar, although only about 10% of the amino acids occur in the same position in different chains.

Now that the tertiary structure of over 50 proteins has been determined by X-ray crystallography, it is possible to draw up tables showing the frequency of appearance of an amino acid residue in, or at the end of, an α-helix, in a β-structure, or a hairpin loop. Thus glutamic acid, methionine, alanine, and leucine residues are often located in helices, whilst valine, isoleucine, tyrosine, and phenylalanine residues are frequently found in β-sheets. Glycine, proline, and aspartic acid residues, on the other hand, are commonly part of hairpin loops.

Using these tables, workers are now attempting to predict locations of secondary structure in a protein, from knowledge of amino acid sequences. Such predictions have considerable success, with up to 80% of helices and β-structures being located correctly.

4.4.4. Quaternary structure—the arrangement of subunits

A higher degree of structural order can occur when the protein molecule is made up of a number of non-covalently linked polypeptide chains. The arrangement of such subunits in the molecule is known as the quaternary structure, and the subunits themselves are often known as 'monomers'. All the 'monomers' of an oligomeric protein may be identical, or there may be more than one type. A 'monomer' can be a single polypeptide chain, or several chains joined by disulphide bridges.

A protein with quaternary structure may be dissociated into subunits by denaturing agents such as urea, guanidine hydrochloride, sodium dodecyl sulphate, or by high salt concentrations. Most proteins of this type contain 2–12 subunits (a few have 3, but no other odd numbers of 'monomers' are found), but multienzyme complexes (see Section 4.5.2, p. 115) may contain over 50 polypeptide chains.

At present, quaternary structure can only be fully investigated by X-ray diffraction techniques. However, much information concerning adjacent groups on different chains can be obtained by the use of chemical cross-linking agents. Such reagents can form covalent bonds between neighbouring groups in the protein molecule whether on the same or different chains; these modified groups can then be identified and conclusions drawn as to which parts of different subunits are close together in the quaternary structure. In addition, if protein subunits are large

FIGURE 4.10 A representation of the haemoglobin molecule showing the folding of the main chain only and not the positions of the side-chains of the amino acid residues. The molecule is built of four subunits, two identical α-chains and two identical β-chains. Black rectangles represent haem groups, each with an iron atom at its centre. Adapted with permission from R. E. Dickerson and I. Geis, *The Structure and Action of Proteins*, Benjamin/Cummings, Menlo Park, Calif. 1969, p. 56. Illustration copyright by Dickerson and Geis

enough, they may be examined by electron microscopy and the quaternary structure determined from electron micrographs.

Figure 4.10 shows the main features of a molecule of *haemoglobin*, a more complex, conjugated protein related to myoglobin. The molecule consists of two pairs of polypeptide chains, each folded in a manner similar to that of myoglobin, and fitted together to form an almost spherical molecule. (Perutz and Kendrew shared the 1962 Nobel Prize for their work on the use of X-ray diffraction to determine this protein structure.)

Many different proteins are known to contain subunits, but the complete three-dimensional structures of most such proteins are still unknown.

4.5 FUNCTION

Proteins have many functions—they act as transport and storage materials, enzymes, hormones, and antibodies, they form the structural elements of living organisms, and they carry out a number of other functions. In this section, a few of the macromolecules, which are representative of each of these functions, will be discussed in detail.

We being with myoglobin and haemoglobin because, historically, these were the first proteins whose detailed molecular architecture was revealed by X-ray crystallography. Some of the interrelations between structure and function learned from these biopolymers are helpful in understanding the interdependence of these properties for other proteins; this is an area of understanding in which much progress is currently being made.

4.5.1. Transport and storage proteins

Some members of this group of proteins have been studied in detail, and indeed, as mentioned above, from a knowledge of the molecular architecture of myoglobin and haemoglobin, general conclusions about protein structure have been drawn.

Myoglobin and *haemoglobin* are responsible for the short-term storage and transportation of oxygen in the body. They are conjugated proteins, with a complex iron–porphyrin prosthetic group, the *haem group* (see Figure 4.11).

The porphyrin ring is flat and has many delocalized π-electrons, whilst the iron—in the iron(II) state under normal physiological conditions—is octahedrally coordinated. The six ligands, arranged octahedrally round the iron atom, are the four nitrogen atoms of the porphyrin ring, a ring nitrogen of a histidine of the polypeptide chain, and the oxygen molecule itself; each ligand donates an electron pair to the vacant orbitals of the iron. In the deoxy forms of myoglobin and haemoglobin, the sixth coordination position is probably occupied by a water molecule. The electronic configuration of the haem group gives the red colour so familiar

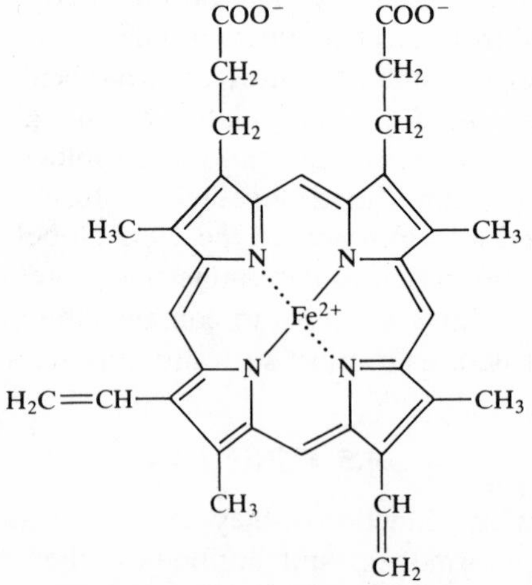

FIGURE 4.11. A haem group: haem b of myoglobin and haemoglobin

in blood; the oxy form of haemoglobin is the brilliant red characteristic of arterial blood, whilst the deoxy form gives venous blood its purplish colour.

The haem group is bound to the protein by coordination with the iron atom and by ionic and hydrophobic bonding; it lies in a hydrophobic pocket in the protein structure with the vinyl side-chains innermost, whilst the propionic acid side-chains project out to the surface of the molecule to form salt bridges with basic amino acid residues. One of the main functions of the protein part of the molecule is to provide the hydrophobic surroundings for the iron atom to help maintain it in the iron(II) state necessary for reversible oxygen binding. [The iron of a free haem group is readily oxidized to iron(III); oxidized myoglobin and haemoglobin containing iron(III) cannot bind oxygen correctly.]

4.5.1.1. Myoglobin.

Myoglobin is the simpler of the two proteins and is found in skeletal muscles, where it receives oxygen from the haemoglobin of the blood and acts as a temporary oxygen store. The molecule consists of one polypeptide chain and one haem group. The amino acid sequence and three-dimensional structure of several myoglobins are known; the folding of the main chain and position of the prosthetic group of sperm whale myoglobin are shown in Figure 4.9. The polypeptide chain contains 153 amino acids and has a molecular weight of 17,000; the shape of the molecule is that of an oblate spheroid, with dimensions of about $44 \times 44 \times 25$Å. The α-helix content of the molecule is high, and almost 80% of the amino acid residues are in helical regions. The 'straight tubes' in Figure 4.9 show these helical regions; there are eight α-helices, all right-handed. The general folding of the molecule results in the non-polar, hydrophobic amino acid side-chains being directed to the interior of the molecule, whilst polar groups point towards, or lie on, the surface. Proline is found at the ends of helical regions, for it cannot form the hydrogen bonds necessary to maintain a helix. The molecule is very compact, i.e. there is little 'empty space' in the interior.

The haem group lies in a pocket in the molecule between two long stretches of α-helix. From one helix, the aromatic ring of a histidine residue projects outwards towards the iron, at right angles to the haem group. A nitrogen atom of this ring provides the fifth pair of electrons donated to the iron orbitals. On the far side of the haem group, another histidine ring is directed towards the iron atom, but is too far from it to coordinate directly. When myoglobin becomes oxygenated, the oxygen molecule fits between the iron and this second histidine, which can then coordinate indirectly with the iron through the oxygen. In addition to the bonding of haem by the first histidine residue, hydrophobic interactions are also important for binding the prosthetic group. There is believed to be

π-bonding between the haem and the ring of a phenylalanine residue which lies close and parallel to the porphyrin structure.

4.5.1.2. Haemoglobin

Haemoglobin is a more complex protein, and transports oxygen in the blood from the lungs to tissues throughout the body; it also facilitates the return of CO_2 to the lungs. Thus, whole blood can take up 21 ml of oxygen per 100 ml of blood, whilst plasma (i.e. without haemoglobin) can carry only 0.3 ml of oxygen per 100 ml in solution.

The molecule consists of four polypeptide chains and four haem groups, i.e. haemoglobin is an oligomeric protein. The molecule is more spherical ($64 \times 55 \times 50$ Å) than myoglobin, and the molecular weight is about 64,000. In the most common form of haemoglobin, there are two pairs of identical chains, the α-chains containing 141 amino acids, and the β-chains containing 146 amino acids. Although the two types of haemoglobin chain and the myoglobin chain have only 21 amino acids in identical positions, it has been found that the folding of the three different chains is very similar. Indeed, all myoglobins and haemoglobins examined so far from a variety of species have the same chain-folding, although the amino acid sequence differs with each species. Each chain has a high helical content, a hydrophobic pocket for the haem group, and is folded in such a way that polar groups are mostly on the surface.

In haemoglobin, the four chains are arranged tetrahedrally (see Figure 4.10) with little contact between like subunits, but close contact between unlike subunits. The haem groups are situated in widely separated pockets on the surface of the molecule such that the iron atoms are 25–30 Å apart. The binding of the prosthetic group in each pocket, however, is very similar to that found in myoglobin. Hydrophobic, ionic, and hydrogen bonding all contribute to the binding holding the subunits together. The molecule has a central cavity lined with polar groups.

Different species produce haemoglobin molecules with the same function and folding, but with different amino acid sequences. However, the presence of certain residues in particular positions in the chains seems essential, for these do not vary from species to species, e.g. the histidine groups on either side of the iron atom. A change in amino acids in other parts of the chain can also have serious consequences for an animal. Substitution of a valine residue for a glutamic acid residue at position 6 of the β-chains of human haemoglobin results in the disease sickle-cell anaemia, whilst replacement by serine of the phenylalanine involved in the π-bonding with haem can be fatal. In general, many amino acid substitutions can occur on the surface of the molecules, without deleterious results, but in functional haemoglobins, a hydrophobic residue in the interior can only be changed to another hydrophobic residue.

4.5.1.3. Function of myoglobin and haemoglobin

The oxygenation of myoglobin and haemoglobin has been studied as a function of oxygen pressure (see Figure 4.12).

The oxygen binding curve for myoglobin is the hyperbola to be expected for a 1:1 association of myoglobin haem and oxygen, i.e. the reversible oxygenation of myoglobin can be explained in terms of a simple equilibrium:

$$\text{Myoglobin} + O_2 \rightleftharpoons \text{Myoglobin—}O_2.$$

At the oxygen pressure of muscle, where haemoglobin must give up oxygen, myoglobin has the high oxygen affinity necessary for the uptake of oxygen. The affinity of myoglobin for oxygen is independent of pH over the range encountered in physiological conditions.

Haemoglobin, on the other hand, gives sigmoid binding curves indicating that a more complex process takes place. [The oxygen binding curves of haemoglobin are similar to the sigmoid kinetic curves obtained with some allosteric enzymes (see Section 4.5.2.2, p. 119), and, indeed, haemoglobin has been taken as a model for allosteric enzymes.] At the oxygen pressure of lungs, haemoglobin has a high affinity for oxygen, whilst at lower pressures, this affinity is decreased, enabling oxygen to be transferred to myoglobin in oxygen-deficient tissues. At lower pH, the oxygen-affinity of haemoglobin is further decreased, again encouraging oxygen release to deficient tissues where an accumulation of CO_2 gives a more acidic

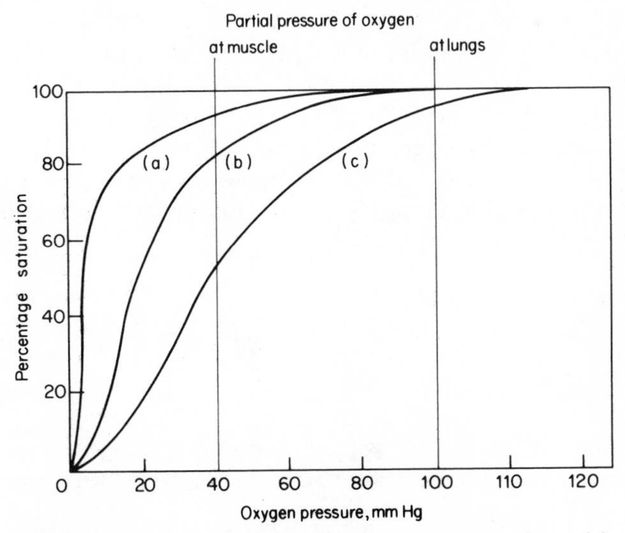

FIGURE 4.12. Oxygen binding curves for (a) myoglobin (invariant with pH); (b) haemoglobin at pH 7.6; (c) haemoglobin at pH 7.0

environment. The tendency to release oxygen at low pH results from the greater ease with which the deoxy form of haemoglobin can bind protons; hence the deoxygenated form is preferred at lower pH values. This change of oxygen affinity with pH is known as the Bohr effect.

It has been proposed that a sigmoid oxygen binding curve is obtained for haemoglobin because of *cooperativity*, i.e. the first oxygen molecule is taken up only with difficulty, but haem–haem interaction during oxygenation then causes the second, third, and fourth molecules to be bound with greater and greater ease. On deoxygenation, the process is reversed, the first oxygen is given up with most difficulty, whilst the fourth is given up most easily, facilitating oxygen release to tissues far removed from the lungs. Because the haem groups are far apart in the haemoglobin molecule, direct haem–haem interaction in this process is not possible.

Perutz has suggested that there is constraint on the deoxy form of haemoglobin resulting in difficulty in uptake of the first oxygen molecule. At some stage during oxygenation, however, small changes in tertiary structure of the polypeptide chains and larger changes in the quaternary structure of the whole molecule take place, giving indirect haem–haem interaction; this releases the constraint on the molecule and increases the oxygen affinity of the unoxygenated haem groups.

Two possible schemes have been put forward to account for the apparent cooperativity. These are the sequential and the symmetry models (see Figure 4.13). Each polypeptide chain can exist in two forms, the low-affinity form (sometimes called the *taut* or T-*form*), and the high affinity form (also called the *relaxed* or R-*form*). In the sequential model, single chains in the molecule can change to the *R*- or *T*-form on oxygenation or deoxygenation; thus molecules can exist containing both *R*- and *T*-subunits. In the symmetry model, on the other hand, all polypeptide chains of one molecule are either in the *R*- or *T*-forms. Evidence from X-ray crystallography favours the second scheme for haemoglobin. It is thought that oxygenation of the α-chains occurs first, which causes conformational changes in the α-chains, rupturing some intrachain and

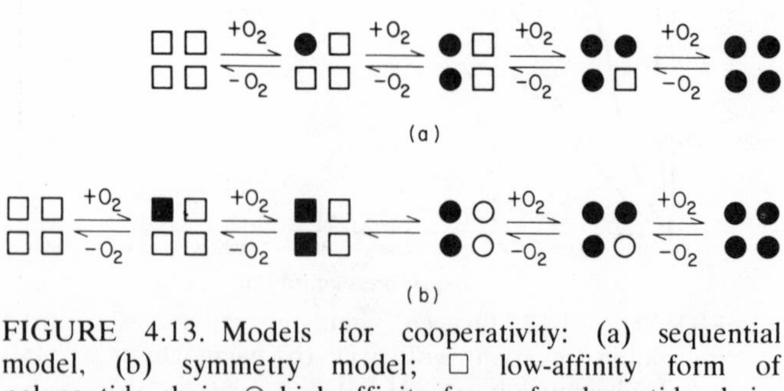

FIGURE 4.13. Models for cooperativity: (a) sequential model, (b) symmetry model; □ low-affinity form of polypeptide chain, ○ high-affinity form of polypeptide chain, ○□ deoxysubunits, ●■ oxygenated subunits

inter-subunit non-covalent bonding. Thus the α–β interfaces are altered, one more than the other (each α-subunit has an interface with each of the two β-subunits), changes occur in the conformations of the β-chains, and the pocket for oxygen in each β-chain becomes enlarged, making binding of oxygen easier.

By X-ray analysis, it has been deduced that the α- and β-subunits rotate with respect to each other on oxygenation. The two α-haem groups move apart by 1 Å, and the two β-haem groups approach each other by 6.5 Å. The atoms at one α–β-interface move 1 Å with respect to each other, whilst at the second interface, the movement is, in general, nearly 2 Å. Thus changes in tertiary structure can be observed, but the order in which they actually occur is not yet known.

The amino acid residues at the *C*-terminal ends of the polypeptide chains are believed to be important for both cooperativity and the Bohr effect at pH values above 6. Removal of the *C*-terminal arginine from the α-chains, and the *C*-terminal histidine from the β-chains greatly inhibits cooperativity and the alkaline Bohr effect. In deoxyhaemoglobin, these residues are involved in interchain salt bridges and thus are anchored in position. The penultimate amino acids in both chains are tyrosine residues and, in deoxyhaemoglobin, are held by hydrogen bonding in pockets between two of the main helices of the polypeptide chain (see Figure 4.14).

On oxygenation, the two helices forming the tyrosine pocket move together and expel the tyrosine residue. The *C*-terminal acid next to the tyrosine is also displaced, and the salt bridges in which it was involved are broken. Figure 4.14 indicates the changes for the α-chains. In the β-chains, a similar type of change can take place, with histidine instead of arginine as the *C*-terminal residue. (The haem pocket of the β-chain is too small in deoxyhaemoglobin to allow direct uptake of oxygen. Thus there must be

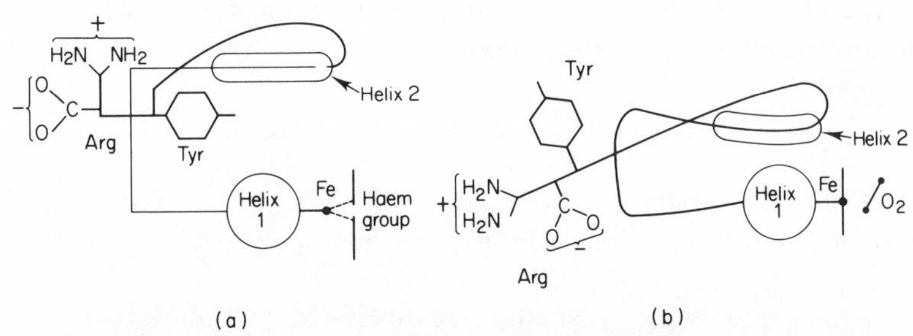

FIGURE 4.14. Schematic representation of spatial changes occurring at the *C*-terminal of the α-chain of haemoglobin in (a) the deoxy form, and (b) the oxy form. In (a) the iron atom lies out of the plane of the porphyrin ring and the tyrosine lies between helices 1 and 2: on oxygenation, the iron atom moves into the plane of the haem group, and helices 1 and 2 move closer together expelling the tyrosine (as in b)

movement in these chains to create a large enough space for the oxygen before oxygenation can take place.)

The freeing of the *C*-terminal acids of the α and β-chains, and rupture of the salt bridges holding them in place in deoxyhaemoglobin, removes the constraints on the molecule, and a change in quaternary structure takes place. The β-chains move further apart, whilst there is little change in the distances between the α-chains. It has been suggested that, in this new form of the molecule, the unoxygenated haem groups have a higher oxygen affinity, thus explaining the sigmoid oxygen binding curve.

In deoxyhaemoglobin, the iron atom lies outside (<0.8 Å) the plane of the porphyrin ring on the side of the nearest histidine residue. On oxygenation, the iron moves into the plane of the ring, and it was first suggested that this movement provided the 'driving force' to move the two helices of the tyrosine pocket together. It is now believed that the movement of the iron may be insufficient by itself to trigger the change in tertiary and quaternary structure, but a satisfactory explanation is not yet available.

The breaking of the anchoring bonds of the *C*-terminal residues on oxygenation is believed to be important for the Bohr effect. In deoxyhaemoglobin, the side-chain of the *C*-terminal histidine of a β-chain interacts with an aspartate residue of the same chain. The histidine ring can thus readily pick up a proton, and the extra positive charge is balanced by the negative charge of the aspartate. The amino group of the *N*-terminal residue of an α-chain can also pick up a proton in deoxyhaemoglobin, and here the $-NH_3^+$ ion is stabilized by ionic bonding to the terminal COO^- group of the opposite α-chain. On oxygenation, these ionic bonds are broken and protons are released. Thus oxygenation is favoured by high pH and deoxygenation by low pH, i.e. oxygen is given up to the tissues, such as muscle, where the pH is lower than in blood.

Carbon dioxide is transported to the lungs by haemoglobin, and it has been found that carbon dioxide can react with the terminal amino groups of all four subunits of haemoglobin:

$$\cdots-NH_2 + CO_2 \rightleftharpoons \cdots-NHCOO^- + H^+$$

Carbon dioxide reacts preferentially with deoxyhaemoglobin, and so is picked up by haemoglobin after oxygen has been given up to peripheral tissues.

In human red cells, a small, negatively-charged molecule is usually found bound to deoxyhaemoglobin. This is diphosphoglycerate ($^{2-}O_3P-OCH_2-CHOPO_3^{2-}-COO^-$), which binds in the central cavity of deoxyhaemoglobin; the cavity is lined by positive charges, which can form ionic bonds to the diphosphoglycerate anion. Thus, diphosphoglycerate (DPG) stabilizes deoxyhaemoglobin, and lowers the oxygen affinity of haemoglobin in blood, making transfer of oxygen from haemoglobin to myoglobin easier. The ionic links between deoxyhaemoglobin and DPG must be broken for oxygenation of haemoglobin to occur, for the central cavity of oxyhaemoglobin is too small for DPG, and the small molecule is expelled on formation of oxyhaemoglobin. Thus, the binding of DPG to haemoglobin at one site affects the binding of oxygen at another--DPG is, therefore, an *allosteric effector* of haemoglobin, i.e. a molecule which binds to a macromolecule and alters the affinity of the macromolecule for a second substance, by stabilizing one conformation of the macromolecule. We shall encounter this type of behaviour again when dealing with allosteric enzymes (see Section 4.5.2.2, p. 119).

Small molecules other than oxygen can also be bound by the haem groups of haemoglobin and myoglobin. Unfortunately, the affinity for carbon monoxide is several times that for oxygen. Thus in an atmosphere containing appreciable amounts of carbon monoxide, this is bound preferentially by haemoglobin. The reaction is not readily reversible, even in excess oxygen, and so tissues far from the lungs are deprived of oxygen, and death can result very quickly.

4.5.1.4. Cytochrome c

Cytochrome c is a haem-containing protein which transports electrons rather than oxygen. The protein has been obtained from a number of different species of animal, and the amino acid sequences have been compared. The molecule contains one peptide chain of 103–114 amino acid residues (depending on the source), has a molecular weight of 12,000–13,000, and is a prolate spheroid (30 × 34 × 34 Å). Over one quarter of the amino acids occur in identical positions in all the cytochrome c chains so far examined, and so these residues must be vital for the function of the protein.

In cytochrome c, the iron does not remain always in the iron(II) state (as is the case under normal conditions for haemoglobin and myoglobin), but changes from the iron(II) to the iron(III) state and back, as the molecule functions in the *electron transport chain*.

[During the oxidation of carbohydrate and fats to provide energy for aerobic organisms, reduced forms of some coenzymes (see Section 4.5.2.2, p. 118) are produced. These are reoxidized whilst molecular oxygen is reduced to water, and at the same time high-energy, chemical intermediates are produced. Oxygen does not act directly as the agent oxidizing the reduced co-enzymes; instead electrons are transferred from the reduced coenzymes via a number of multimolecular complexes to molecular oxygen—this sequence is the electron transport chain.]

It is believed that structural changes take place on oxidation and reduction of cytochrome c, but X-ray work on the three-dimensional structures of the oxidized and reduced forms has failed to show major conformational changes.

X-ray studies of cytochrome c show there is some α-helix in the molecule, and the polypeptide takes the form of an extended chain which is folded round the haem group (see Figure 4.15). The molecule as a whole provides an excellent example of chain-folding to give a hydrophobic interior and a charged surface. The haem group enveloped by hydrophobic side-chains lies in a crevice formed by the first 47 residues on one side and the next 43 on the other, with the remaining residues stretching across the top in an α-helix. One edge of the haem is exposed to solvent, and a propionic acid side-chain of the prosthetic group lies in a polar environment at the surface of the molecule, whilst the other is buried in the hydrophobic interior and is involved in a network of hydrogen bonds. The haem in cytochrome c is covalently bonded to the polypeptide chain

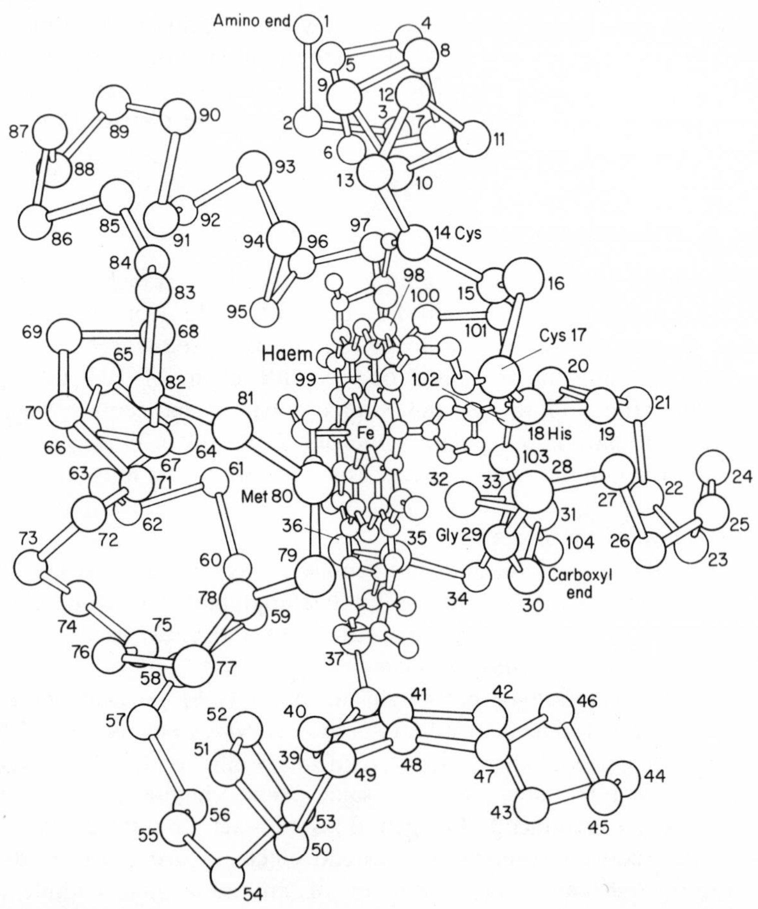

FIGURE 4.15. α-Carbon skeleton of horse heart oxidized cytochrome c. Only α-carbon atoms are shown, with —CONH— amide groups represented by straight bonds. Only those side-chains are shown which are bonded to the haem group, which is seen here nearly in an edge view. Reproduced with permission from 'The Structure and History of an Ancient Protein' by R. E. Dickerson, *Scientific American*, April 1972. Illustration copyright by Dickerson and Geis

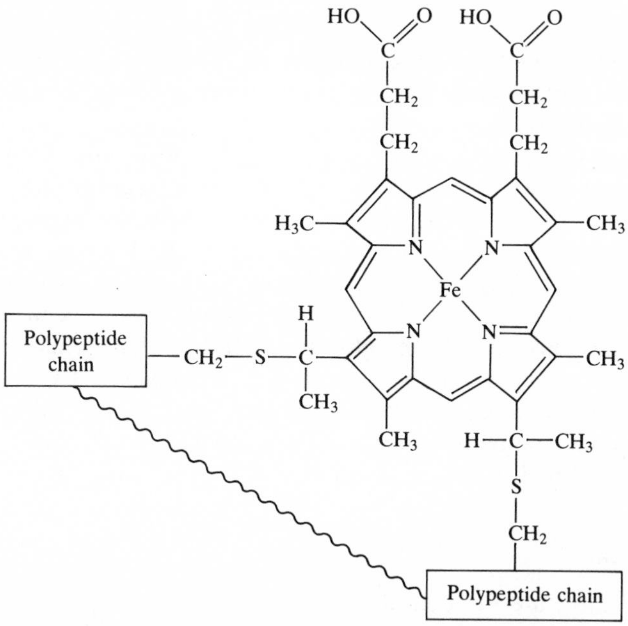

FIGURE 4.16. Binding of haem to the polypeptide chain in cytochrome c.

through interaction of two cysteine residues with the vinyl groups of haem (see Figure 4.16), and the iron is again octahedrally coordinated, with four electron pairs coming from the porphyrin ring structure, the fifth from a histidine, and the sixth from a methionine residue of the protein which projects into the crevice.

Two 'channels' filled with hydrophobic side-chains lead out from either side of the haem group to the surface of the molecule, and in these, parallel pairs of aromatic rings are found.

As cytochrome c functions in the electron transport chain, it acquires electrons from a membrane-bound, haem-containing complex, *cytochrome c reductase*, and donates electrons to a second membrane-bound, haem-containing complex, *cytochrome c oxidase*. Both reductase and oxidase are believed to bind to sites (not identical, but close together) on the cytochrome c molecule near the exposed edge of the haem. Positive charges distributed about the perimeter of the haem crevice are believed to be important for this binding.

A satisfactory explanation for the mechanism of electron transfer from the reductase to cytochrome c, and thence to the oxidase, has not yet been found, but the exposed edge of the haem group is believed to be very important. Close and parallel alignment of the haem groups of cytochrome c and its reductase or oxidase, during binding of cytochrome c to the respective complex, may facilitate electron transfer. Information about the mechanism involved would likely be gained from X-ray diffraction studies of cytochrome c bound to the oxidase or reductase, but because these complexes are membrane-bound, it has not been possible to obtain crystals suitable for such work.

Comparison of the amino acid sequence of cytochromes from different species has allowed conclusions to be made concerning molecular evolution. Thus, species containing cytochromes with the largest number of differences in primary structure are assumed to be furthest apart on the evolutionary tree, whilst species containing very similar cytochromes are thought to be more closely related. Similar studies have been carried out with other proteins, e.g. haemoglobin and myoglobin, where the amino acid sequences from a number of species are known. Eventually, it is thought that biological classifications may be facilitated by protein-sequence comparisons, rather than relying, as at present, solely on external characteristics.

4.5.2. Enzymes

Enzymes constitute the extremely important group of proteins which act as biological catalysts, and increase the rates of reactions by a factor of 10^8 to 10^{20} without altering the equilibrium positions of those reactions. They are highly specific; each one usually catalyses only one type of reaction, and thus, in a living organism, where many hundreds of chemical reactions are occurring, many hundreds of different enzymes exist.

Although many enzymes have now been purified and studied, the exact mechanisms by which these proteins greatly increase reaction rates are not yet fully known, although plausible postulates have been put forward for several enzymes. In general, a reaction is speeded up in the presence of an enzyme because the activation energy for the reaction is decreased. Several ways in which an enzyme might bring about this decrease have been suggested.

An enzyme forms an unstable complex with one or more of its substrates. [A reactant with which an enzyme interacts is called a *substrate*.] In this enzyme–substrate complex, there may be geometric or electronic strain of the substrate, which lowers the activation energy required to transform the substrate to a transition state. Also, immobilization of the substrate, by binding to the enzyme in an orientation favourable for reaction, will result in a loss of entropy, which can again decrease the activation energy for the reaction. Furthermore, for some enzymes, covalent bonds may be formed between enzyme and substrate; if each chemical step has a lower activation energy than the uncatalysed reaction, then the overall rate will be increased. Enzymes can provide functional groups capable of donating or receiving protons, and such groups, in the enzyme–substrate complex, are brought close to the bond to be broken, so that interaction can readily take place. Where two substrates are involved, the rate of a reaction is increased by the substrates being brought together on the enzyme in the correct orientation for reaction.

Enzyme molecules may be relatively simple, consisting of a single polypeptide chain, or more complex, being made up of several non-covalently bonded, polypeptide chains. Historically, the simpler enzymes were the first to be studied in great detail, but now three-dimensional structures of oligomeric enzymes (whose molecules consist of several subunits) have been determined by X-ray diffraction, and changes in tertiary and quaternary structure connected to enzyme function are beginning to be understood.

Several enzymes involved in catalysing different steps of one biochemical pathway are often found in association in Nature as *multienzyme complexes*. Ultimately, it is hoped that the reactions taking place on the complex will be understood at the molecular level in terms of the structure of each component of the complex. At present, however, work on such complexes has only reached the stage of determining the number of polypeptide chains in the complex, the possible function of each and the areas of contact between each component chain.

4.5.2.1. Active sites

The tertiary and quaternary structure of an enzyme is all-important for its activity, and there is a specific region on the surface of each enzyme where the substrate is bound, and where reaction takes place—this region is known as the *active centre*, or *active site*. The amino acid residues at the active centre may be widely spaced in the primary sequence of the protein, but the folding of the tertiary structure of the enzyme brings these residues together. The precise geometry of the active centre plays a large part in determining the specificity of an enzyme, for the only substances accepted as substrates by the enzyme will be those whose molecules contain the necessary stereochemical features to react correctly with the active site.

The residues involved at the active site may constitute a very small fraction of the whole molecule. Much of the remainder of the polypeptide chain exists to hold the active-site residues in the correct arrangement in space and chemical environment for catalysis. In many enzymes, however, parts of the molecule may also be involved in regulation of enzyme activity (see next section).

As indicated earlier, reagents such as strong acid or alkali, or 8 M urea, or extremes of temperature, which cause large changes in tertiary structure (i.e. cause denaturation) usually destroy activity.

The structures of the enzymes investigated to date have several features in common. In general, the α-helix content of the molecules is not as high as in myoglobin, but stretches of β-sheet are not unusual. Water-soluble enzymes possess a large number of charged groups on the surface; those not on the surface are usually involved in the active site. Most of the hydrogen-bond donor groups not available to surface water are found in the interior of the enzyme molecule in a convenient orientation to a hydrogen-bond acceptor. Large parts of the interior, however, are hydrophobic in character. The active site is situated in either a definite cleft in the molecule, or a shallow depression on the surface. These structural features, which have been investigated in the solid crystalline state, are in general maintained in solution. An enzyme molecule in a crystal is 'trapped' in one conformation, whereas molecules in solution are likely to be more flexible. Current evidence suggests, however, that the overall folding of an enzyme molecule, as determined by X-ray diffraction methods, is maintained in solution. In a macromolecule there may be regions where local conformational changes can take place in solution; this is particularly common in the area of the active site.

The high specificity of enzymes as catalysts was originally explained by the hypothesis that the substrate fitted exactly into the enzyme surface like a key in a lock. A later hypothesis suggested that binding of the substrate induces conformational changes at the active site, and X-ray investigation of complexes related to enzyme–substrate complexes has shown that this 'induced fit' theory is more correct, for amino acid side-chains in enzymes may move their position by a distance of several Ångstroms when the substrate is bound.

The nature of the amino acid side-chains at the active centre may be investigated by chemical modification of the enzyme. A reagent relatively specific for one type of amino acid side-chain is allowed to react with the enzyme; if enzymic activity is lost, it is possible that the amino acid involved in the reaction forms an important part of the active site. Care must be taken, however, to eliminate the possibility that the enzyme modification has altered the conformation of the macromolecule so that the spatial arrangement of the active site is affected, instead of specific residues at the active site. For example, iodoacetate reacts with sulphydryl groups:

$$\cdots\!-SH + ICH_2COO^- \longrightarrow \cdots\!-S-CH_2-COO^- + HI$$

and it can, in some cases, be deduced that these groups are at the active sites of enzymes where iodoacetate destroys the enzymic activity. However, reagents such as iodoacetate are not completely specific (iodoacetate can also react with amino groups and histidine), but the use of non-specific reagents can be improved by difference labelling. Here, an enzyme is treated with a reagent in the presence of substrate and the positions of modified residues in the amino acid sequence subsequently determined. Another sample of enzyme is treated in the same way in the absence of substrate. Modified amino acid residues found in the enzyme after the second, but not the first, reaction are assumed to have been 'protected' by the substrate, and are therefore considered to be located close to the active site.

More recent investigations have involved the use of substances of higher specificity; in many cases these substances are closely related to the substrates of the enzymes. Such studies have been made on chymotrypsin, an enzyme catalysing the hydrolysis of esters and peptide bonds near amino acids with aromatic side-chains. p-Nitrophenyl acetate has been used as a pseudo-substrate which modifies the enzyme, and an acid-stable, acetyl chymotrypsin has been found. On hydrolysis of the enzyme, the acetyl group was found to be attached to a serine residue (serine 195 in the amino acid sequence). Other reagents such as diisopropyl phosphofluoridate also react only with this particular serine residue in the enzyme to give an inactive diisopropyl phosphoryl enzyme:

$$i\text{-}Pr-O-P(=O)(O-i\text{-}Pr)-F + HO-CH_2-CH \longrightarrow i\text{-}Pr-O-P(=O)(O-i\text{-}Pr)-O-CH_2-CH$$

Thus this serine must play an important part in the active centre.

Reagents related structurally to the substrate, but possessing an extra reactive group, can also be useful in investigations of the active site. Again, substances of this type have been used on chymotrypsin—the reagent is bound to the active site, and the reactive group can attack an amino acid side-chain in, or near, the active centre. Using a chloromethyl ketone derived from *N*-tosylphenylalanine (the phenylalanine part of the reagent resembles the normal substrate; see Figure 4.17), it has been found that a histidine residue (57 in the amino-acid sequence) lies in or very close to the active centre of chymotrypsin:

$$\cdots-CH_2-(\text{imidazole, N-H}) + Cl-CH_2-C(=O)-CH(NH-SO_2-C_6H_4-CH_3)-CH_2-(\text{phenyl}) \longrightarrow$$

$$\cdots-CH_2-(\text{imidazole, N}-CH_2-C(=O)-CH(NH-SO_2-C_6H_4-CH_3)-CH_2-\text{phenyl})$$

Partial hydrolysis of the enzyme to yield a modified peptide shows where the modified amino acid residue lies in the primary sequence of the polypeptide chain.

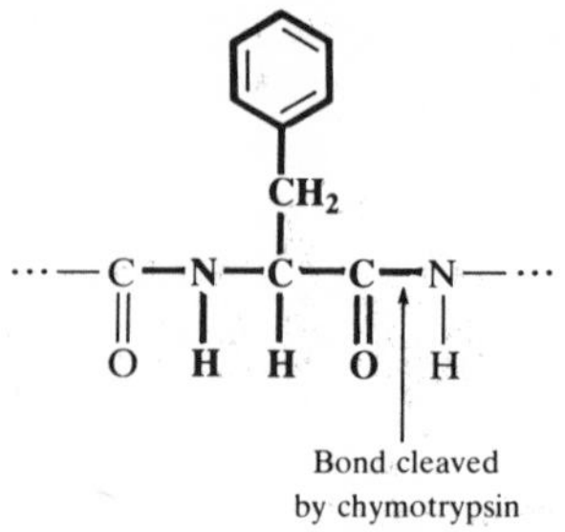

(a)

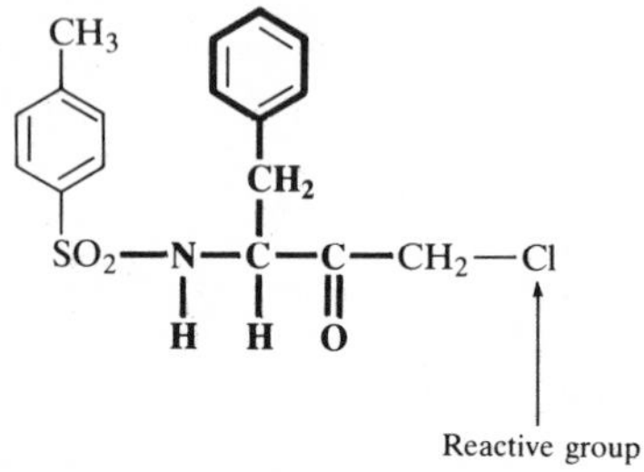

(b)

FIGURE 4.17. Labelling of active site of chymotrypsin: (a) part of natural substrate; (b) reactive molecule for labelling active site. Common structural features are shown in heavy type

4.5.2.2. *Activators and inhibitors*

Enzyme activity can be affected by the presence of other small molecules; these molecules are known as *activators* or *inhibitors*, depending whether they increase or decrease the rate of an enzyme-catalysed reaction. Many enzymes, in fact, require the presence of non-protein components, known as *cofactors*, before they can act as catalysts. These cofactors may be metal ions, or small organic molecules, called *coenzymes*. The metal ion can act as a catalytic centre, can stabilize the active conformation of the enzyme, or can act as a bridging group to help bind the substrate to the enzyme. (Because metal ions form part of many enzymes, traces of these metals are required in the human diet.)

Coenzymes, on the other hand, can be covalently bound to enzyme molecules (in which case they are considered prosthetic groups), or loosely bound. In this second instance, the coenzyme is actually a specific substrate of the enzyme. Coenzymes generally function as intermediate carriers of functional groups, or single atoms, or electrons. In many cases, parts of their structures come from vitamins. (Again lack of certain vitamins in the human diet leads to deficiency diseases because some enzymes cannot function properly.)

A coenzyme may participate in many reactions, i.e. several different enzymes may be capable of binding and utilizing one coenzyme. X-ray diffraction studies have shown that the region of each enzyme binding a particular coenzyme is folded in a similar way (even when there are large differences in amino acid sequence), although other regions of the enzyme molecules may vary widely from enzyme to enzyme depending on the reaction catalysed. This, and other, evidence has given rise to the idea of 'domains' in a protein molecule. Thus a long polypeptide chain may be folded-up to give several distinct regions or 'domains' within the molecule. In some cases, different functions can be assigned to different domains of an enzyme molecule, e.g. one domain may contain a coenzyme binding site, and another may contain, say, an inhibitor binding site.

Inhibition of an enzyme may be reversible or irreversible, depending on the nature of the inhibitor. Many of the reagents, used in the investigation of the nature of the amino acids at the active site of an enzyme, can be classed as irreversible inhibitors. Irreversible inhibition of one enzyme may have serious, sometimes fatal, effects on a living system, emphasizing the importance of each individual enzyme. For example, cyanide preferentially inactivates cytochrome oxidase, an enzyme involved in the utilization of atmospheric oxygen in the body (see Section 4.5.1.4, p. 113); death results in minutes. Diisopropyl phosphofluoridate inhibits not only proteases, such as trypsin and chymotrypsin (see Section 4.5.2.4, p. 126), but also acetylcholinesterase, an enzyme essential for the transmission of nerve impulses. This inhibitor belongs to the family of nerve gases, and causes paralysis, and rapid death. Severe skeletal deformities result from the ingestion of another specific enzyme inhibitor, β-aminopropionitrile (see Section 4.5.5.1, p. 146), which blocks the action of a lysine oxidase, necessary for the formation of cross-links in the structural protein, collagen.

In contrast to this type of action, natural regulators of enzyme action in a living organism usually function by reversible inhibition (or activation). These inhibitors may be related in structure to the substrate, in which case they compete with the substrate for the binding position at the enzyme active site, and so slow down the reaction rate. Some enzymes, however, are inhibited by molecules which are structurally unrelated to the substrate.

These molecules often bind at a specific site on the enzyme molecule distinct, and removed, from the catalytic active site. Enzymes, whose activity can be modulated by non-covalent binding of specific molecules at sites other than the active sites, are known as *allosteric enzymes*. Allosteric enzyme activity is often regulated by inhibitors and activators in living organisms. (Inhibitors are sometimes the end-products of a metabolic pathway in which the enzyme is involved, and this type of regulation is known as *feed-back inhibition*). Often these enzymes have complicated molecular structures consisting of several subunits, i.e. they are oligomeric proteins. For some allosteric enzymes, the rate of the enzyme-catalysed reaction varies sigmoidally with substrate concentration in a manner similar to the oxygen binding of haemoglobin (compare Figure 4.12). However, not all enzymes showing kinetics of this type are allosteric.

Where allosteric enzymes show sigmoid kinetics and have molecules composed of subunits, it has been postulated that the subunits can exist in two main conformations, the *T*-(or low-affinity) and *R*-(or high-affinity) conformations similar to those suggested for haemoglobin (see Section 4.5.1.3, p. 108). Again, two models for interconversion of *T*- and *R*-forms on substrate binding have been proposed, the sequential and symmetry models (see Section 4.5.1.3, p. 108, and Figure 4.13), to account for the cooperativity indicated by sigmoid kinetics.

It has also been suggested that the inhibitors and activators cause changes in the quaternary structures of these enzymes, thus affecting the catalytic activity. Inhibitor and activator binding probably takes place at sites distant from the active centres, often even at sites on separate subunits. An allosteric inhibitor would bring about a change in tertiary structure, causing the enzyme to 'lock' in a *T*-, or low-affinity, state. An activator, in contrast, would 'lock' the enzyme in a *R*-, or high affinity, state (see Figure 4.18). In this case, sigmoidal kinetics would no longer be observed.

Some enzymes can exist in multiple forms in an organism; such forms are known as *isozymes*. Isozymes catalyse the same reaction, but may differ

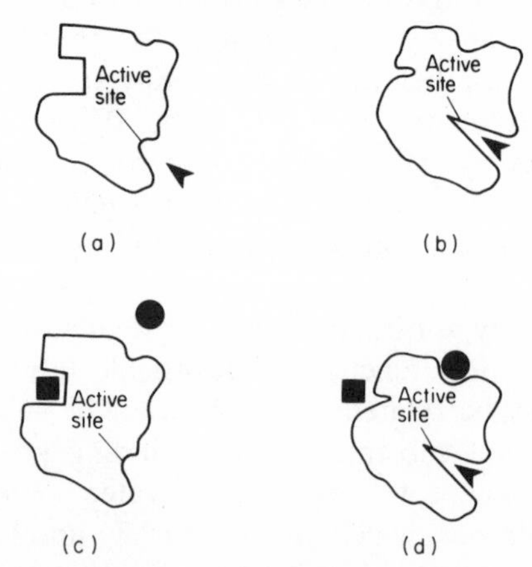

FIGURE 4.18. Regulation of an allosteric enzyme: (a) enzyme in *T*-state unable to bind substrate; (b) enzyme in *R*-state able to bind substrate; (c) enzyme 'locked' in *T*-state by binding inhibitor; (d) enzyme 'locked' in *R*-state by binding activator; ➤ substrate, ■ inhibitor, ● activator

slightly in amino acid sequence, and in affinity for substrate and regulatory molecules. The presence of such isozymes allows for differential control of reactions catalysed by the enzyme in different parts of the organism.

4.5.2.3. Lysozyme

The first tertiary structure of an enzyme to be elucidated by X-ray work was that of lysozyme, an enzyme consisting of one polypeptide chain of molecular weight 14,600, with 129 amino acid residues, which catalyses the hydrolysis of a glycosaminoglycan (see Section 6.5.2.6, p. 311). The molecule contains some short α-helical portions and a region of antiparallel pleated sheet, but the most striking feature of the structure is a large cleft in the surface, which contains the active site (see Figure 4.19). The

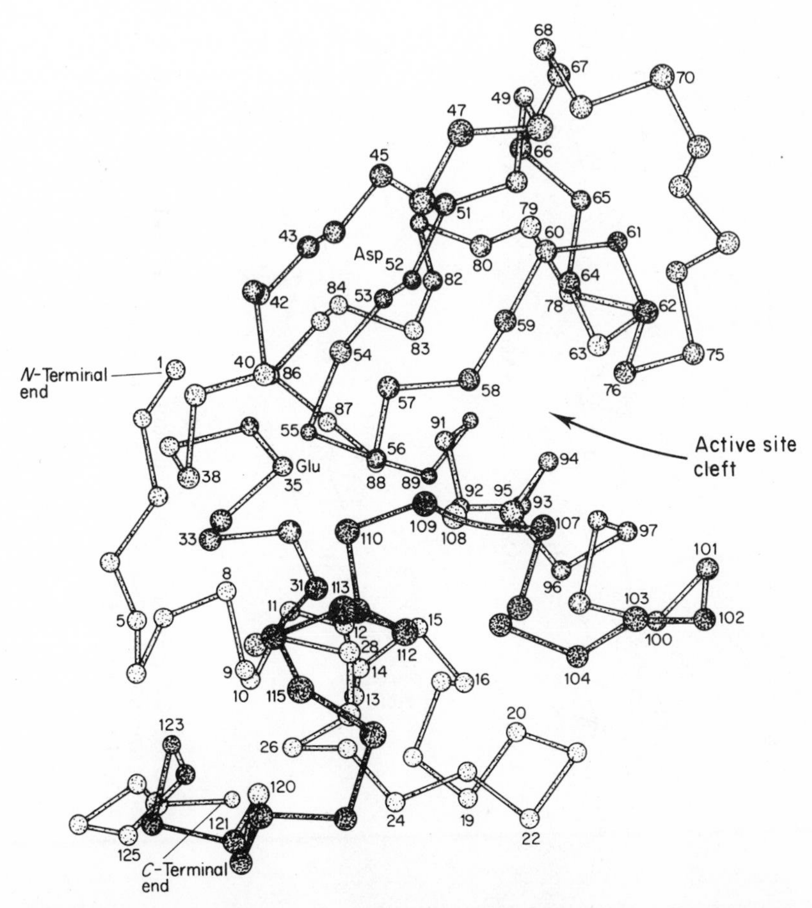

FIGURE 4.19. Folding of the polypeptide chain of lysozyme. Reproduced with permission from R. E. Dickerson and I. Geis, 'The Structure and Action of Proteins', Benjamin/Cummings, Menlo Park, Calif., 1969, p. 71

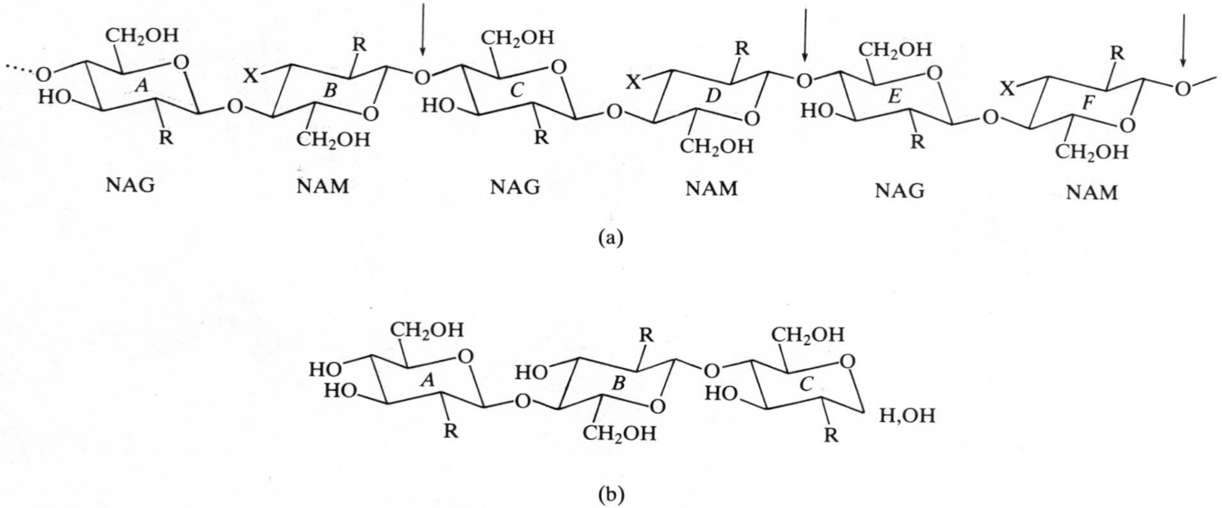

NAG NAM NAG NAM NAG NAM

(a)

(b)

FIGURE 4.20. Substrate and inhibitor of lysozyme: (a) part of natural substrate (for further details see Section 6.5.2.6), (b) an inhibitor, tri-*N*-acetylglucosamine; R = $—NHCOCH_3$, X = $—OCH(CH_3)COOH$, ↑ = possible sites of bond scission catalysed by lysozyme

majority of hydrophobic side-chains are buried in the interior of the molecule, whilst the charged polar groups lie on the surface.

X-ray studies have been made on a complex of the enzyme with tri-*N*-acetylglucosamine, a substance which, although closely related to the substrate of lysozyme (see Figure 4.20), is not hydrolysed by the enzyme. These investigations showed that the tri-*N*-acetylglucosamine is found in the cleft, and a tryptophan residue (62 in the amino acid sequence) moves about 0.75 Å towards one sugar ring of the substrate. The entire crevice closes down slightly showing that the induced fit theory for substrate binding is correct in this case.

A hypothesis for the reaction mechanism of lysozyme has been proposed: The substrate consists of alternating residues of *N*-acetylglucosamine (NAG) and *N*-acetylmuramic acid (NAM) (see Figure 4.20). X-ray diffraction studies of the enzyme –tri-*N*-acetylglucosamine complex showed how three sugar rings could be fitted into the cleft, and gave information on possible hydrogen bonding between amino acid residues of the enzyme and the inhibitor.

It is known that efficient catalysis occurs with a substrate containing six sugar rings, and that a bond between NAM and NAG is broken (see Figure 4.20). Model building, with the knowledge of how the three sugar rings of the inhibitor bind to the enzyme, indicated how six sugar rings may be accommodated at the active site of the enzyme. Possible interactions are shown in Figure 4.21.

Rings *B*, *D*, and *F* contain bulky side-groups (X in Figure 4.21) which cannot fit into the crevice in lysozyme; thus these rings must be bound in such a way that the side-groups project away from the enzyme surface. Ring *C* must not contain this side-group, for it would clash with an isoleucine residue of the enzyme which extends into the cleft. Thus the ring binding at *C* must be NAG, and other rings must be fitted in accordingly. It is then found that ring *D* cannot bind to the enzyme in the normal chair conformation, for the $—CH_2OH$ group would come too close to a main-chain carbonyl and a tryptophan residue of the enzyme, and to the $—NH.CO.CH_3$ group of ring *C*. It is therefore believed that ring *D* is forced into a half-chair conformation such that carbons 1, 2, and 5 and the ring-oxygen lie in a plane. It is now known that the $—CH_2OH$ group on ring *D* is essential for bond scission to take place. This forced change in conformation introduces considerable strain into the substrate.

As indicated in Figure 4.20, scission may occur between rings *B* and *C*, or between *D* and *E*. But the inhibitor tri-*N*-acetylglycosamine contains the equivalent of rings *B* and *C*, and yet is not hydrolysed. The site of catalysis must therefore be between rings *D* and *E*.

Two amino acid residues of lysozyme are believed to be involved in the catalytic step—an aspartate (52 in the amino acid sequence) and a glutamic acid residue (35 in the amino acid sequence). These residues are known to lie in the vicinity of the bond to be hydrolysed, and, indeed, an irreversible

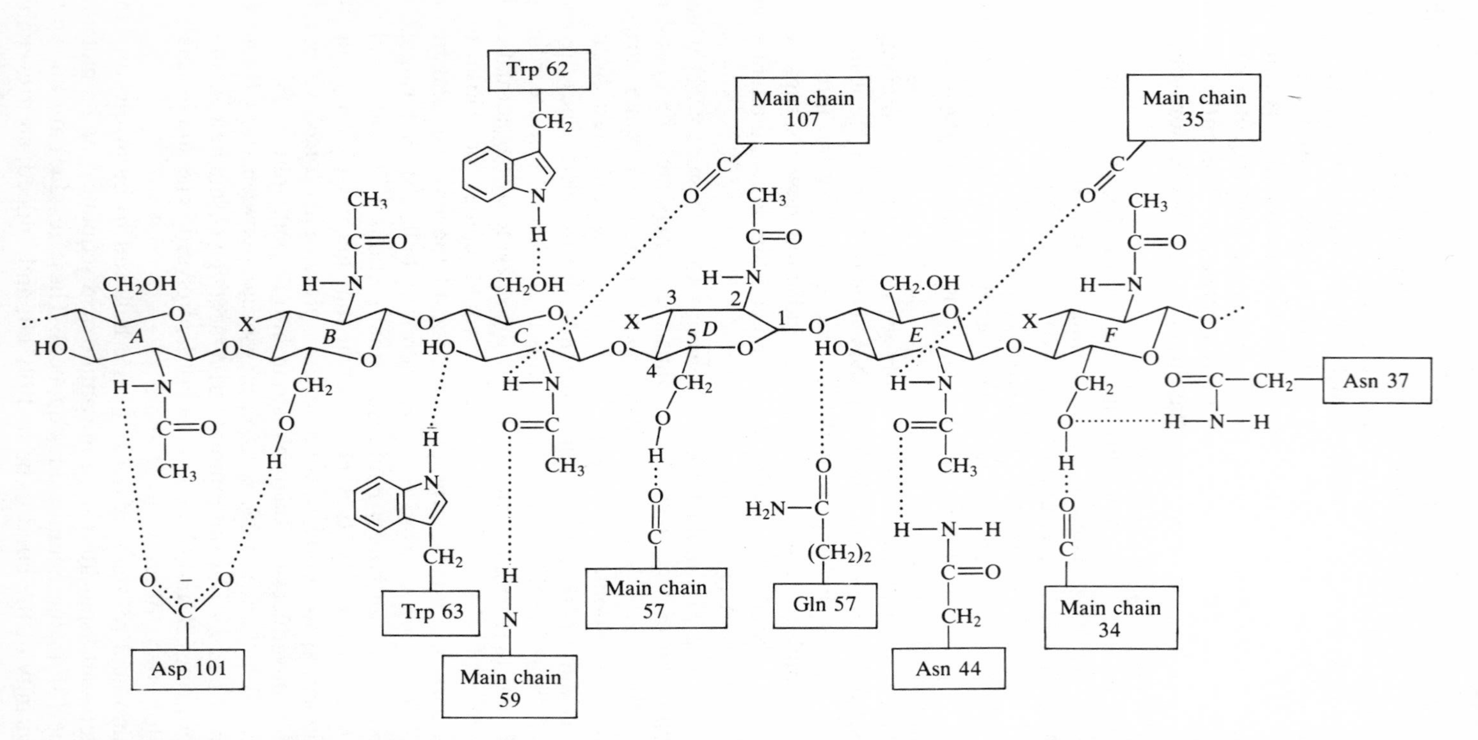

FIGURE 4.21. Schematic representation of the active site of lysozyme: X = —OCH(CH_3)COOH. Amino acid residues are numbered from the *N*-terminal end of the polypeptide chain

inhibitor of lysozyme (an epoxypropyl, $-CH_2-CH-CH_2$, derivative of di-*N*-acetylglucosamine), which binds to the *B* and *C* ring sites, attacks aspartate 52. The aspartate residue is surrounded by a polar environment and so is probably ionized, whilst the glutamic acid lies near hydrophobic groups and is most likely to remain unionized.

The proton of glutamic acid attacks the linking oxygen between rings *D* and *E* (see Figure 4.22a) and the $C_{(1)}-O$ bond breaks, giving a carbonium ion (see Figure 4.22b). Under such conditions, the ring of the carbonium ion would take up a half-chair conformation with carbons 1, 2, and 5 and the ring-oxygen in a plane, i.e. the conformation already favoured by the

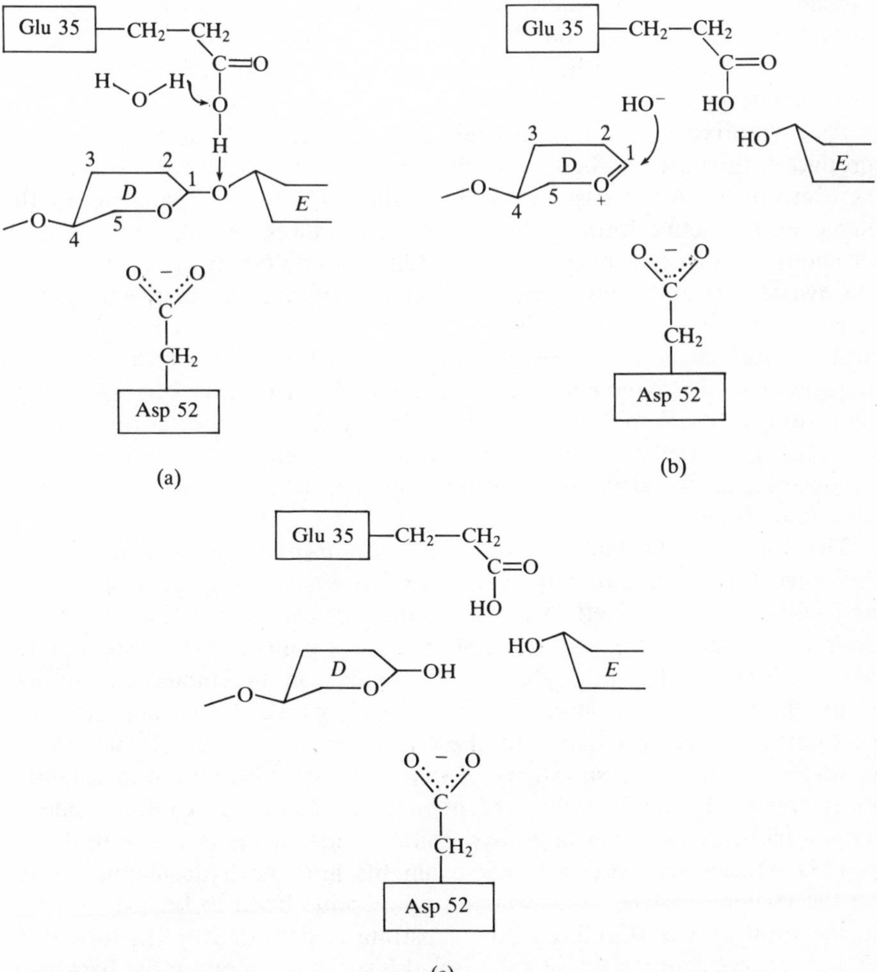

Glu 35 —CH₂—CH₂ C=O O H O D E Asp 52

(a)

(b)

(c)

FIGURE 4.22. Proposed mechanism of cleavage of the polysaccharide bond by lysozyme (see text for details)

enzyme. The ionized aspartic group helps to stabilize the carbonium ion, and these factors lower the activation energy for bond-breaking. A proton from an ionizing water molecule replaces the proton on glutamic acid, and the remaining hydroxyl group attacks the carbonium ion to complete the reaction (see Figure 4.22c).

4.5.2.4. Chymotrypsin and trypsin

Detailed reaction mechanisms have also been proposed for the enzymes, chymotrypsin and trypsin, whose structures are known. These proteolytic enzymes, which have already been mentioned (see Section 4.4.1.3, p. 90) as useful tools in the investigation of protein sequence, are found in the human digestive tract, where they accelerate the breakdown of proteins during the digestion of food. The resultant amino acids, or small peptides, can then be used to synthesize new protein constituents of the body, or to provide energy.

As chymotrypsin and trypsin could hydrolyse the cells in which they are produced, they are synthesized in the pancreas in an inactive form, known as a *zymogen*. After passage of the zymogen from the pancreas to the intestine, the active form of the enzyme is produced by the action of acid, or another proteolytic enzyme, which splits peptide bonds in the zymogen.

Chymotrypsinogen, the zymogen of chymotrypsin, becomes activated by a complex series of steps. First, trypsin catalyses the hydrolysis of an arginine–isoleucine bond (between residues 15 and 16 of the zymogen sequence) in the zymogen to give an active form of chymotrypsin, which then attacks itself to lose two dipeptides, and leave the 'normal' active form of the enzyme. This form consists of three polypeptide chains, produced from the single chain of the zymogen, which are held together by disulphide bonds.

The three-dimensional structure of chymotrypsin shows that there is extensive β-structure, that there are only two short segments of α-helix, and that there is no cleft in the molecule. Instead, a histidine (57 in the amino acid sequence) and a serine residue (number 195), essential for activity, are situated in a shallow depression on the molecular surface. Near these is an aspartic acid residue (102), which is also important for catalysis. In close proximity to these residues is a hydrophobic pocket believed to define the specificity of chymotrypsin. Chymotrypsin catalyses most effectively the hydrolysis of peptide bonds on the carboxyl side of amino acid residues with large hydrophobic side-chains (see Figure 4.17a, p. 118). During catalysis, this side-chain fits into the hydrophobic pocket on the enzyme surface, thus bringing the peptide bond to be broken close to the catalytic site. Binding of the substrate is stabilized by the formation of a short transient stretch of antiparallel β-structure between the backbone of the substrate on the amino-terminal side of the bond to be cleaved and the backbone of the enzyme near the specificity pocket.

Present evidence suggests that chymotrypsinogen, and the different forms of chymotrypsin, have basically the same chain-folding, and the removal of two dipeptides on activation causes little change in structure, certainly in the area of the active site. The main change on activation is believed to occur after the first scission of the arginine–isoleucine bond, when the newly formed amino group of the isoleucine residue attracts the carboxyl group of an aspartic acid residue (194), causing the main polypeptide chain in that region to move. This movement changes the conformation at residues 187, 192, and 193, and forms the 'specificity pocket' of the active site. (One side of the specificity pocket is lined by residues 189–192.)

Detailed studies of the interaction of inhibitors and 'poor' substrates with chymotrypsin and related enzymes have allowed a reaction mechanism to be postulated.

Electronic strain in the enzyme is believed to be an important factor in the mechanism of catalysis. An aspartic acid residue (102), buried in a non-polar environment, interacts with a histidine residue (57) which in turn shares a proton with a serine residue (195) (see Figure 4.23), producing electronic strain. This hydrogen-bonding network produces a serine residue which is strongly nucleophilic, and which can react with the substrate. Stable acyl-chymotrypsin compounds have been obtained using modified substrates, indicating the formation of an acyl enzyme at serine 195 as intermediate step.

The reaction is thought to take place as shown in Figure 4.24. The reactive serine (195) attacks the substrate (see Figure 4.24a), whilst the histidine (57)–aspartate (102) system acts as a proton acceptor. A transient tetrahedral intermediate is formed (Figure 4.24b), and is believed to be stabilized by hydrogen bonding of the oxygen of the substrate (O^- in Figure 4.24b) to two —NH groups of the backbone of the enzyme polypeptide chain (at positions 193 and 195 on the enzyme).

X-ray diffraction studies on complexes of similar enzymes with peptide

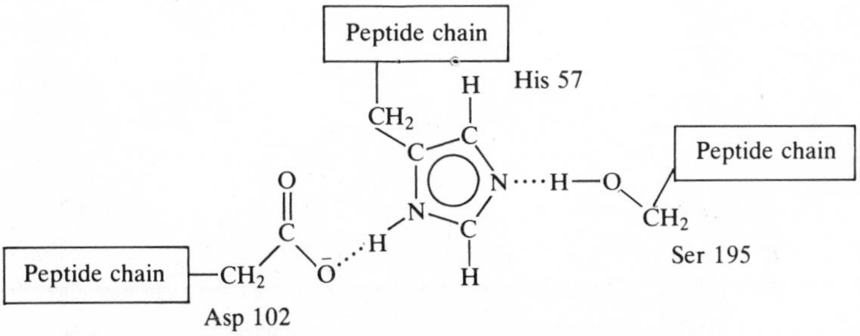

Peptide chain
His 57
Peptide chain
Ser 195
Peptide chain
Asp 102

FIGURE 4.23. Proposed catalytic site for chymotrypsin: interaction of a serine residue (number 195 in amino acid sequence) with histidine residue (number 57) and aspartic acid residue (number 102) makes the serine strongly nucleophilic

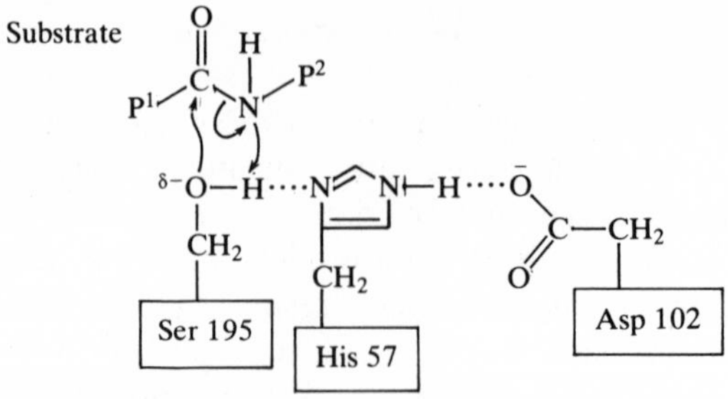

Substrate

O
H
P[1] C N P[2]
δ− O—H····N N—H····O
C—CH₂
CH₂ CH₂ O
Ser 195 His 57 Asp 102

(a)

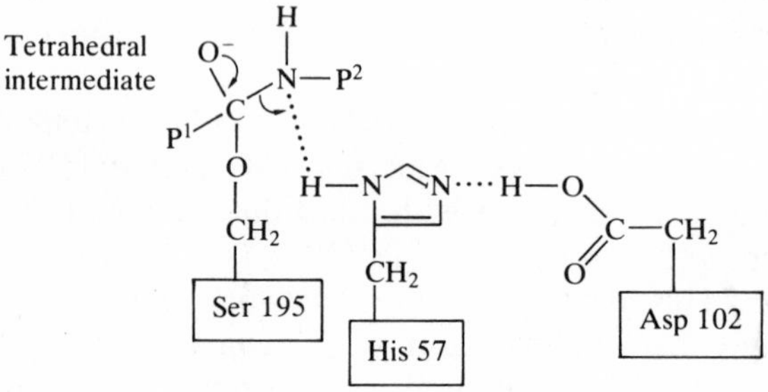

H
Tetrahedral intermediate
O⁻ N—P[2]
P[1] C
O
H—N N····H—O
CH₂ C—CH₂
CH₂ O
Ser 195 His 57 Asp 102

(b)

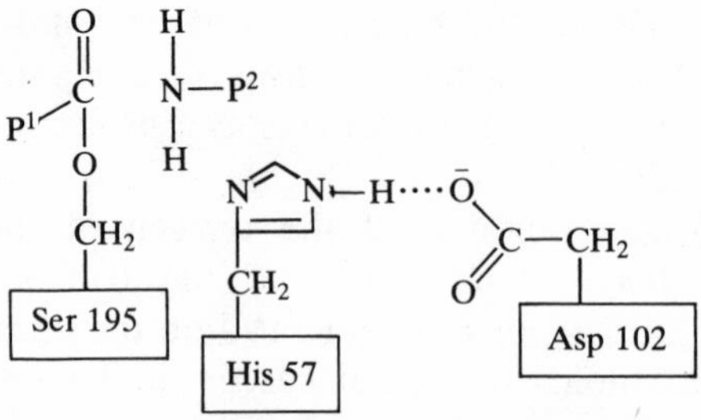

O H
P[1] C N—P[2]
O H
N N—H····O
CH₂ C—CH₂
CH₂ O
Ser 195 His 57 Asp 102

(c)

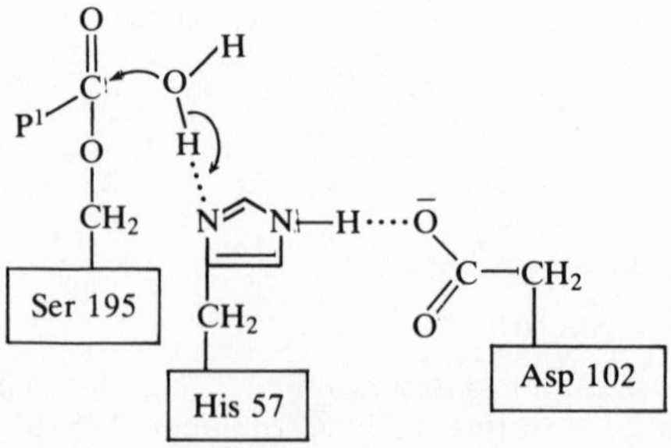

O
H
P[1] C O
O H
CH₂ N N—H····O
C—CH₂
CH₂ O
Ser 195 His 57 Asp 102

(d)

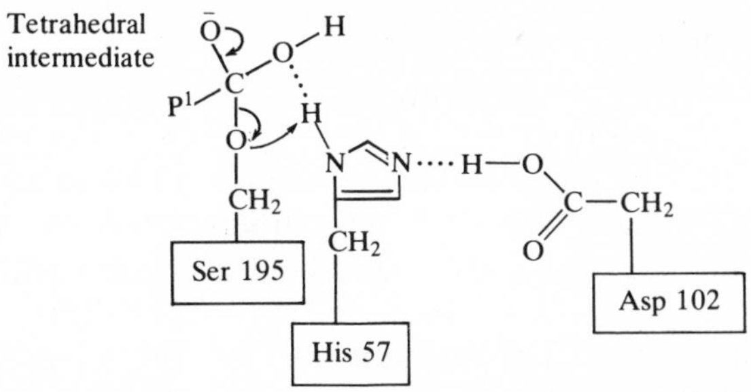

(e)

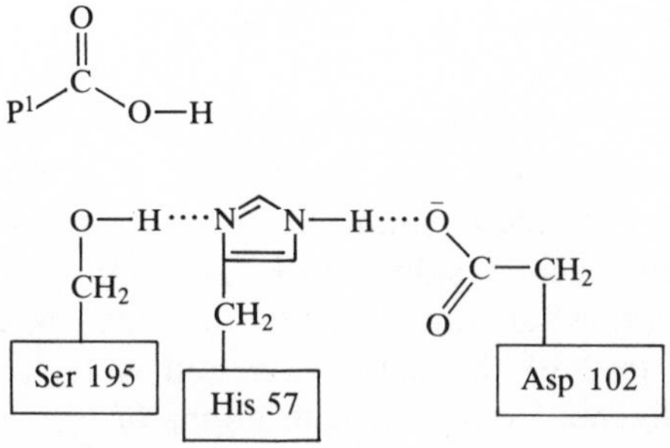

(f)

FIGURE 4.24. Proposed mechanism of cleavage of peptide bond by chymotrypsin (see text for details). P^1 and P^2 represent the remainder of the polypeptide chain of the substrate

inhibitors have indicated the presence of a tetrahedral arrangement at the reactive serine. This arrangement is exactly that which would be expected for a transition state in the formation of an acylated serine residue. Thus, as for lysozyme, the enzyme acts by stabilizing the transition state of the reaction.

The tetrahedral intermediate then breaks down to give an acyl enzyme and a fragment of the original substrate (P^2—NH_2, see Figure 4.24c). This fragment, P^2—NH_2, then diffuses away from the enzyme.

The histidine (57)–aspartate (102) system then promotes transfer of the acyl group from serine (195) to the second substrate, water (see Figure 4.24d, e, and f). Again a tetrahedral intermediate is formed.

The arrangement of amino acids at the active site of *trypsin* is believed to be very similar to that described for chymotrypsin. However, the different specificity of trypsin, i.e. hydrolysis of peptide bonds on the carboxyl side of basic amino acid residues, can be explained by the presence of an aspartic acid residue in the specificity pocket. This aspartic acid residue facilitates the binding of basic side-chains, such as those of lysine and arginine.

4.5.2.5. Carboxypeptidase A

The molecule of carboxypeptidase A—an enzyme catalysing the hydrolysis of a protein from the *C*-terminal end of a chain, but acting most readily when the last residue has an aromatic side-group (see Section 4.4.1.1, p. 88)—contains a sizeable proportion of α-helix, and also some β-sheet in the interior of the molecule. The β-structure involves 15% of the molecule's 307 amino acid residues, and consists of eight parallel or anti parallel extended chains. The molecule also contains a zinc atom, which is essential for enzymic activity.

This enzyme also is synthesized as a zymogen, *procarboxypeptidase A*, but is hydrolysed to give a single polypeptide chain of 307 amino acid residues as its active form.

The active site is again believed to be situated in a shallow depression on the surface, close to a mainly hydrophobic pocket capable of binding the aromatic side-chain of the last residue of the substrate. This enzyme provides a further example of 'induced fit', for X-ray studies have indicated movement of amino acid residues 145, 248, and 270 involved in binding and catalytic action when the enzyme–substrate complex is formed. The zinc atom forms part of the active site and is tetrahedrally coordinated, three of its four ligands being nitrogen atoms of histidines 69 and 196 and an oxygen atom of glutamic acid 72 of the enzyme, whilst the fourth is the carbonyl oxygen of the peptide bond of the substrate which is being split. The carboxy terminus of the substrate is held in place by interaction with an arginine residue of the enzyme (see Figure 4.25). Tyrosine (248) and

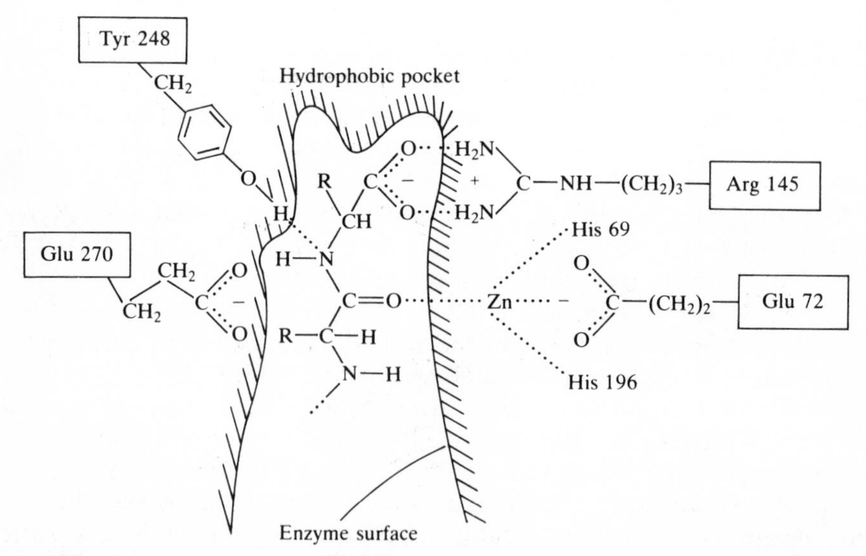

FIGURE 4.25. Binding of substrate to carboxypeptidase A; R = hydrophobic side-chain of substrate

glutamate (270) residues are believed to be important during the hydrolysis reaction.

The zinc atom is believed to make the C=O bond of the substrate more polar, and thus render the carbon atom more susceptible to nucleophilic attack. The presence of glutamate 270 also favours polarization of the C=O bond, i.e., the enzyme is believed to induce electronic strain in the substrate. Glutamate 270 itself is thought to attack the carbonyl carbon, or to activate a water molecule to attack the substrate (see Figure 4.26, p. 132).

4.5.2.6. Pancreatic ribonuclease

The enzyme, catalysing the hydrolysis of ribonucleic acid (see Section 5.4.1.2, p. 189), can exist in two active forms, both of which have been synthesized chemically. The first form, synthesized by Merrifield (see Section 4.6.1) consists of one polypeptide chain of 124 amino acids. In the second form, one peptide bond has been cleaved to give two chains of 20 and 104 residues. Although there is no covalent bond between the two chains, hydrophobic bonding is sufficient to hold them together and give an active enzyme. Hirschmann and his colleagues (see Section 4.6.2, p. 161) have synthesized the 104-residue chain of this form of ribonuclease, and by combination with a natural 20-unit chain (which itself exhibited no catalytic activity) have obtained enzymic activity characteristic of natural ribonuclease. X-ray studies have shown that the molecular structure round the active site is identical in both forms of ribonuclease, and that they differ only in an outer chain not involved in the catalytic activity.

The molecule of ribonuclease contains a large cleft, like lysozyme, where the active centre is situated. The outline of this crevice is defined by a double-stranded, V-shaped stretch of β-sheet, and there is relatively little α-helix in the biopolymer. The polypeptide chain is folded so that the interior of the molecule is hydrophobic.

Molecules of ribonuclease have been chemically cross-linked whilst in the crystalline form. This prevents them going into solution. But these molecules retain catalytic activity, indicating that there must be extensive similarity between the conformations of the enzyme in the crystalline form and whilst active in solution.

Ribonuclease has four disulphide bridges. If these are reduced and the molecules denatured with 8 M urea, all catalytic activity is lost. However, on removal of the urea by dialysis, some catalytic activity returns. This demonstrates that the molecule can refold spontaneously to the active conformation, and reform the correct disulphide bridges (although some 'wrong' disulphide bonds also form). The result is all the more remarkable when one realises that four disulphide bridges can be formed in 105 different ways from eight cysteine residues. This work was the first demonstration that the tertiary structure of a protein is determined by its amino-acid sequence.

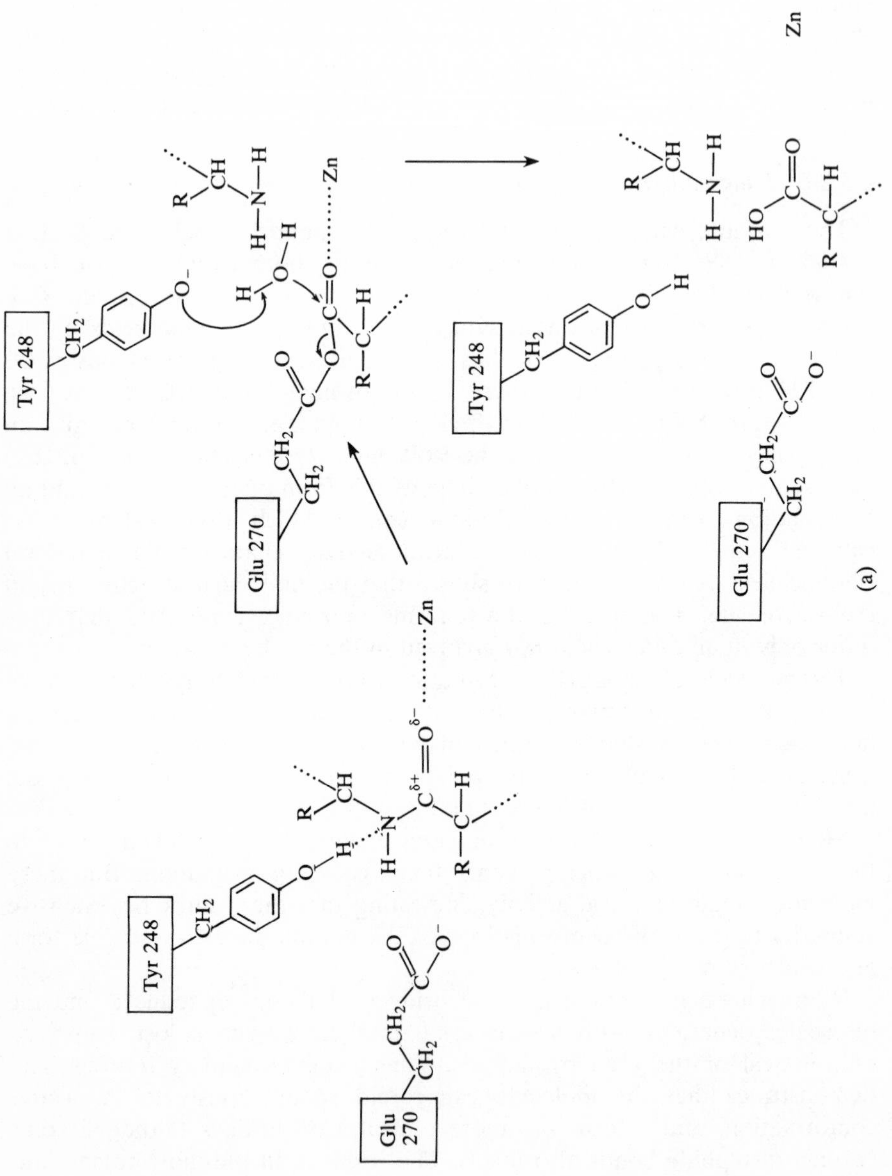

Tyr 248

Glu 270

Zn

Tyr 248

Glu 270

Zn

Tyr 248

Glu 270

Zn

(a)

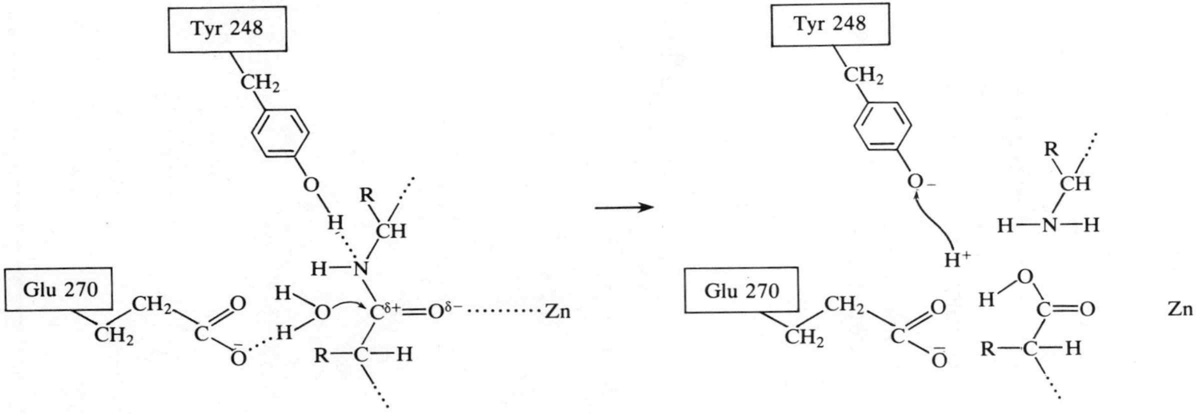

(b)

FIGURE 4.26. Proposed mechanisms of cleavage of peptide bond by carboxypeptidase A: (a) direct attack by the side-chain of glutamate (270) residue; or (b) activation of water molecule by glutamate (270) residue. Other mechanisms are possible which do not involve ionization of tyrosine

Three amino acid residues have been identified as being important for the catalytic activity of ribonuclease. These are two histidine residues (12 and 119) and a lysine residue (41). Histidine 12 is located in the small 20-unit peptide which had to be added to the 104-unit peptide synthesized by Hirschmann (containing residues 41 and 119) to restore catalytic activity. [In 1972, Anfinsen, Moore, and Stein received the Nobel Prize for their work on the amino acid sequence of ribonuclease, and the identification of residues at the active site.]

The enzyme is believed to act by providing a proton acceptor and a proton donor in close proximity to the bond of the substrate to be cleaved. The two histidine residues (12 and 119) may act as the acceptor and donor, whilst lysine (41) may help to stabilize a cyclic phosphate intermediate (see Figure 4.27).

Further synthetic work on ribonuclease is now in progress. Attempts are being made to synthesize an analogue of ribonuclease, with several amino

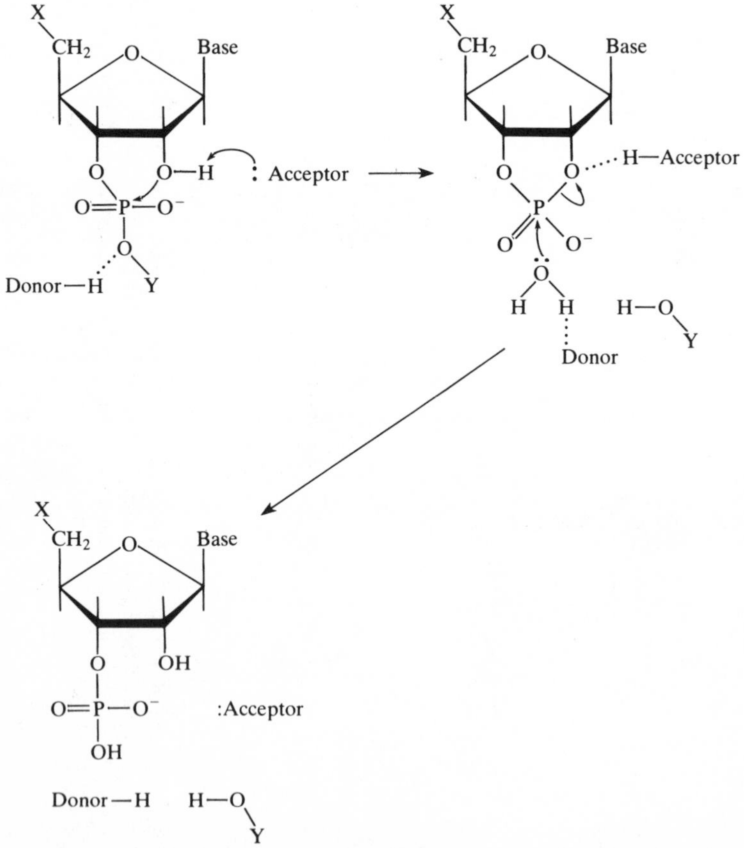

FIGURE 4.27. Proposed mechanism of action of ribonuclease: X, Y = remainder of ribonucleic acid chain

acid substitutions and some deletions, in the hopes of investigating in greater detail the influence of structure on catalytic activity.

4.5.3. Hormones

Hormones are substances which act on specific cells in plants or animals, and have a profound effect on the metabolism of these target cells. Some hormones are small proteins or large polypeptides, and again it is thought that the biochemical information carried by these molecules is defined by their three-dimensional structures. For some hormones, the mechanism of action is understood, but for most, much work remains to be carried out on the relation between three-dimensional structure and function. Several protein hormones have been subjected to partial hydrolysis by proteolytic enzymes, and it has been found that smaller sections of the whole molecule may retain biological activity. Thus, the remainder of the molecule may serve merely to modify or enhance the activity, or stabilize the hormone during transport from the point of synthesis to the target cells.

4.5.3.1. Insulin

Insulin, the most widely-studied protein hormone, is secreted by the pancreas, and plays a major rôle in the regulation of carbohydrate, fat, and amino acid metabolism. It stimulates glycogen (see Section 6.5.1.1, p. 278) and fatty acid synthesis, and inhibits their degradation. This hormone also promotes uptake of glucose and amino acids from blood. The most important target cells are in liver, muscle, and fatty tissues. A deficiency of this hormone causes one type of diabetes.

The insulin molecule contains two polypeptide chains, linked together by disulphide bridges. The molecular weight is 6,000, but the molecules readily dimerize. Bovine insulin, with an *A*-chain of 21 amino acids and a *B*-chain of 30 residues, was the first protein to have its complete amino acid sequence determined (see Figure 4.6, p. 98).

Insulin and modified analogues have now been synthesized chemically so that investigations have been made of the relation between structure and function. The free amino and hydroxyl groups in the molecule are not necessary for activity, but the disulphide bridges are important for the maintainance of the correct conformation for biological action—in fact, the intact three-dimensional structure is believed to be essential. The tripeptides at the *N*- and *C*-terminal ends of the *B*-chain are not necessary, but the amino acid at the *C*-terminal of the *A*-chain is, however, essential. The acid at the *N*-terminal of the *A*-chain is also important.

It is now thought that residues 1, 5, 19, and 21 (numbered from the *N*-terminal end) of the *A*-chain, and residues 12, 16, and 22–26 of the *B*-chain are involved in binding to the insulin receptor of target cells (see below). Many of these residues are invariant in insulin molecules from different species.

The three-dimensional structure of a complex of the insulin dimer with zinc has been studied using X-ray diffraction. This technique has shown that the *A*-chain is compact, and contains two short stretches which are almost completely α-helical. The longer *B*-chain is wrapped round the shorter chain, and has an α-helical central portion. The remainder of the *B*-chain is extended, and in the dimer, parts of the two *B*-chains lie close together and appear to form an antiparallel pleated sheet. Hydrophobic residues in the interior of the molecule stabilize the three-dimensional structure by hydrophobic interactions. It is believed that some of the hydrophobic residues on the molecular surface, which are important for dimerization, may be the residues involved in interaction with the receptor site on the target cell. It has not yet been possible to explain the biological activity of insulin in terms of this three-dimensional structure, but residues believed important for the activity are on the surface of the molecule.

Like several other protein hormones, insulin is believed to bind to specific protein receptors on the outer membranes of its target cells. This binding probably triggers a change in concentration of one or more chemical messengers within the cell, and as a result, the activities of several enzymes of that cell are changed. The detailed mechanism is not yet understood, but as a number of processes, e.g. glucose uptake into cells, glycogen synthesis and breakdown, fatty acid synthesis and oxidation, are affected, there may be more than one mechanism of action in operation. Changes in level of cyclic nucleotides (see Figure 5.25, p. 226) and movement of calcium within cells may be involved in bringing about the observable effects of insulin action.

Insulin is believed to be synthesized as a single polypeptide chain, *preproinsulin*, with 104–109 amino acid residues, depending on the species involved. A block of 23 residues is then hydrolysed from the *N*-terminal end, to give *proinsulin* (see Figure 4.28). Proinsulin, like insulin, readily forms dimers, but is biologically inactive.

Biosynthesis of the precursor as one long chain facilitates formation of the correct disulphide bonds in the insulin molecule. Limited proteolysis of proinsulin cleaves the molecule at the *C*-terminal end of the *B*-chain and the *N*-terminal end of the *A*-chain to yield the active form of insulin (see Figure 4.28).

4.5.3.2. Other hormones

Several protein hormones are secreted by the pituitary, or hypophysis, a gland at the base of the brain, and are transported round the body to stimulate the action of other glands. Like insulin, many of these hormones are synthesized as a larger precursor, which undergoes partial hydrolysis to give the active form of the hormone. *Thyrotropin*, a glycoprotein with a molecular weight 28,000, stimulates the thyroid, whilst *adrenocorticotropin* (ACTH) acts on the adrenal cortex to stimulate production of certain steroids. Thyrotropin consists of two subunits, and the amino acid

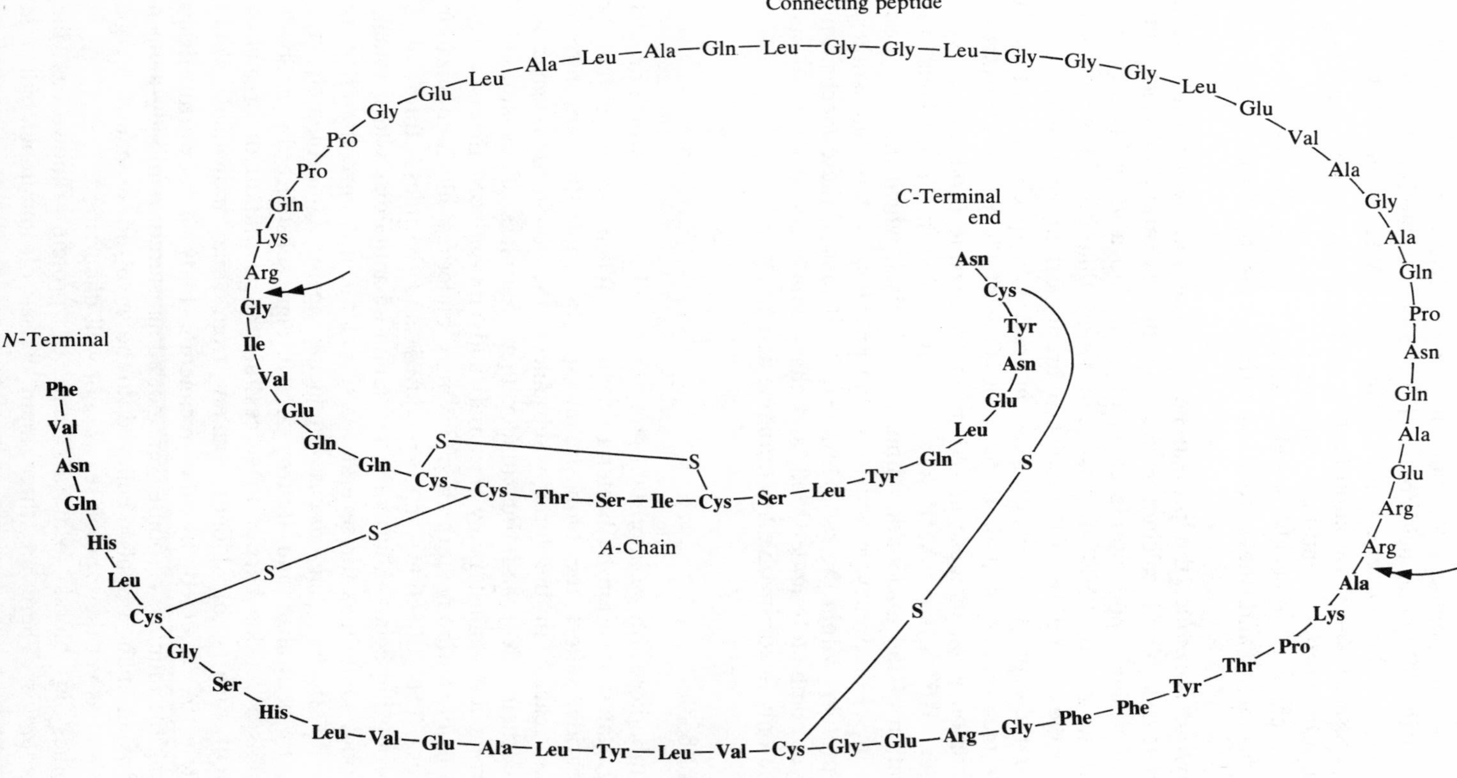

FIGURE 4.28. Amino acid sequence of porcine proinsulin. Residues of the *A*- and *B*-chains of insulin are in bold type; →→ sites of limited proteolysis for the conversion of proinsulin to insulin

sequences of both polypeptide chains have been determined. The primary sequence of the 39 amino acids of ACTH has been established, and it has been found that the sequence of 24 acids at the *N*-terminal is active. In particular, residues 6–19 are important for activity and receptor binding.

Human *somatotropin*, another pituitary hormone, consists of a single polypeptide chain of molecular weight 22,000. The primary sequence has been determined, and the secondary structure is believed to be 50% α-helix.

Parathyroid hormone, the protein hormone of the parathyroid gland, which increases blood calcium levels and the excretion of phosphate, consists of a single polypeptide chain of 84 amino acid residues, with a molecular weight of 9500. It has been found that the 34 amino acid residues at the *N*-terminal of the molecule are important for activity.

Most of these polypeptide hormones interact with specific receptors on the outer membranes of their target cells. One of the primary results of this interaction is an increase in activity of an enzyme producing a cyclic nucleotide, 3′,5′ cyclic AMP (see Figure 5.25, p. 226). Thus the concentration of the nucleotide within the cell rises, which in turn causes increases in selected enzyme activities, and probably preferential synthesis of some proteins within the cell. However, the details of these mechanisms are not yet completely understood, and changes in cyclic AMP level may not be the only factor involved in hormone action.

4.5.4. Antibodies

The antibodies form an important group of blood proteins, which protect the body against invasion by bacteria or viruses. When foreign protein, or polysaccharide, enters the bloodstream, specific antibodies are formed which can react with the foreign substances, the so-called *antigens*, to neutralize their effect and thus render them harmless. Thus antibodies afford protection against many bacterial- and viral-induced diseases, and stimulation of antibody formation in man underlies all immunization procedures. The injection of a modified and harmless form of a disease-producing agent induces the formation of antibodies which remain in the blood, sometimes for several years, and can minimize the effect of contact with the virulent disease-producing agent. Unfortunately, the tendency to neutralize and destroy any foreign substance by antibody formation is a serious barrier to successful transplantation of organs—a replacement heart, or kidney, taken from one individual would immediately be rejected by the recipient's body if powerful drugs suppressing the immune response were not administered. Antibody–antigen interactions in man are also believed to be involved in many allergic reactions, but as yet these processes are not well understood.

Antibodies in blood form a group of proteins known as the *immunoglobulins*. There are three main classes of immunoglobulins in adult human blood, and within each class there are a large number of different, but closely related, antibodies, for each antigen stimulates the synthesis of specific antibodies. All immunoglobulins are glycoproteins, and the main class contains 3% carbohydrate. The molecules of this class, the *IgG* antibodies, consist of four polypeptide chains, two identical 'heavy' chains of molecular weight about 50,000, and two identical 'light' chains of molecular weight about 25,000. The four chains are held together by disulphide bridges and non-covalent interactions, giving a molecule of weight about 150,000 (see Figure 4.29). The proline content of antibodies is fairly high, and the molecules are believed to possess little α-helical structure, although β-structure is present.

As blood contains so many closely related antibodies, differing in only a few amino acid residues per molecule, it was extremely difficult to obtain pure, homogeneous antibody for structural studies. In some cancerous conditions, however, one globulin is produced in excess, and it has been an

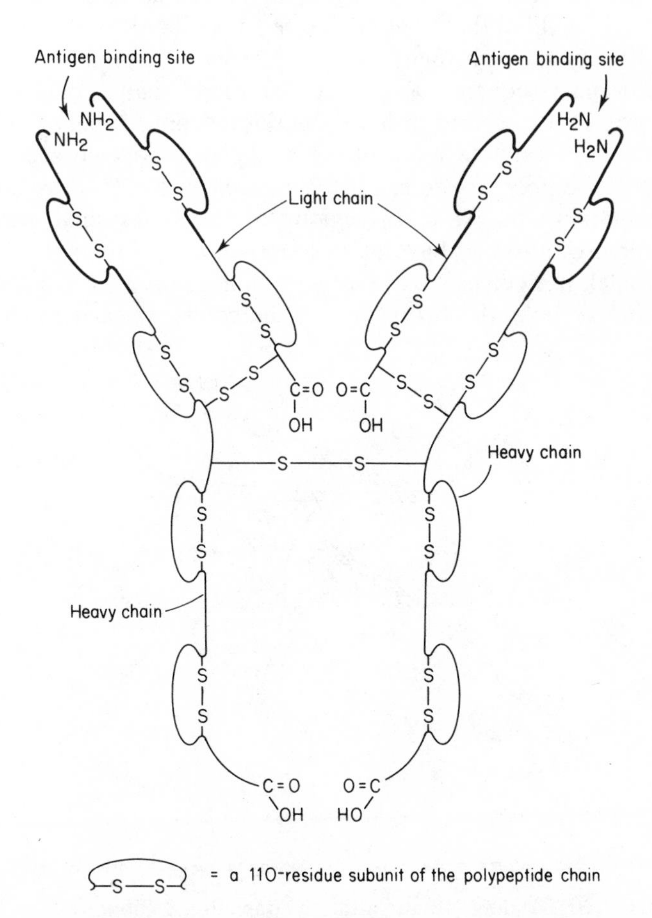

FIGURE 4.29. Schematic representation of an antibody molecule. Variable regions of the heavy and light chains are shown in heavy type

easier task to isolate and purify this protein. Amino acid sequence studies and electron microscopy have been carried out on these myeloma globulins (and also on other antibodies). Although these immunoglobulins are produced by a diseased animal, it is believed that they are normal proteins synthesized in abnormal amounts, and so the results of these structural studies are believed to be valid for all antibodies of this class.

Primary sequence studies have shown that each 'heavy' and 'light' chain may be considered to consist of a constant and a variable region. In the heavy chains of different antibodies, the sequences of the first 330 amino acids from the *C*-terminal end of the chain are very similar, and may be thought of as three sections of 110 residues of similar sequence. Many differences are found in the remaining 110 residues nearest the *N*-terminal end of the chain. A similar situation is found for the light chains. In this case the sequences of the 110 amino acids nearest the *C*-terminal end are similar, whilst the sequences of the remaining 110 acids of the chains can differ widely; within these variable regions there are hypervariable sequences. Thus, the specificity of each antibody is determined by the *N*-terminal primary sequence of its heavy and light chains. Each 110 amino acid sequence of the chains can be considered as a type of subunit, or domain, and the complete molecule is made up of twelve such domains. The folding of the polypeptide chain within each 'subunit' is similar, but the subunits containing the variable regions of the *N*-terminal ends of the heavy and light chains may have extra loops. In the subunits, two irregular, roughly parallel, β-sheets surround a hydrophobic core (see Figure 4.30). A disulphide bridge links the two sheets in the centre of each subunit, whilst

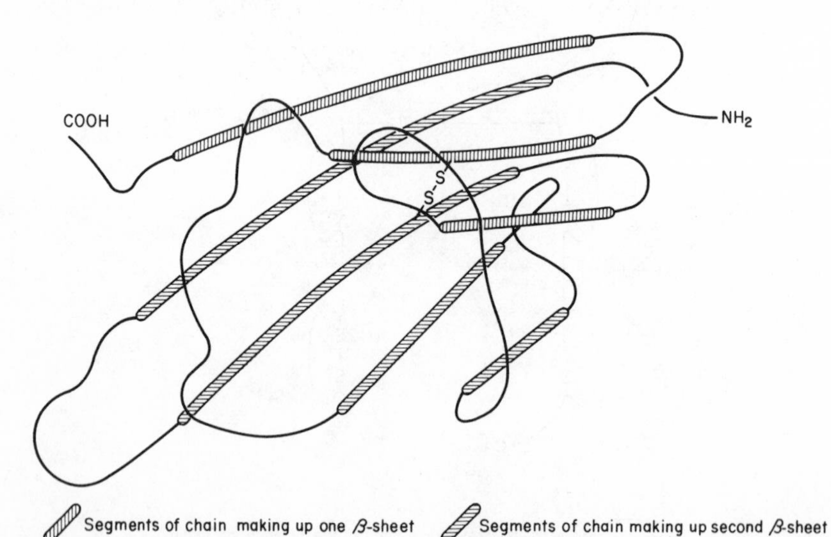

FIGURE 4.30. Folding of an antibody domain or subunit. The figure shows a constant domain. Variable domains are similar, but contain longer, or extra, loops joining the segments of β-structure

the subunits, or domains, are linked together in the whole molecule by segments of extended peptide chain.

Electron microscope and X-ray studies have indicated the the molecules can adopt a flexible Y-shape (see Figure 4.29) with the antigen binding sites near the *N*-terminal ends of the chains. The overall shape of the molecule may change, e.g. from a Y to a T, when antigen is bound, but there is probably little change at the antigen binding site itself. The overall shape change can take place as a result of the flexibility of the segments of chain which link individual domains.

There are two antigen-binding sites per molecule, each involving parts of one heavy and one light chain. X-ray studies suggest that the binding site is situated in a cleft formed by the hypervariable portions of a light and a heavy chain, i.e. the arrangement of domains in the molecule is such that the hypervariable regions of one heavy and one light chain are brought into close proximity at one end of the molecule to give a cleft, or groove, into which the antigen can fit. The exact shape of this groove varies according to the specificity of the antibody.

The nature of the binding sites has been studied by methods similar to those used for investigating enzyme active sites. The work has been complicated, however, by the difficulty of obtaining a homogeneous antibody specific for one antigen. Modification of certain antibodies has provided evidence that tyrosine and tryptophan residues form an important part of some binding sites, indicating that hydrophobic interactions are probably involved in antibody–antigen interactions. Other studies suggest that carboxyl groups may be involved in binding positively-charged antigens. In general, non-covalent forces are believed to be important in antigen–antibody binding, but the nature of these forces will depend on the nature of each particular antigen.

If an antigen is large, then antibody–antigen interaction can result in the formation of a precipitate. This effect 'neutralizes' the antigen so that harmful effects are not produced in the animal which it has invaded. The antigen has probably more than one antibody-combining site so that a three-dimensional, cross-linked network of alternating antigen–antibody can be formed, which will precipitate when it reaches a sufficient size. This precipitate does not contain a fixed proportion of antigen and antibody, but the composition depends on the ratio of components originally present.

One of the major unsolved problems involving antibodies concerns the mechanism by which antigen can stimulate the formation of antibody. It is known that antibodies to a new antigen are synthesized *de novo* from amino acids, and are not produced by the modification of existing proteins. It is also believed that an animal possesses the information necessary to make the antibody specific for any antigen before the introduction of the latter, and that rearrangement and processing of nucleic acids (see Chapter 5) must take place for new antibodies to be synthesized. Many of the details, however, of new antibody synthesis remain unknown.

4.5.5. Structural proteins

Structural proteins, because of their function, must be insoluble in water, a fact which makes investigations of their structure difficult. Although they are synthesized as soluble precursors, they often become incorporated into complicated, insoluble aggregates containing more than one type of polymer. The components of such aggregates can become cross-linked to give huge *supermolecular complexes*. Ultimately, it is hoped that the details of the assembly of these complexes will be well understood, but as yet we can often do little more than identify the components, and put forward hypotheses about the interactions between them.

Several of the methods used to solubilize these proteins for structural studies bring about chemical reactions, and hence changes to the protein molecules. Thus it is extremely difficult, in some cases, to draw conclusions about the properties of these proteins and their interactions in the native, insoluble state.

4.5.5.1. Collagen

Collagen is probably the most abundant protein in the animal kingdom, and may constitute 25% of the total protein of a mammal. It is the major component of skin, tendon, cartilage, bone, and teeth; it is important in the healing of wounds; and it becomes involved in a variety of diseases affecting joints, the skeleton and other connective tissue. Although inert and water-insoluble in adult animals, the collagen of young animals is more soluble. Also, a fraction of adult collagen is soluble in mineral acids, and structural studies have been made mainly on these soluble materials.

The fundamental structural unit of collagen is the *tropocollagen* molecule, 2800 Å long, 15 Å in diameter, and with a molecular weight of about 300,000. The amino-acid composition is unusual, for the protein contains about 33% glycine, whilst proline and hydroxyproline constitute another 20–30% of the total. Another unusual acid, hydroxylysine, is also found in smaller quantities. In addition, small amounts of carbohydrate (0.5–1.5% of glucose and galactose) are attached to the hydroxylysine residues in some collagen molecules, the amount and nature of this carbohydrate varying with the source of the protein.

The tropocollagen molecule consists of three left-handed helical polypeptide chains wound round each other to give a triplex helix (see Figure 4.31). The three chains are held together in helical conformation by hydrogen bonding between the hydrogen of the —NH group of glycine on one chain and the oxygen of the —C=O group of a residue on another chain. The formation of these bonds and the triple helix is possible only when every third residue in the chain is glycine. Breaking of these bonds by heat, or reagents such as urea, gives a denatured product with random structure known as *gelatine*. When a gelatine solution is cooled, some molecules can renature to give collagen-like molecules, whilst the

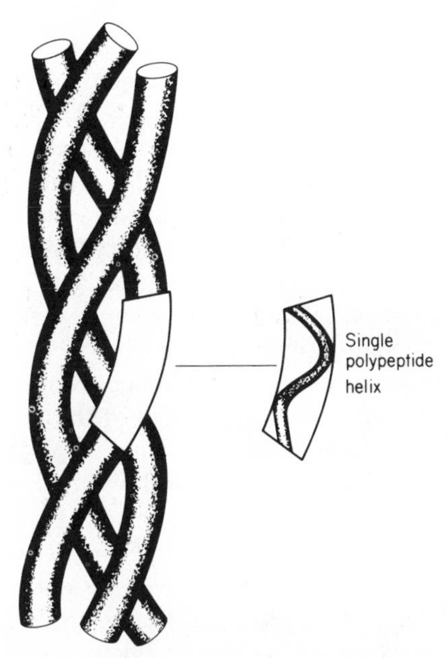

FIGURE 4.31. Schematic representation of the triple helix of the tropocollagen molecule

remainder give non-orientated, high molecular weight aggregates. The resultant increase in viscosity (or gel formation at high concentrations) are properties widely used in the food industry.

In addition to the hydrogen bonds, the tropocollagen molecules are stabilized by covalent, interchain cross-links, which involve lysine groups near the *N*-terminal ends of the molecule. The free $—NH_2$ groups of the lysine residues become oxidatively deaminated to aldehyde groups; the reaction is catalysed by an enzyme, lysine oxidase. The aldehyde groups can then take part in an aldol condensation reaction:

$$
\begin{array}{ccccc}
\cdots\!-\!NH\!-\!CH\!-\!CO\!-\!\cdots & & \cdots\!-\!NH\!-\!CH\!-\!CO\!-\!\cdots & & \cdots\!-\!NH\!-\!CH\!-\!CO\!-\!\cdots \\
| & & | & & | \\
(CH_2)_3 & & (CH_2)_3 & & (CH_2)_3 \\
| & & | & & | \\
CH_2\!-\!NH_2 & & CHO & & CH\!-\!OH \\
 & \longrightarrow & & \longrightarrow & | \\
CH_2\!-\!NH_2 & & CHO & & CH\!-\!CHO \\
| & & | & & | \\
(CH_2)_3 & & (CH_2)_3 & & (CH_2)_2 \\
| & & | & & | \\
\cdots\!-\!NH\!-\!CH\!-\!CO\!-\!\cdots & & \cdots\!-\!NH\!-\!CH\!-\!CO\!-\!\cdots & & \cdots\!-\!NH\!-\!CH\!-\!CO\!-\!\cdots
\end{array}
$$

This is one of several different cross-links between lysine and modified lysine residues found in collagen (see also Section 4.5.5.4, p. 150).

These cross-linkages are heat-stable, and so they are present in gelatine. In fact, the gelatine molecules in which the three polypeptide chains are held together covalently are those which renature most easily to the triple helix form. More complex tri- and tetra-functional cross-links are believed to exist in insoluble collagen (see Figure 4.32). These may involve the side-chains of histidine residues as well as of lysine residues.

In most mammalian collagen, two of the chains of one molecule are identical in amino acid composition—these are called α_1-chains—whilst the third chain, the α_2-chain, is different. Some collagens, however, are made up of three identical chains. The complete amino acid sequence of some collagen chains has been established. This investigation was a major undertaking for each chain contains approximately 1000 amino acids. The sequences so far studied indicate that glycine occupies every third position throughout almost the whole length of each of the polypeptide chains; only near chain-ends is this regularity not found. Regions relatively rich in non-polar amino acids, such as proline and hydroxyproline, alternate along the chain with regions richer in polar amino acids. The non-polar regions contain high amounts of the tripeptide, —Gly—Pro—Hyp. On each chain of the triple helix there are three residues per turn. Where —Gly—Pro—Hyp— repeating units are found, the side-chain rings of both proline and hydroxyproline are on the outside of the triple helix, whilst the side-chain hydrogen of glycine can pack inside the triple helix.

The amino acid sequences near the *N*-terminal ends of the chains are highly polar, and do not contain as much glycine as the remainder of the molecule. Thus the chain-ends may not be capable of taking up the coiled

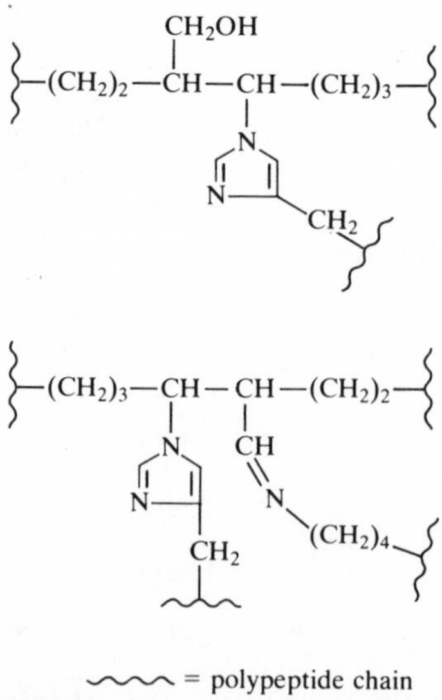

~ = polypeptide chain

FIGURE 4.32. Possible intermolecular cross-links in collagen, joining three or four polypeptide chains

helix structure of the main part of the molecule. There is no free amino group at the *N*-terminal ends of most collagen molecules for the terminal acid is pyrrolidone-5-carboxylic acid. This is probably formed by cyclization of a terminal glutamic acid residue:

$$\begin{array}{c} CH_2 \\ H_2C \quad\quad C{=}O \\ HO \\ HOOC{-}CH{-}NH{-}\cdots \end{array} \longrightarrow \begin{array}{c} CH_2 \\ H_2C \quad\quad C{=}O \\ HOOC{-}CH{-}N{-}\cdots \end{array}$$

Glutamic acid residue — Pyrrolidone-5-carboxylic acid

In living organisms, numbers of tropocollagen molecules are found closely packed together parallel to each other, forming *collagen fibrils*. The molecules are probably not linked end-to-end, but there may be a gap of approximately 400 Å between the end of one molecule and the start of the next. The molecules are, however, bonded side-by-side, and may be regularly staggered by approximately a quarter of their length (see Figure 4.33). This arrangement is thought to produce the characteristic banding of collagen fibrils seen in the electron microscope. This banding normally has a repeat distance of 600–700 Å. The bands are believed to correspond to the alternating polar and non-polar regions of the molecules.

The known amino acid sequences show that an approximate quarter-stagger of molecules brings stretches of polar residues close together on adjacent molecules. A similar situation arises for hydrophobic residues. Thus molecules are held together by non-covalent interactions as well as by cross-links. The situation is not clear, however, and the suggestion has been made that the quarter-stagger hypothesis is insufficient to explain all the properties of the fibrils. It has further been suggested that the staggered tropocollagen molecules are grouped together in multistranded ropes. Controversy exists concerning the numbers of molecules in these 'ropes' for 2, 4, 5, and 8 have been suggested.

Electrostatic attraction between molecules in the fibril is, by itself, insufficient to account for the high tensile strength so characteristic of collagen, and indeed covalent links are believed to hold molecules together in the mature, insoluble fibrils. These intermolecular bonds are closely related to the interchain covalent bonds within one tropocollagen molecule, for they involve the aldehyde derived from the side-chain of a lysine residue. However, the intermolecular bonds may differ from the

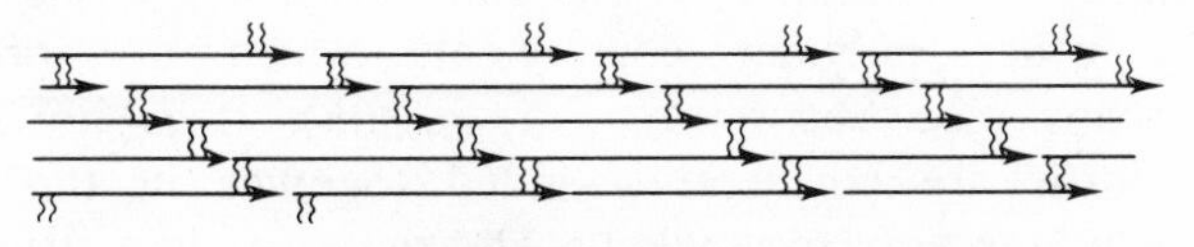

~~= intermolecular cross-links

FIGURE 4.33. Schematic representation of arrangement of molecules in a collagen fibril

intramolecular links, as here the lysine aldehyde may condense with the side-chain of hydroxylysine, or the corresponding aldehyde. If tropocollagen molecules are lined up head-to-tail and staggered by a quarter of their length, a lysine and hydroxylysine residue of one molecule are in register with a hydroxylysine and a lysine residue of the adjacent molecule. Thus intermolecular links could readily be formed:

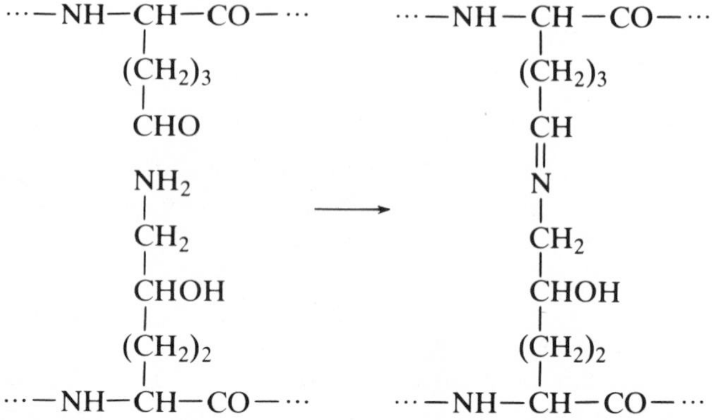

$$\begin{array}{ccc}
\cdots-NH-\underset{|}{CH}-CO-\cdots & & \cdots-NH-\underset{|}{CH}-CO-\cdots \\
(CH_2)_3 & & (CH_2)_3 \\
| & & | \\
CHO & & CH \\
 & & \| \\
NH_2 & & N \\
| & \longrightarrow & | \\
CH_2 & & CH_2 \\
| & & | \\
CHOH & & CHOH \\
| & & | \\
(CH_2)_2 & & (CH_2)_2 \\
| & & | \\
\cdots-NH-CH-CO-\cdots & & \cdots-NH-CH-CO-\cdots
\end{array}$$

a Schiff base

The importance of these cross-links in determining fibril strength has been shown by a study of the disease, lathyrism, which results in skeletal deformities and loss of fibril strength. This disease can be caused by substances such as β-aminopropionitrile, which are now known to inhibit oxidation of lysine to the aldehyde, and thus prevent cross-link formation. β-Aminopropionitrile acts as an inhibitor of lysine oxidase, for it closely resembles the natural substrate, the side-chain of lysine:

$$N{\equiv}C-CH_2-CH_2-NH_2 \qquad -(CH_2)_2-CH_2-CH_2-NH_2$$

β-Aminopropionitrile — Side-chain of lysine residue

The extra reactive $-C{\equiv}N$ group attacks the enzyme giving a covalently modified polypeptide which is no longer active.

The solubility and stability of collagen is also believed to be related to the numbers of cross-links in the fibrils. The tanning process, which converts raw hide to leather, involves introducing additional cross-links into the collagen by using aldehydes.

In living tissues, the collagen fibrils are packed parallel to one another in the form of fibres. Formation of fibrils and fibres of a tissue is probably influenced by other polymer components which are present, e.g. glycosaminoglycans (see Section 6.5.2.5, p. 306). It is not yet known whether the fibrils are covalently bonded together in the fibres, but extensive cross-linking appears unlikely. The rigid, rod-like structure of the tropocollagen molecules, and their arrangement in the fibrils, gives the collagen fibres their characteristic, high, tensile strength and resistance to stretch, which is so important to their function in multicelled organisms.

The arrangement of fibres in a particular tissue is related to the function of that tissue. Thus, in tendon, the fibres are closely packed parallel to one another, giving the tensile strength of a light steel wire; in skin, the fibres are randomly orientated and woven into a feltwork to produce a structure with some flexibility in all directions; and in the cornea of the eye, laminated sheets of parallel fibres are found, the direction of the fibres in adjacent sheets being at 90° to one another. (It is not yet known how this gives the transparent nature of the cornea, but where the organization breaks down, as in corneal scars, the resultant tissue is opaque.) The addition of calcium salts to the collagen fibre network of bones and teeth confers the necessary rigidity for these structures.

The biosynthesis of collagen is complex. Collagen is synthesized as a larger precursor, *procollagen*, which undergoes specific cleavage at both the *N*- and *C*-terminal ends to yield tropocollagen, after exportation from the synthesizing cell. Hydroxyproline and hydroxylysine are not incorporated directly into the growing polypeptide chains (see Chapter 5), but proline and lysine residues are hydroxylated after their insertion into the chain. This hydroxylation is facilitated in the presence of ascorbic acid, vitamin C. Some hydroxylysine residues may be further modified by the attachment of galactose or glucosylgalactose residues. After the addition of carbohydrate, the procollagen molecules are secreted into the extracellular space and are hydrolysed to give tropocollagen, which then aggregates to form insoluble fibrils. At this stage, some lysine residues may be oxidized to the aldehyde, and intra- and inter-molecular cross-links can be formed to give mature collagen fibres.

Removal of the 'extra' peptides from procollagen to give tropocollagen is necessary for the correct formation of fibrils and fibres. In animals where this process is blocked by, for example, a genetic defect, disorders of connective tissue are found, and stretchable, or fragile, skin or hide and easily dislocated joints are observed.

The decrease in solubility of collagen with age is probably due to changes in the intermolecular bonds, but the reason for the deleterious changes in collagen in old age are not yet known.

4.5.5.2. Keratin

Keratin is the main structural protein of the protective outer covering of higher animals. It is the chief component of hair, skin, fur, wool, horn, and feathers. Reinforced with calcium salts it is also found in hooves, nails, claws, and beaks. Keratin is extremely insoluble, and so structural studies have again proved to be difficult. It is characterized by a high sulphur content, and, indeed, soluble protein fragments can be obtained only after reduction or oxidation of disulphide bonds.

The structure of keratin varies widely with the source of the material. For example, the keratin of wool consists of many different polypeptides which may be linked together by disulphide bonds. Three groups of polypeptide can be distinguished in wool fibres: the so-called low-sulphur, high-sulphur, and high-glycine proteins. Wool fibres are made up of microfibrils embedded in an amorphous matrix; the low-sulphur proteins form the microfibrils, whilst the other two groups of protein are located in the matrix.

The low-sulphur proteins have molecular weights of 40,000 to 58,000, and the amino acid sequences of segments of the polypeptide chains have been determined. In some cases, a 7-residue periodicity of non-polar residues has been found. X-ray diffraction data indicate that these proteins take up a right-handed α-helical secondary structure (see Figure 4.7a, p. 99). In wool, protofibrils probably consist of three α-helices wound round each other to give a superhelix analogous to the collagen triple helix, but winding in the opposite sense (compare Figure 4.31, p. 143). The 'coiled coils' may be held together by disulphide bonds. The basic unit is 160 Å long, but can form aggregates more than 2000 Å long. Indeed, several protofibrils become bonded together side-by-side, as well as end-to-end, to form microfibrils, 70 Å in diameter.

The high-sulphur proteins can contain up to 30% cystine, and human hair, with a fibril and matrix arrangement similar to wool, is one of the richest known sources of cystine. The molecular weights of high-sulphur proteins are generally in the range 10,000 to 28,000, and several amino acid sequences have been determined. Repeating sequences such as -Pro-Ser-Cys-Cys-Gln-Pro-Ile-Cys-Cys-Asp- have been found. There is no evidence of an α-helical conformation in these, or the high-glycine, proteins.

The high-glycine proteins are small, with molecular weights less than 10,000. The content of aromatic amino acid residues is also high, so that glycine and aromatic residues may make up 50% of the polypeptides. Some sequences have been determined, but as yet, little is known of the arrangement of these proteins in the matrix, or of the molecular interactions of the different groups of protein in a wool or hair fibre.

Feather keratin appears to be very different. The three groups of proteins of wool have not been found in feather, and no amino acid sequence homology has been discovered between feather and hair keratins. The micro fibrils of feather seem to contain β-sheet rather than α-helical structures.

Hair and wool are extensible, and can return to their original length after moderate stretching. Stretching of the fibril probably breaks some hydrogen bonds stabilizing the α-helix, and indeed stretched wool appears to have a pleated sheet rather than an α-helical structure. However, the helices are believed to be cross-linked by disulphide bonds, and when the stress is removed, these bonds restore the original α-helical structure, and so shorten the stretched fibre. The retention of hair fibre structure by disulphide bonds is utilized in the hairdressing industry, where the disulphide bonds are broken by chemical reduction, and reformed after the hair fibres have been shaped as desired. In human hair there is, on average, one disulphide bond per six amino acid residues. Keratin fibres can also assume an almost permanent change in shape or length when steamed under stress—a property much used in the clothing trade. The keratins of horn, claw, and hooves are much less extensible than wool or hair.

4.5.5.3. Fibroin

One of the main constituents of silk is fibroin, which again consists of multiple units cross-linked to give a strong fibre. The protein is extremely insoluble, and consequently its structure is not yet well characterized. The amino acid composition of fibroin depends on the species of origin, but the fibroin of the commercial silkworm has a high proportion of serine, alanine, and glycine. Limited sequence studies have shown that glycine occupies every second position along parts of the polypeptide chain, and sequences such as

$$\text{Gly-Ala-Gly-Ala-Gly-(Ser-Gly-[Ala-Gly]}_2)_8\text{-Ser-Gly-Ala}$$

have been found. The structure of the polypeptide chains is believed to be partly crystalline and partly amorphous. The crystalline regions contain mostly glycine, alanine and serine, and consist of antiparallel, pleated sheets (see Figure 4.7c, p. 99). The chains are held together within one sheet by hydrogen bonds, and in silkworm fibroin all the serine and alanine residues are situated on one side of the sheet and the glycine on the other. Small side-chains on amino acid residues are necessary for this type of regular arrangement. The amorphous regions contain more of the amino acids with bulky side-chains. The molecular weight is believed to be 365,000, and the molecule may consist of three small components of weight 26,000 and one large component of weight 280,000. These components are probably cross-linked together.

Silk fibres are very strong, for resistance to tension is borne by the covalent bonds of the backbone of the polypeptide chains. As these chains in the pleated sheets are already fully extended, the fibres are relatively inextensible. The amorphous regions account for their very limited extensibility. The sheets are held together by relatively weak van der Waals' forces, and so silk is quite flexible.

Fibroin is synthesized in a soluble form, and is secreted as a thick solution. The protein then solidifies to an insoluble thread as the silk is spun by the insect.

4.5.5.4. Elastin

Elastin is another important and interesting structural protein, for it has many of the properties of a natural rubber (see Chapter 7). It is one of the major proteins of the elastic fibres of higher animals, and gives elasticity to arterial walls and ligaments, whilst a little is found in tendons and skin.

Again, the insolubility of the polymer has made structural studies difficult. Elastin is synthesized as a soluble precursor, *tropoelastin*, with a molecular weight of about 70,000. The tropoelastin is later covalently cross-linked to form insoluble elastin. Precipitation of insoluble elastin may be influenced by the presence of other polymers in the tissue, e.g. glycosaminoglycans (see Section 6.5.2.5, p. 306). The protein contains a high proportion of glycine, valine, alanine, and proline, and also a little hydroxyproline, but only a small proportion (5%) of polar amino acids, i.e. over 90% of the amino acid residues are non-polar. Some work has been done on the amino acid sequence, and multiple copies of sequences such as —Lys—Ala—Ala—Lys— and —Lys—Ala—Ala—Ala—Lys— have been found. These sequences are believed to come from cross-linking regions. Sequences such as —(Pro—Gly—Val—Gly—Val—Ala)$_n$— are thought to be situated between the cross-links. The nature of the cross-links has been investigated extensively, and these are formed by modification of lysine residues within the tropoelastin. Two main types of cross-link have been found. The first is similar to those found in collagen, i.e.:

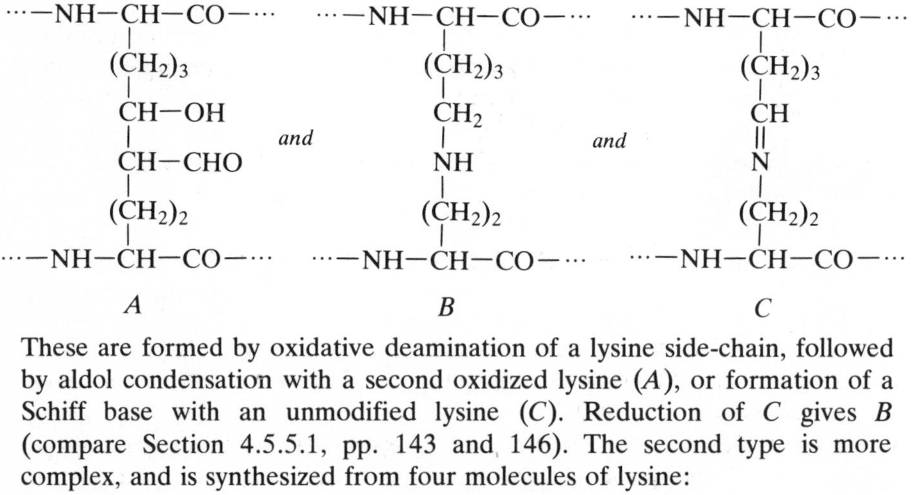

$$\begin{array}{ccccc}
\cdots\text{—NH—CH—CO—}\cdots & & \cdots\text{—NH—CH—CO—}\cdots & & \cdots\text{—NH—CH—CO—}\cdots \\
| & & | & & | \\
(CH_2)_3 & & (CH_2)_3 & & (CH_2)_3 \\
| & & | & & | \\
\text{CH—OH} & & CH_2 & & CH \\
| & \textit{and} & | & \textit{and} & \| \\
\text{CH—CHO} & & NH & & N \\
| & & | & & | \\
(CH_2)_2 & & (CH_2)_2 & & (CH_2)_2 \\
| & & | & & | \\
\cdots\text{—NH—CH—CO—}\cdots & & \cdots\text{—NH—CH—CO—}\cdots & & \cdots\text{—NH—CH—CO—}\cdots \\
A & & B & & C
\end{array}$$

These are formed by oxidative deamination of a lysine side-chain, followed by aldol condensation with a second oxidized lysine (*A*), or formation of a Schiff base with an unmodified lysine (*C*). Reduction of *C* gives *B* (compare Section 4.5.5.1, pp. 143 and 146). The second type is more complex, and is synthesized from four molecules of lysine:

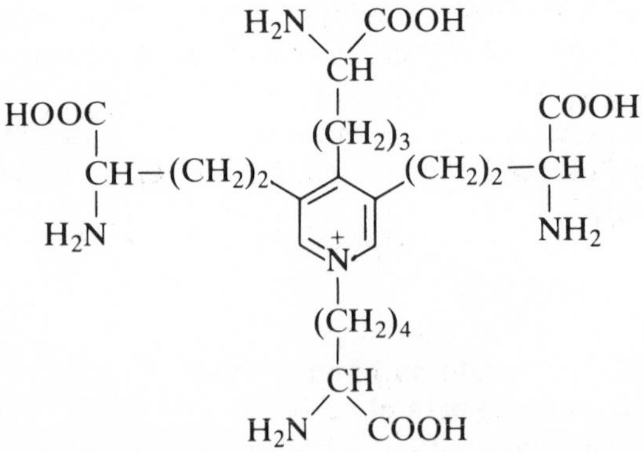

$$\begin{array}{c}
H_2N\;\;COOH \\
\diagdown\;\diagup \\
CH \\
| \\
(CH_2)_3 \\
HOOC{-}CH(NH_2){-}(CH_2)_2{-}\text{[pyridinium ring, }N^+\text{]}{-}(CH_2)_2{-}CH(NH_2){-}COOH \\
| \\
(CH_2)_4 \\
| \\
CH \\
\diagup\;\diagdown \\
H_2N\;\;COOH
\end{array}$$

Desmosine

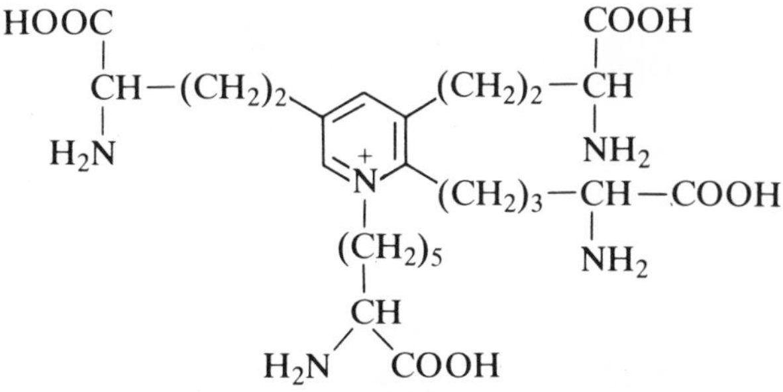

HOOC — CH(H_2N)—$(CH_2)_2$— ; COOH — CH(NH_2)—$(CH_2)_2$— ; $(CH_2)_3$—CH(NH_2)—COOH ; $\overset{+}{N}$—$(CH_2)_5$—CH(H_2N)(COOH)

Isodesmosine

Such cross-links are obviously capable of joining up to four polypeptide chains together. The importance of these cross-links in giving strength and elasticity to elastin has been deduced from studies of animals where cross-linking has been inhibited.

Little detail is known of the secondary structure, which X-ray studies have indicated may be mostly random coil, but which is believed to contain a small proportion of α-helix. Whilst elastin is hard and inextensible when dry, it behaves like a rubber when swollen in water. As the elastic properties depend on the nature of the solvent, it has been suggested that the structure of elastin may be regarded as a number of globular molecules, or possibly flexible coils, with some hydrophilic groups on the surface, and hydrophobic groups in the interior. During stretching, the hydrophobic groups may be forced into water, and the energy for contraction would then come from the return of these groups to a non-polar environment. The various cross-links join neighbouring globules, or coils, to give a network assembly resembling a three-dimensional mattress spring. This assembly can then behave as a rubber (see Chapter 7).

4.5.5.5. Actin and myosin

The contraction of striated muscles is believed to involve structural proteins, the most important of which are actin and myosin. Actin and myosin are also found in many non-muscle cells. Together they are probably important for cell movement. Where actin occurs alone, the biopolymer is most likely involved in defining, or maintaining, the shape of a cell.

Striated muscles are made up of parallel fibres, which in turn consist of striated fibrils, called *myofibrils*. In the electron microscope, the striations can be seen as a pattern of light and dark bands, which consist of overlapping thick and thin filaments in the myofibrils. The thick filaments are mainly myosin, whilst actin forms an important part of the thin filaments. When the muscle contracts, these filaments slide past one another (see Figure 4.34).

It is believed that each thick filament contains 200–400 molecules of myosin. The molecules are about 1600 Å long, with a molecular weight of

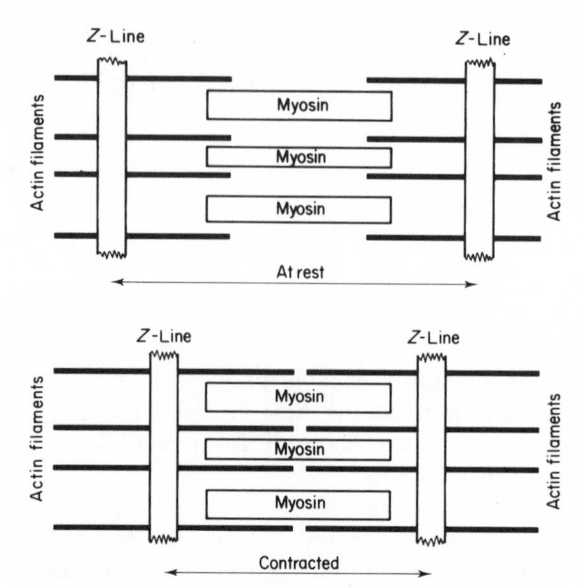

FIGURE 4.34. Schematic representation of the movement of actin and myosin filaments during the contraction and relaxation of muscle. The *Z*-line is a structural element of striated muscle, itself made up of protein components

470,000. They consist of two large polypeptide chains of molecular weight about 200,000, and four small chains of molecular weight 15,000–20,000. The molecules have a rod-like portion 1340 Å long, consisting of the *C*-terminal stretches of the large or 'heavy' chains. These chains have an α-helical structure, and are wound round each other to give a double-stranded coiled coil. Attached to this is a less helical head made up of the small or 'light' chains, and the *N*-terminal regions of the heavy chains; this head actually consists of two globules (see Figure 4.35a). All the proline in the structure is found in this 'head', whilst the 'tail' has a low content of amino acids with small hydrocarbon side-chains. Unusual amino acid residues, such as 3-methylhistidine and *N*-methyllysine, are found in the head. Amino acid sequences of some fragments of myosin are known.

The heads of the molecules project outwards from the thick filaments to form cross-bridges with the actin filaments, and have enzymic activity which is important in muscle contraction. Each globule of the myosin head carries one enzymic site, and one site for binding actin.

The enzymic activity catalyses hydrolysis of a nucleotide, adenosine triphosphate (ATP, see Figure 5.2b, p. 171), to adenosine diphosphate (ADP) and inorganic phosphate. The reaction is universal in living organisms as an energy-yielding reaction. In this case, the energy produced is used to slide the thick and thin filaments past each other to give muscle contraction. Although the details of the sliding mechanism are not

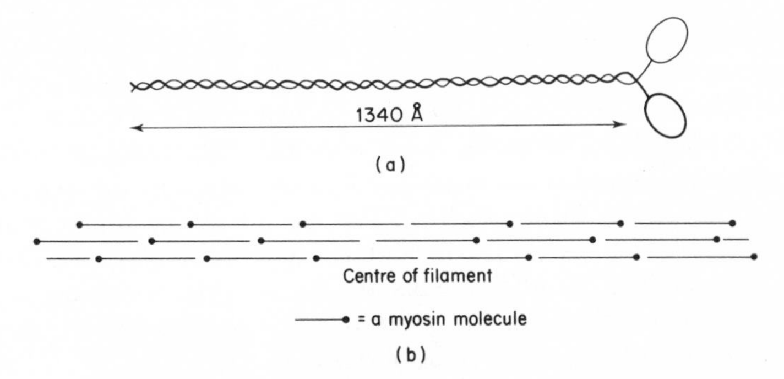

FIGURE 4.35. (a) Schematic representation of myosin molecule; (b) arrangement of myosin molecules in a thick filament

completely understood, the ATP hydrolysis is thought to provide energy for a conformational change in myosin, which then results in a movement of filaments relative to each other.

The myosin molecules lie antiparallel to each other at the centre of a thick filament, and parallel to each other in a regularly staggered arrangement in the rest of the filament (see Figure 4.35b).

Actin can be obtained in a globular form consisting of a single polypeptide chain with a molecular weight of 42,000. The molecules contain a fairly high amount of proline, and so the α-helix content is low. The amino acid sequences of several actins are now known, and the sequence varies with the source of the protein. Striated-muscle actin, like myosin, contains 3-methylhistidine. As yet the function of this unusual residue in muscle is unknown. In the presence of KCl and Mg^{2+}, the globular protein polymerizes to a fibrous form known as *F-actin*, and it is this form which is found in muscle. The structure of *F*-actin consists of two helically intertwined strands of spheres, where each sphere is a molecule of globular actin (see Figure 4.36).

The exact details of how the muscle proteins bring about contraction are not yet known. However, it is thought that the cross-bridges from myosin to actin can exert a longitudinal force on the actin fibres when muscle contracts, thus causing the thin filaments to slide past the thick ones.

At low calcium levels ($<10^{-6}$M), striated muscle is in the relaxed state, and no cross-bridges between myosin and actin are seen. Adenosine triphosphate can be bound to the myosin head and hydrolysed to adenosine diphosphate, which is not released at this stage. The energy

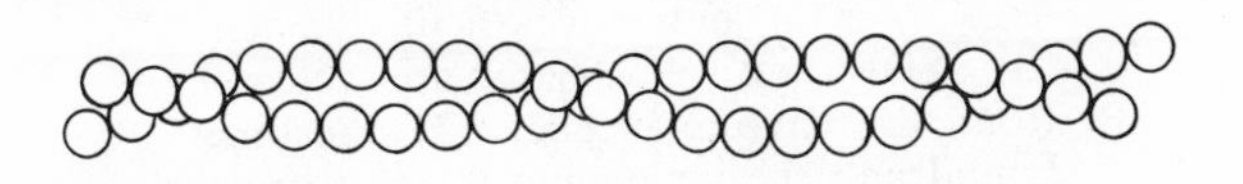

FIGURE 4.36. The helical structure of *F*-actin

provided by ATP hydrolysis is believed to be stored in some way in the myosin head.

In the presence of calcium, the myosin head combines with actin, i.e. forms a cross-bridge between thick and thin filaments in muscle. A conformational change in the myosin alters the angle of the head to actin and moves the actin filament about 100 Å. Adenosine diphosphate then dissociates from the myosin head and another molecule of ATP can be bound. The binding of ATP breaks the cross-bridge between actin and myosin, so that another 'round' of movement can take place (see Figure 4.37). In the meantime, other cross-bridges have formed between the actin and myosin filaments so that no backslip takes place. Calcium uptake by the membrane structures of the muscle cell brings about relaxation.

Other proteins of the thin filaments, *troponin* and *tropomyosin*, are involved in muscle relaxation, preventing the formation of the

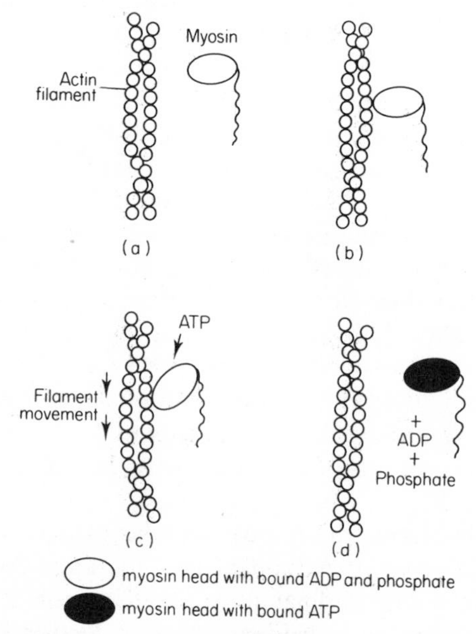

FIGURE 4.37. (a) Muscle in relaxed state: myosin head contains bound ADP and phosphate. (b) After increase in Ca^{2+} concentration, myosin head becomes bound to actin. (c) Change in conformation of myosin alters angle of head to actin and moves actin filament. (d) ADP and phosphate are released from the myosin head, ATP becomes bound and the actin–myosin cross-bridge breaks. ATP can then be hydrolysed to ADP and phosphate and the cycle of steps (a)–(d) is repeated

cross-bridges. Tropomyosin consists of two coiled α-helical chains which fit into the grooves in *F*-actin, whilst troponin is made up of three subunits—troponin C which binds calcium, troponin T which binds to tropomyosin, and troponin I which binds to actin and inhibits the interaction of actin and myosin. At low calcium levels, the troponin–tropomyosin inhibits cross-bridge formation and muscle is relaxed. When a nerve signal releases calcium into the muscle cell, this calcium is bound by the troponin C subunit. The binding brings about a conformational change in troponin which, in turn, is relayed to tropomyosin. The tropomyosin molecules change their position in the actin grooves, allowing the myosin-binding sites on actin to become exposed. Thus myosin–actin interaction can take place, and the muscle contracts.

4.5.6. Other functions

Nucleoproteins are found in chromosomes and ribosomes (see Chapter 5). The chromosomes, which contain the genetic information of a living organism, contain nucleic acid, and basic and acidic proteins. As yet, less is known of the acidic proteins than of the basic proteins, the *histones*.

Histones contain high amounts of lysine and arginine, and are capable of binding negatively-charged nucleic acids by electrostatic attractions. It has been found that histones can be divided into five main types: two arginine-rich, two moderately lysine-rich, and one lysine-rich. The five types are now known as histones *H*1, *H2A*, *H2B*, *H*3, and *H*4; histone *H*3 has the lowest lysine:arginine ratio (0.72), whilst for histone *H*1, the ratio can be as high as 20. Histones are single polypeptide chains of molecular weight about 10,000–20,000; the molecular weights are: *H*1 = 21,000, *H2A* = 14,000, *H2B* = 13,800, *H*3 = 15,300, and *H*4 = 11,300. These polymers have a low content of sulphur-containing and aromatic amino acids, but may contain methylated and acetylated lysine and phosphorylated serine residues. Histidine and threonine residues can also become modified. The modification of these residues may be directly linked with the function of the histones on the chromosomes. Indeed, phosphorylation of serine can be observed in response to some hormones.

Sequence studies indicate that the distribution of residues along the chains is irregular—very basic sequences may be found at one end of a molecule, whilst other regions contain more acidic, aromatic, and apolar residues. Thus most histones have a concentration of positive charge in the *N*-terminal half of the polypeptide chain; only for *H*1 are there more positively-charged residues in the *C*-terminal half. The clusters of basic side-chains are probably important for binding negatively-charged deoxyribonucleic acid, DNA (see Chapter 5). In the arginine-rich histones, the primary sequences are very similar in different animals. Thus two residues of *H*4 have changed in 1200×10^6 years of evolution. This conservation of amino acid sequence throughout evolution suggests that the complete molecular structure must be importnat for its function. The sequence of histone $H1$ is the most variable from species to species.

Little work has so far been carried out on the three-dimensional structure of histones. However, it is believed that these proteins have a definite tertiary structure when complexed with DNA, although they behave as random coils in solution, because of the large number of positive charges along the chains. The detailed structure of DNA–histone complexes is, at present, unknown, but it is believed that the histones are bound in the grooves of the DNA helices (see Chapter 5). Histones may enhance supercoiling of the DNA, and are believed to be important for the correct packing of DNA in the chromosomes.

It is now known that the five histones are complexed to DNA in chromosomal material in a repeating pattern, which is nearly invariant amongst eukaryotic (higher) organisms and cell types. The pattern looks like beads on a string and the repeating unit, called a nucleosome, consists of a 'bead' or core particle, with a link to the adjacent core particle. The core particle is about 100 Å in diameter, and contains two molecules each of histones $H2A$, $H2B$, $H3$, and $H4$ and a stretch of DNA (140 base-pairs long; see Chapter 5). Histones $H3$ and $H4$ form a core around which the DNA is wound, and to which histones $H2A$ and $H2B$ are attached. Thus $H3$ and $H4$ probably define the folding of the DNA molecule, whilst $H2A$ and $H2B$ stabilize it. The linker region is a shorter stretch of DNA on which one molecule of $H1$ is less-tightly bound. The linker itself is highly folded in the chromosome.

The binding of histones to DNA may prevent formation of RNA or new DNA; thus (as will be described in Chapter 5), histones are probably involved in the complex mechanisms of controlling protein and DNA synthesis. Histones are likely non-specific repressors of protein synthesis, whilst the more acidic proteins of chromosomes may be specific agents of control.

Very many other functions for proteins are known, e.g. as regulators of protein synthesis, and in the ribosomes as parts of the 'machinery' for this synthesis (see Chapter 5), but in the space of this chapter it has only been possible to describe some of those which are best understood.

4.6 CHEMICAL SYNTHESIS

The chemical synthesis of a peptide can be achieved in a general way by blocking all but one of the functional groups of an amino acid, activating this remaining group (which must be $-NH_2$ or $-COOH$), and then allowing it to react with a suitably protected derivative of a second amino acid. Very great problems had to be overcome in this field of classical organic synthesis in the choice of the blocking groups and the type of condensation reactions. Synthesis of small peptides was first successfully achieved by the pioneering work of Fischer, but more recently methods have been evolved for the larger products and proteins.

This work will be illustrated here by a brief description of two separate approaches to the chemical synthesis of the enzyme, ribonuclease A, as well as general comments on some problems of protein synthesis.

4.6.1. Solid-phase synthesis

An important advance in techniques for synthesizing peptides and proteins was made by Merrifield when he devised the method of attaching the *C*-terminal end of an amino acid to an *insoluble resin*. Further amino acids are then added in a stepwise manner, according to the known sequence in the protein to be synthesized, so that the growing polypeptide chain, covalently bonded to the resin, remains insoluble. Excess reagents may thus be washed out easily, and at the end of the reaction sequence the polypeptide chain can be obtained in a fairly pure form by cleavage from the resin support.

A common resin used for protein synthesis is a chloromethyl derivative of a styrene–divinylbenzene polymer, and the first amino acid is attached to the resin by an esterification reaction between the resin and a salt of the acid. Other derivatives of polystyrene resins are now available, where the completed polypeptide can be cleaved from the resin more easily or, in some cases, where the resin–peptide link is more stable. Different solid supports are now being developed, e.g. porous glass beads, to which amino acids can be attached; these allow the various reactions of stepwise synthesis to be carried out with very high yields.

To prevent side-reactions, the amino group of the amino acid must be protected, and the most common protecting group is a tertiary butyloxycarbonyl (Boc) group:

$$(CH_3)_3CO-\overset{\overset{\displaystyle O}{\|}}{C}-NH-CHR-COO^- + ClCH_2-C_6H_4-RESIN \longrightarrow$$

Salt of butyloxycarbonyl amino acid — Chloromethyl resin

$$(CH_3)_3CO-\overset{\overset{\displaystyle O}{\|}}{C}-NH-CHR-COOCH_2-C_6H_4-RESIN.$$

Boc-amino acyl resin

(It is extremely important for peptide synthesis that blocking groups be chosen to minimize the risk of racemization of the amino acids, i.e. conversion of L- to D- amino acids. In general, a protected amino acid can racemize according to the scheme:

$$R^1-\overset{\overset{O}{\|}}{C}-\underset{\underset{H}{|}}{N}-\underset{\underset{H}{|}}{\overset{\overset{R^2}{|}}{C}}-\overset{\overset{O}{\|}}{C}-X \underset{\substack{+H^+ \\ \text{D,L-product}}}{\overset{-H^+ \text{ (basic conditions)}}{\rightleftharpoons}} H-\overset{\overset{O=C-X}{|}}{C}(-R^2)-N=C(R^1)-O^- \underset{\substack{+X \\ \text{D,L-product}}}{\overset{-X}{\rightleftharpoons}}$$

$$\text{oxazolone: } R^2-C(H)-C(=O)-O-C(R^1)=N \underset{\substack{+H^+ \\ \text{racemization}}}{\overset{-H^+}{\rightleftharpoons}} R^2-C^- \text{ (ring) } -C(=O)-O-C(R^1)-N$$

An oxazolone

The oxazolone can lose a proton reversibly, and so can racemize to give a D, L-mixture. The process depends partly on the nature of the R^1 group, and moieties of the type $R^1 = R^3$—O, e.g. the Boc group, tend to suppress racemization.)

The Boc group is then removed so that the amino group of the acid may form a peptide bond with the second acid of the chain. The relatively mild conditions necessary for the removal of the Boc group may leave the covalent bond between the first acid and the resin intact, and do not racemize the amino acid residue:

$$(CH_3)_3CO-\overset{\overset{O}{\|}}{C}-NH-CHR-COOCH_2-C_6H_4-RESIN \xrightarrow[\text{acid}]{\text{HBr/acetic}}$$

$$Br^-NH_3^+-CHR-COOCH_2-C_6H_4-RESIN + (CH_3)_2C=CH_2 + CO_2$$

$$\xrightarrow[\text{dimethylformamide}]{\text{ethylamine}} NH_2-CHR-COOCH_2-C_6H_4-RESIN$$

The second Boc-amino acid can then be coupled to the first using dicyclohexyl carbodiimide (DCC); $C_6H_{11}-N=C=N-C_6H_{11}$ as activator:

$$NH_2-CHR^1-COOCH_2-C_6H_4-RESIN + Boc-NH-CHR^2-COOH$$

$$\xrightarrow[\text{(in } CH_2Cl_2 \text{ or dimethylformamide)}]{\text{DCC}} Boc-NH-CHR^2-CO-NH-CHR^1-COOCH_2-C_6H_4-RESIN$$

The free incoming amino acid probably reacts with the DCC to give an intermediate as follows:

$$Boc-NH-CHR^2-COOH + C_6H_{11}-N=C=N-C_6H_{11} \longrightarrow$$

$$Boc-NH-CHR^2-C(=O)-O-C(=N-C_6H_{11})-NH-C_6H_{11}$$

This intermediate then reacts with the amino acid on the resin, forming the dipeptide on the resin

$$C_6H_{11}-NH-CO-NH-C_6H_{11}$$

Peptide bond formation can also be brought about without the use of DCC, by employing 'active' esters of amino-protected amino acids. These can react directly with a free amino group of a resin-bound residue to form a new peptide bond. 'Active' esters are esters of acids with alcohols having electronegative substituents, e.g. *p*-nitrophenyl, 2,4,5-trichlorophenyl, or *N*-hydroxysuccinimidyl

$$\begin{array}{l} H_2C-C=O \\ \quad\;\; | \qquad \backslash \\ \quad\;\; | \qquad\; N-O- \\ \quad\;\; | \qquad / \\ H_2C-C=O \end{array}$$

esters. It is indeed difficult to incorporate glutamine or asparagine residues into growing peptide chains by coupling with DCC, and these two residues are usually added via their *p*-nitrophenyl esters.

Where the side-chain of the amino acid used contains a reactive group, this must also be protected by a group which is not removed under conditions of cleavage of the Boc—NH bond. In the synthesis of ribonuclease A, benzyl groups were used to protect the reactive side-chains of aspartic and glutamic acids, cysteine, serine, threonine, and tyrosine, whilst the benzyloxycarbonyl derivative of lysine, the nitro derivative of arginine, and the sulphoxide of methionine were also used.

Other side-chain-protecting groups are available; for example tosyl groups (*p*-toluenesulphonyl) can be used to protect side-chains of lysine, arginine, histidine, and tyrosine. In general, protecting groups must be stable in the conditions used for peptide bond formation, and for removal of Boc moieties from the residues at the amino-terminal end of the peptide chain. Many side-chain-protecting groups are removed by the reaction which cleaves the completed peptide from the resin, and *must* be removable by a procedure which does not hydrolyse peptide bonds, or racemize amino acid residues.

Peptide bond formation may be repeated, as above, several times to give the complete amino acid sequence desired. The steps involved in chain synthesis do not appear to cause racemization of the amino acids, so that use of L-amino acid derivatives yields a poly-L-amino acid chain. This process of polypeptide synthesis on a resin using Boc-amino acids can now be mechanized and automated. A metering pump draws a precise volume of reagent into the reaction vessel, where it is agitated to facilitate reaction. After mixing, excess reagent is removed by filtration, and the processes of washing, deprotection, neutralization, and coupling are automatically programmed.

In the Merrifield procedure, it is extremely important that each step of the synthesis proceeds with high yield and gives little, or no, side-reactions. Otherwise, as the steps are repeated, the concentrations of unwanted side-products increase, and low yields of the desired product are obtained. For example, if coupling efficiency is 99.9%, a 50-residue peptide prepared by solid-phase synthesis can be expected to be just under 95% pure. But if coupling efficiency is 99.0%, then the final product will be approximately 60% pure.

Cleavage of the completed chain from the resin can be brought about with hydrogen fluoride, or hydrogen bromide in trifluoroacetic acid. In the synthesis of ribonuclease, this treatment also removed all side-chain-protecting groups. The liberated polypeptide, now soluble, can be purified by conventional methods such as gel permeation chromatography and electrophoresis.

The polypeptide chain of ribonuclease A, consisting of 124 amino acids, was synthesized by Merrifield using this method, and disulphide bridges were subsequently introduced into the chain by reduction followed by air oxidation. A fraction of the crude product, after purification, showed catalytic activity and specificity similar to that of natural ribonuclease A, although the activity per unit weight was less for the synthetic than the natural protein. This lowered activity was due to the presence of closely related, inactive molecules containing the 'wrong' disulphide bonds.

Further purification gave a product with 80% of the activity of the natural enzyme.

4.6.2. Synthesis in solution

A large fragment of ribonuclease A (containing 104 of the 124 amino acids) has also been synthesized chemically by Hirschmann, using a technique involving reactions carried out in solution. Peptide bonds are formed using *N*-carboxyanhydrides of the amino acids:

$$\text{R}^1\text{-CH-C(=O)-O-C(=O)-NH (ring)} + \text{H}_2\text{N-CH(R}^2\text{)-COOH} \xrightarrow[\text{borate, } 0^\circ\text{ C}]{\text{pH 10.2}}$$

$$\text{R}^1\text{-CH(NH-COO}^-\text{)CO-NH-CH(R}^2\text{)-COO}^- \xrightarrow{\text{H}_3\text{O}^+}$$

$$\text{CO}_2 + \overset{+}{\text{N}}\text{H}_3\text{-CH(R}^1\text{)-CO-NH-CH(R}^2\text{)-COOH}$$

$$\text{R-CH-C(=O)-S-C(=O)-HN (ring)}$$

A *N*-thiocarboxy anhydride

$$\text{Boc-NH-CHR-COON}\langle\text{C(=O)-CH}_2\text{-CH}_2\text{-C(=O)}\rangle$$

A Boc-hydroxysuccinimide ester

N-Thiocarboxy anhydrides and Boc-hydroxysuccinimide esters can also be used. In this way several small fragments of the ribonuclease molecule have been synthesized, and these were coupled together using azide derivatives:

$$\cdots\text{-CO-NH-CHR}^1\text{-CON}_3 + \text{H}_2\text{N-CHR}^2\text{-CO-NH-}\cdots$$

$$\longrightarrow \cdots\text{-CO-NH-CHR}^1\text{-CO-NH-CHR}^2\text{-CO-NH-}\cdots + \text{HN}_3$$

(Azides can be formed by the reaction of hydrazine, then nitrous acid, on a carboxylic acid ester.)

The reaction conditions involved in this method are such that only the side-chains of cysteine and lysine had to be protected. After synthesis of the 104-unit-long polypeptide chain, the protecting groups were removed, the product was purified, and disulphide bonds were formed by air oxidation. By addition of the peptide (obtained from natural ribonuclease) containing the sequence of 20 amino acids of ribonuclease missing from the synthetic product, catalytic activity characteristic of ribonuclease was obtained, again indicating the importance of primary structure in determining activity.

Several other proteins have been synthesized in solution, but *N*-carboxy anhydrides are not now used because of the fairly high occurrence of side-reactions.

In general, for solution synthesis, long fragments are first produced. The carboxyl group of the residue which will form the C-terminal of a fragment can be protected as the methyl or ethyl ester. This residue, with a free amino group, is coupled to an *N*-protected amino acid, either by using DCC or an 'active' ester of the second amino acid. The *N*-protecting group is removed from the second acid residue and a third acid is added to the growing peptide (see Figure 4.38). These steps are repeated to give the required peptide fragment. Two adjacent fragments of the desired protein can be condensed using the azide method outlined above, as this minimizes racemization of amino acid residues. Racemization is also reduced if the *C*-terminal residue of a fragment is glycine or proline (where racemization cannot take place). Continued condensation of large fragments, in the correct order, gives, finally, the desired protein.

One of the greatest problems of protein synthesis is the formation of the correct disulphide bonds in the completed molecule. Incorporation of cysteine residues into the polypeptide chain, followed by oxidation, allows

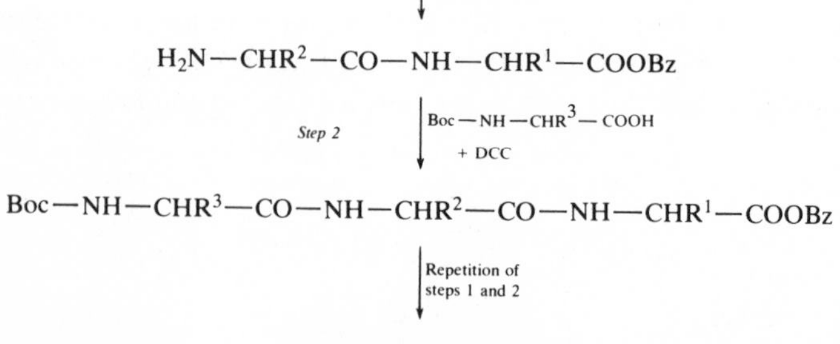

$$\text{Boc—NH—CHR}^2\text{—COOH} + \text{H}_2\text{N—CHR}^1\text{—COOBz}$$

$$\Big\downarrow \text{DCC}$$

$$\text{Boc—NH—CHR}^2\text{—CO—NH—CHR}^1\text{—COOBz}$$

Step 1 $\Big\downarrow$ HBr in acetic acid

$$\text{H}_2\text{N—CHR}^2\text{—CO—NH—CHR}^1\text{—COOBz}$$

Step 2 $\Big\downarrow$ $\text{Boc—NH—CHR}^3\text{—COOH}$ + DCC

$$\text{Boc—NH—CHR}^3\text{—CO—NH—CHR}^2\text{—CO—NH—CHR}^1\text{—COOBz}$$

$\Big\downarrow$ Repetition of steps 1 and 2

Completed polypeptide

FIGURE 4.38. Peptide synthesis in solution; Bz = benzyl group

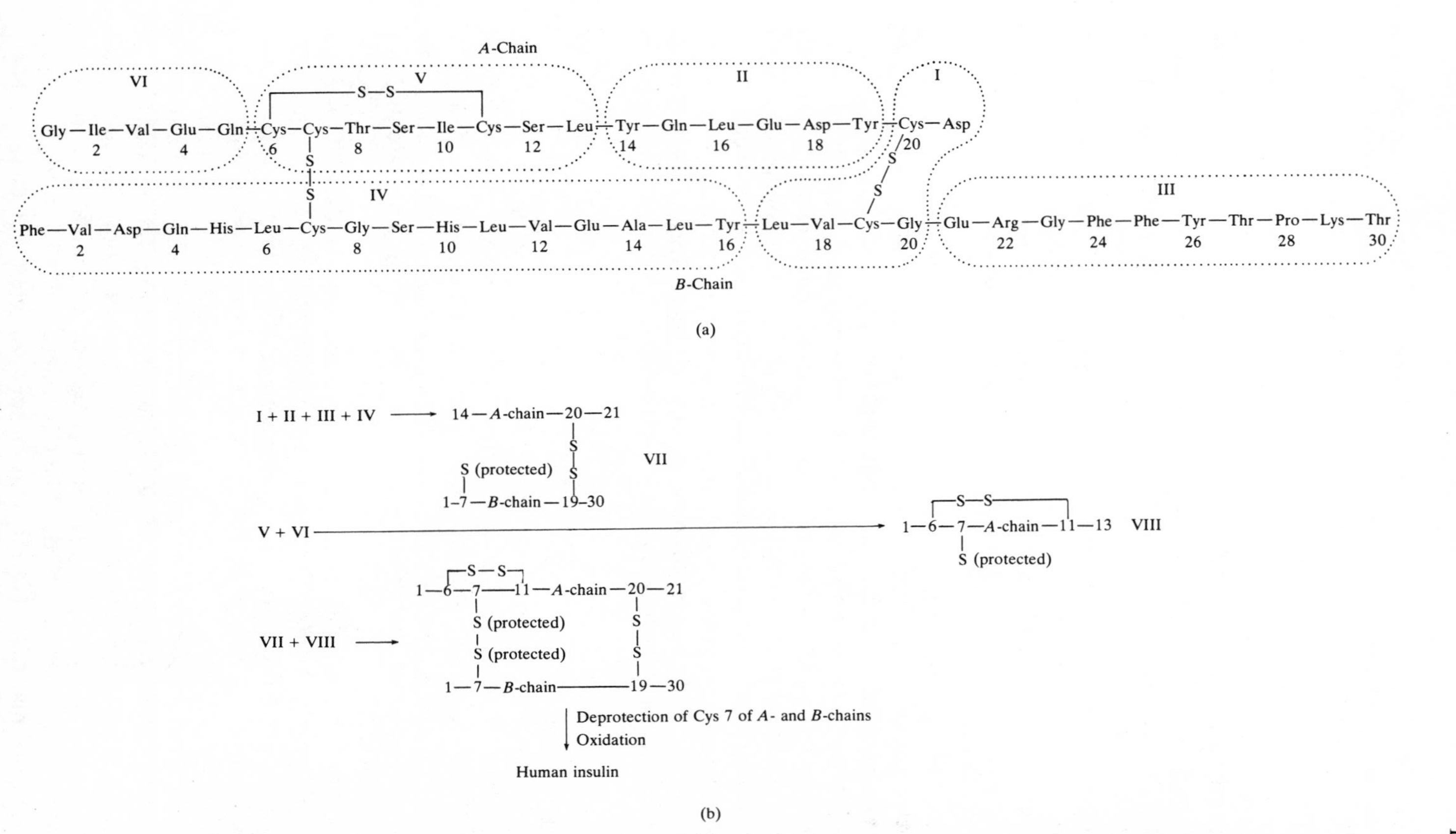

FIGURE 4.39. Chemical synthesis of human insulin: (a) human insulin molecule; (b) coupling of fragments during synthesis

random formation of disulphide bridges. Thus many unwanted side-products, as well as the required molecules, result. This problem may be surmounted, however, by synthesizing a protein from fragments containing preformed disulphide bridges. In this way fully active human insulin (see Figure 4.39a) was synthesized. First, a fragment containing *A*-chain residues 20 and 21 and *B*-chain residues 17 to 20 (I) was prepared, with the *A*20–*B*19 disulphide bridge already formed. To this were coupled an *A*-chain fragment (II), residues 14 to 19, and two *B*-chain fragments, residues 21–30 (III) and 1–16 (IV, with the cysteine residue at *B*7 protected). A small fragment, *A*6–13 (V), was then synthesized with the cysteine residue at *A*7 protected and the *A*6–*A*11 disulphide bridge intact. To this was coupled fragment *A*1–5 (VI), and finally the two large fragments (VII) and (VIII) were coupled together. Their cysteine groups (*A*7 and *B*7) were unblocked and oxidized with iodine in acetic acid (see Figure 4.39b).

Chemical synthesis of ribonuclease, and other enzymes such as lysozyme, as well as insulin and other peptide hormones, has confirmed the relation between structure and activity, and will be of great importance in the future in the preparation of modified proteins differing only in the nature of one amino acid residue in several hundred. Such modifications may enable the rôle of individual amino acids in the functioning of a large protein molecule to be established. Such investigations can be facilitated by a 'semisynthetic' approach where fragments, prepared by partial hydrolysis of a natural protein, are coupled to new synthetic fragments containing altered amino acid sequences.

4.7 USES

In addition to the natural functions of proteins in living organisms, proteins are also essential as a foodstuff for man and the higher animals. These animals cannot synthesize their own amino acids by the fixation of atmospheric nitrogen, or by the conversion of inorganic nitrates, and thus require an external supply of amino acids, i.e. protein is an essential part of their diets.

For man, this protein must contain certain amino acids—the so-called *essential amino acids* (Val, Leu, Ile, Thr, Lys, Met, Phe, and Trp)—because these cannot be synthesized in the body. The other amino acids, however, are not essential in the diet, for they can be produced by modification of other substances. Thus the quality of a human diet is determined not necessarily by total protein content, but by the amount of the essential amino acid present in the lowest proportion. In areas of the world where cereals, such as wheat, rice, or maize, form the staple diet, lysine and tryptophan *may* be limiting factors, for cereals are relatively deficient in these amino acids. Plant breeders are, therefore, producing new types of maize and wheat with higher proportions of these acids.

In this brief outline, it is possible to mention only a few of the very many properties of proteins that man has made use of for centuries. Thus, the enzyme system of yeast, which catalyses fermentation, i.e. the conversion of starch or glucose to alcohol, has been employed for thousands of years. More recently, enzymes have been used industrially—mainly in the food industry, but also in the detergent industry. As already mentioned, denatured collagen, i.e. gelatine, is widely used in the food industry. The properties of several structural proteins such as wool and silk have enabled them to be used, after weaving, as the basis of clothing. Hormones are used therapeutically, e.g. the administration of insulin to diabetics, whilst antibodies can be used for immunochemical assays, often of great importance in medicine, and for affinity chromatography (see Section 2.3.1.5, p. 20).

4.8 ADDITIONAL READING

The Proteins (Ed. H. Neurath), 3rd ed., Academic Press, New York; Vol. 1 (1975), Vol. 2 (1976), Vol. 3 (1977).

The Primary Structure of Proteins W. A. Schroeder, Harper & Row, New York, 1968.

The Structure and Action of Proteins R. E. Dickerson and I. Geis, Harper & Row, New York, 1969.

Chemistry and Molecular Biology of the Intercellular Matrix (Ed. E. A. Balazs), Academic Press, London and New York, 1970.

Atlas of Protein Sequence and Structure (Ed. M. O. Doyhoff), National Biomedical Research Foundation, Silver Spring, Maryland.

General accounts in books such as:

Biochemistry D. E. Metzler, Academic Press, London and New York, 1977.

Biochemistry L. Stryer, W. H. Freeman & Co., San Francisco, 1975.

Recent developments can be found in:

Methods in Enzymology

Advances in Enzymology

Advances in Protein Chemistry

Annual Reviews of Biochemistry

Nature

The Enzymes (Ed. P. Boyer), 3rd ed., Academic Press, London and New York.

Scientific American

'Amino acids, Peptides and Proteins', *Specialist Periodical Reports of the Chemical Society, London*.

(If no date is given after a book, this is because a new volume is brought out regularly, in some cases every year.)

Chapter 5

Nucleic Acids

5.1. INTRODUCTION

This chapter describes the nucleic acids, which are remarkable polymers carrying and transmitting the genetic information of a living organism. Cell nuclei contain structures known as chromosomes, and genetic information is located, in the form of a code, in the *deoxyribonucleic acids* (the DNAs) present in the chromosome.

The genetic information specifies protein structure, and the proteins determine in turn the characteristics of each cell, and, consequently, that of the whole organism. Furthermore, this genetic information is passed on from one generation to the next.

Deoxyribonucleic acids contain the coded information for controlling protein structure, but other nucleic acids, the *ribonucleic acids* (RNAs), are involved in the mechanism of protein biosynthesis. Most RNAs are synthesized in the cell nuclei, but carry out their functions in the cytoplasm. One type of RNA acts as a *messenger* (*m*-RNA) to carry the coded information from the DNA in the nucleus of the cell to the protein-synthesizing sites in the cytoplasm. Two further RNAs, *ribosomal* (*r*-RNA) and *transfer* (*t*-RNA), take part in the actual polypeptide synthesis.

Methods exist for studying the structures of nucleic acids, and although this knowledge is not yet complete, many aspects of the relation between structure and function of the nucleic acids can now be understood.

5.2 THE MONOMERS

5.2.1. Structure and properties

Nucleic acids are formed by the condensation of *nucleoside triphosphates*, complex small molecules consisting of a heterocyclic base, a sugar, and a triphosphate group (see Figure 5.1).

Only four major bases are found in each type of nucleic acid. In DNA, the bases are *adenine* (A), *guanine* (G), *thymine* (T), and *cytosine* (C). The corresponding bases in RNA are adenine, guanine, *uracil* (U), and cytosine. Adenine and guanine are purines, whilst thymine, cytosine, and uracil are pyrimidines. The sugar is a pentose in the furanose ring form (see Section 6.2); in deoxyribonucleic acid it is 2-*deoxy-D-ribose*, whilst in ribonucleic acid, the sugar is *D-ribose* (hence the names of the polymers).

The combination of a base and a sugar is known as a *nucleoside*. The nucleoside consisting of ribose and adenine is *adenosine*; with guanine, uracil, or cytosine, the corresponding nucleoside is *guanosine, uridine* or *cytidine*. The less common nucleoside containing ribose and thymine is known as *thymidine*. For the *deoxynucleosides*, the prefix deoxy- is added to the name of the nucleoside.

A nucleoside phosphorylated on the sugar is called a *nucleotide*;

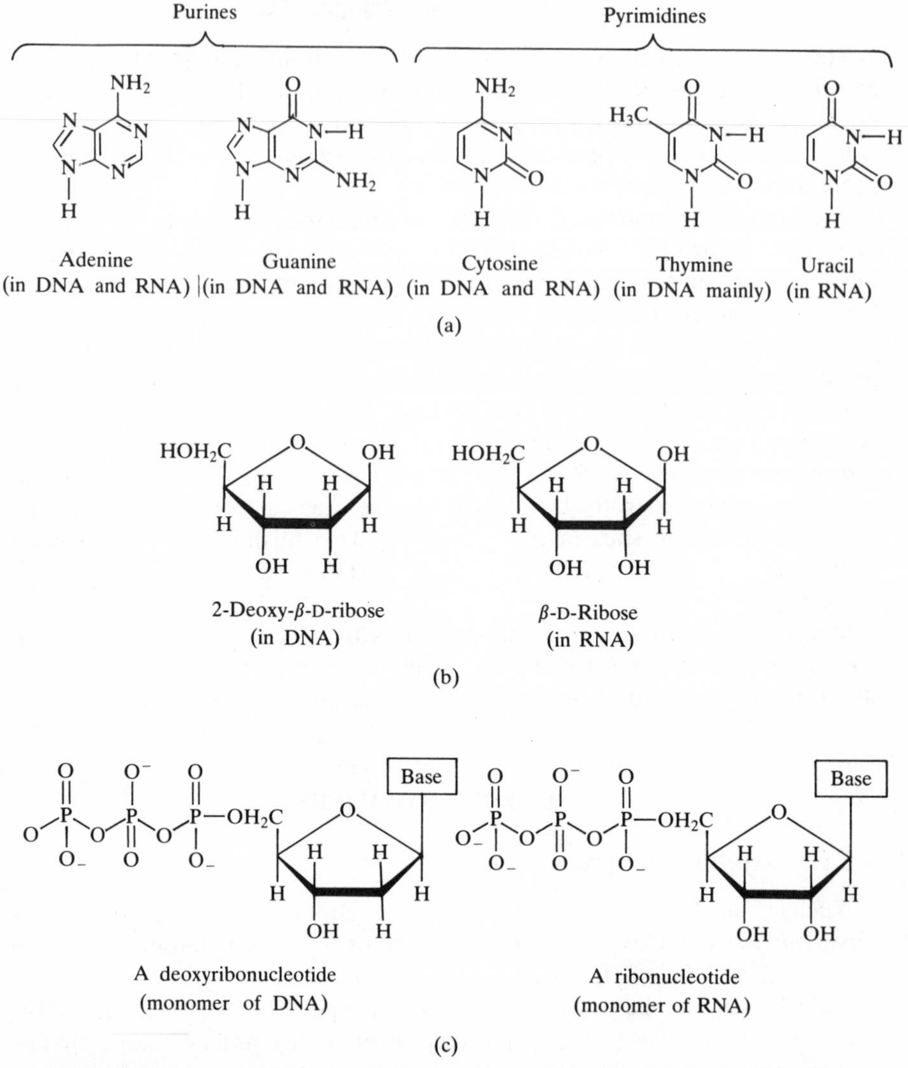

Purines | Pyrimidines

Adenine (in DNA and RNA) | Guanine (in DNA and RNA) | Cytosine (in DNA and RNA) | Thymine (in DNA mainly) | Uracil (in RNA)

(a)

2-Deoxy-β-D-ribose (in DNA) | β-D-Ribose (in RNA)

(b)

A deoxyribonucleotide (monomer of DNA) | A ribonucleotide (monomer of RNA)

(c)

FIGURE 5.1. (a) The heterocyclic bases of nucleic acids; (b) the sugars of nucleic acids; (c) nucleoside triphosphates

nucleotides may contain mono-, di-, or tri-phosphate groups. In the nucleotides shown in Figure 5.2, the phosphate group is esterified to the C-5′ of the sugar ring, and the base is linked to the sugar by a β-glycosidic link at C-1′ of the monosaccharide. (For further details of glycosidic linkages, see Section 6.2.2, p. 254.

As primary phosphate groups are strong acids, both the nucleotides and the polymers are acidic. Hydrolysis of nucleic acids may yield nucleoside monophosphates containing the two acidic groups of a substituted

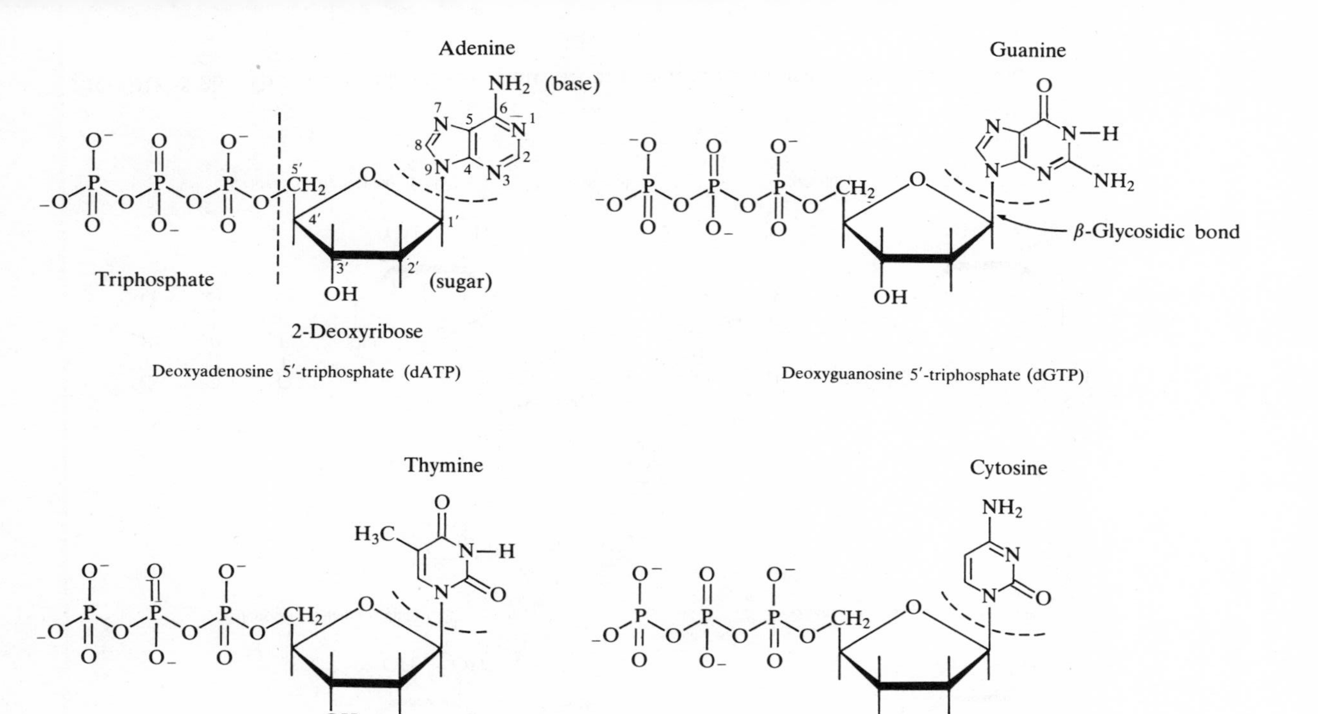

FIGURE 5.2a. The nucleoside triphosphates from which are synthesized deoxyribonucleic acid. The numbering of the ring atoms in sugar and base is shown for deoxyadenosine triphosphate

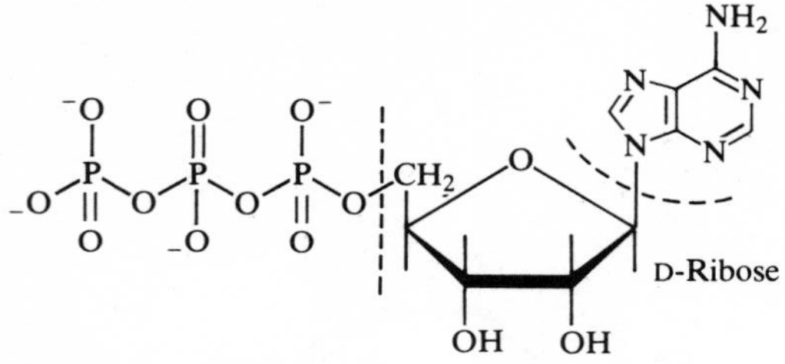

Adenosine 5′-triphosphate (ATP)

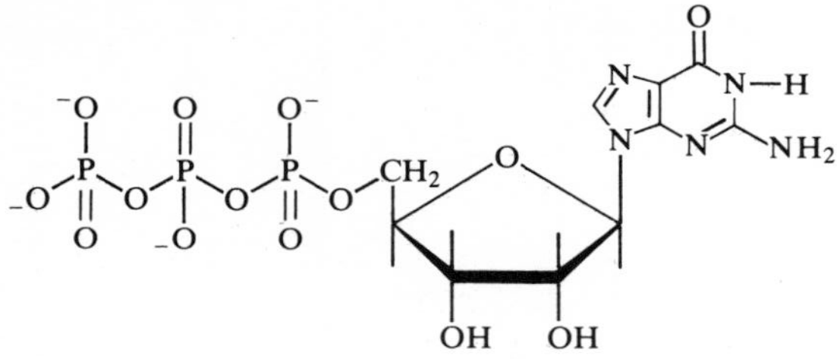

Guanosine 5′-triphosphate (GTP)

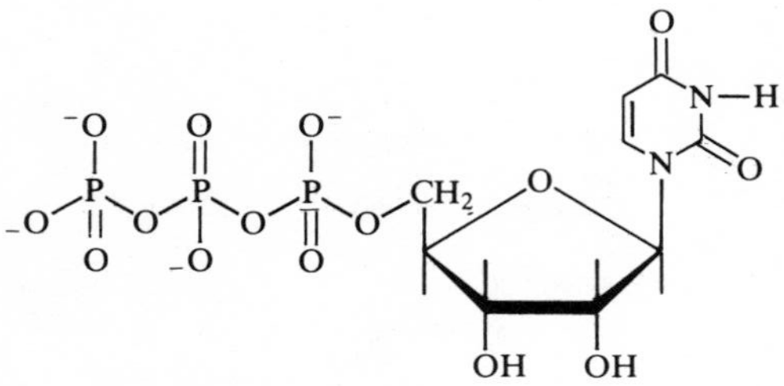

Uridine 5′-triphosphate (UTP)

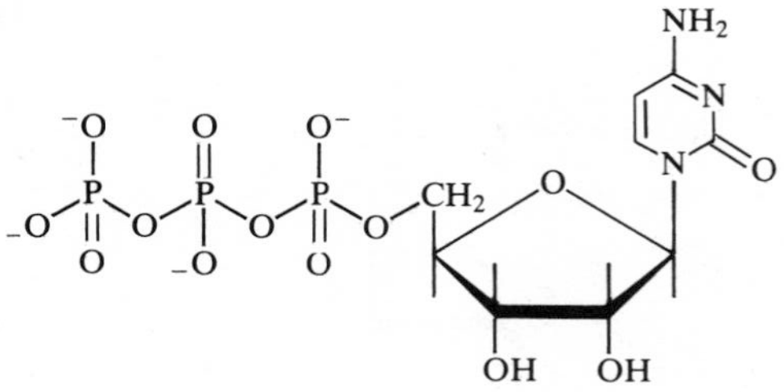

Cytidine 5′-triphosphate (CTP)

FIGURE 5.2b. The nucleoside triphosphates from which are synthesized ribonucleic acid

phosphoric acid. The dissociation of these groups is characterized by a pK value near 1 for the primary phosphate ionization, and a pK value near 6 for the secondary dissociation.

The bases guanine, thymine, cytosine, and uracil can show lactam–lactim tautomerism, as shown in Figure 5.3, but it is believed that the lactam predominates at neutral pH.

The enolic hydroxy groups on guanine, thymine, cytosine, and uracil (in the lactim form) can also ionize, with pK values in the range 10–12.5, whilst the amino groups of adenine, guanine and cytosine have pK values in the range 3–4.5. The presence of such ionizable groups in nucleotides, nucleosides, and the bases, allows these substances to be separated by ion-exchange chromatography and electrophoresis.

Pyrimidine rings are flat, whilst those of the purines are slightly puckered. In both nucleotides and nucleic acids, these rings are almost at right angles to the sugar rings.

The conjugated ring system of the purine and pyrimidine bases results in marked absorption of UV light, with a maximum in the absorption spectrum of each base near 260 nm. This property is useful for locating nucleic acids within cells, and also for locating the products of polymer hydrolysis, such as nucleotides, nucleosides, or free bases, after chromatographic separations.

The amino groups of the bases, and the hydroxyl groups of the sugars can be methylated, and indeed many methyl derivatives of the common nucleotides are found in the naturally occurring nucleic acids.

Other derivatives of the amino groups of the bases can be formed, and these are widely used to protect the amino groups during chemical synthesis of nucleic acids (see Section 5.6, p. 232). For example, the amino group of adenine is frequently benzoylated:

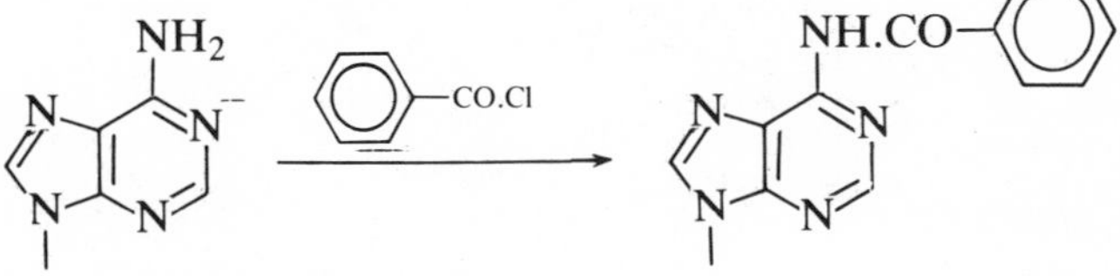

NH_2 — CO.Cl → NH.CO—

Bases containing amino groups are oxidatively deaminated by nitrous acid, and thus are converted to other bases, e.g. cytosine is converted to uracil:

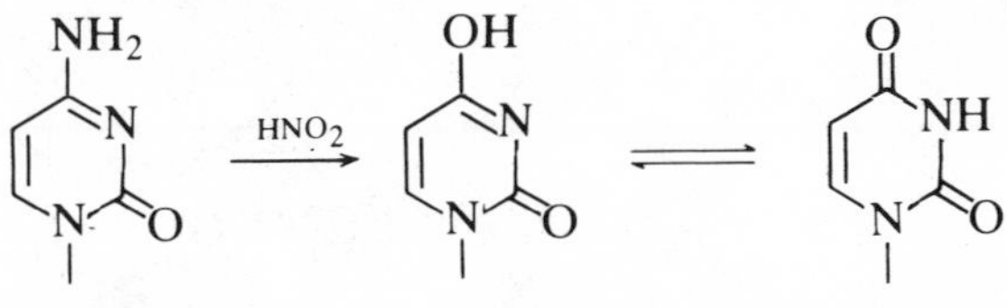

NH_2 —HNO_2→ OH ⇌ O, NH

Cytosine Uracil

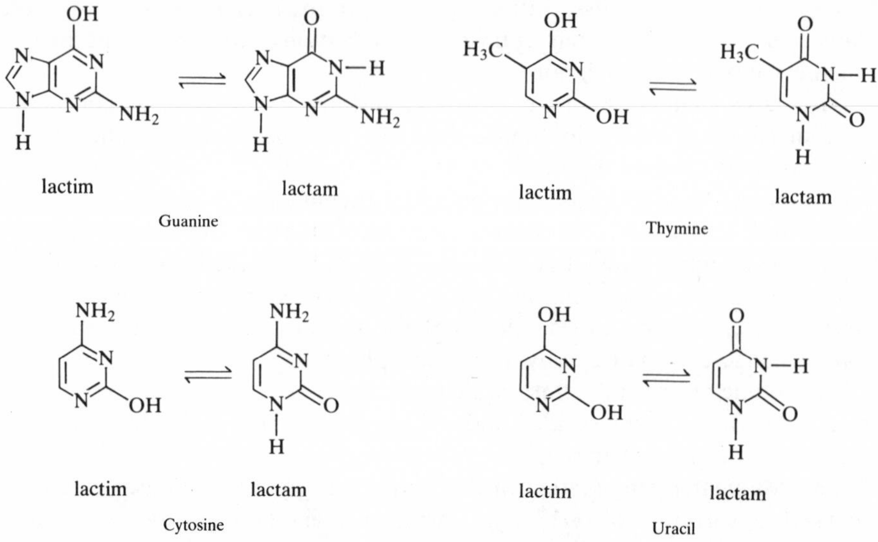

lactim ⇌ lactam

Guanine

lactim ⇌ lactam

Thymine

lactim ⇌ lactam

Cytosine

lactim ⇌ lactam

Uracil

FIGURE 5.3. Lactam–lactim tautomerism of bases of DNA and RNA

Such a change can seriously affect the biological function of the nucleic acid containing the altered base (see Section 5.5.7, p. 228).

Several derivatives of the sugars of nucleotides can be formed, and again some of these are important in the chemical synthesis of nucleic acids (see Section 5.6, p. 232). Thus free hydroxyl groups of the sugars can be acetylated:

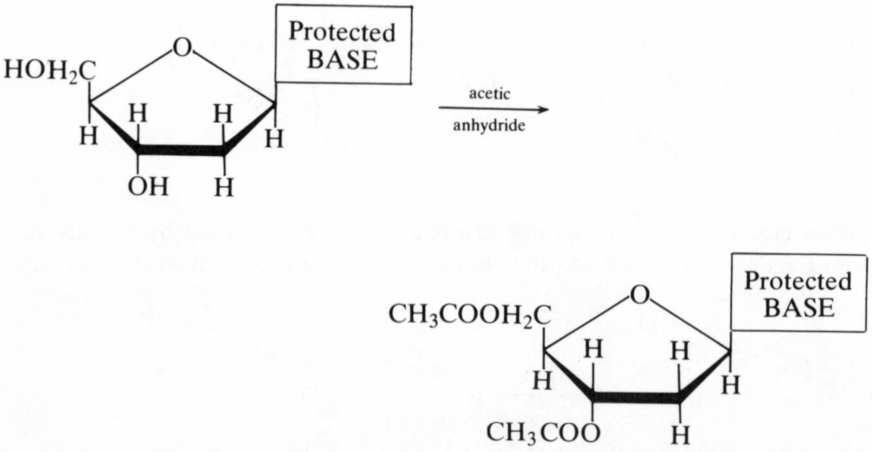

HOH_2C — Protected BASE — $\xrightarrow{\text{acetic anhydride}}$ CH_3COOH_2C — Protected BASE; CH_3COO

(For further details of sugar derivatives see Sections 5.6 and 6.2.1, p. 248.)

In the ribonucleotides, the diol group of the sugar ring can be oxidized by periodate to yield a dialdehyde. The product can then be reduced using sodium borohydride (see Section 6.2.1, p. 248). Often tritiated borohydride is used to give a radioactive diol, which can then be separated and identified more readily from a reaction mixture. This technique can be useful in the end-group analysis of ribonucleic acids.

Phosphate ester—O—CH_2 — Base — OH OH $\xrightarrow{KIO_4}$

Phosphate ester—O—CH_2 — Base — CHO CHO $\xrightarrow{\text{borohydride}}$

Phosphate ester—O—CH_2 — Base — HOH_2C CH_2OH

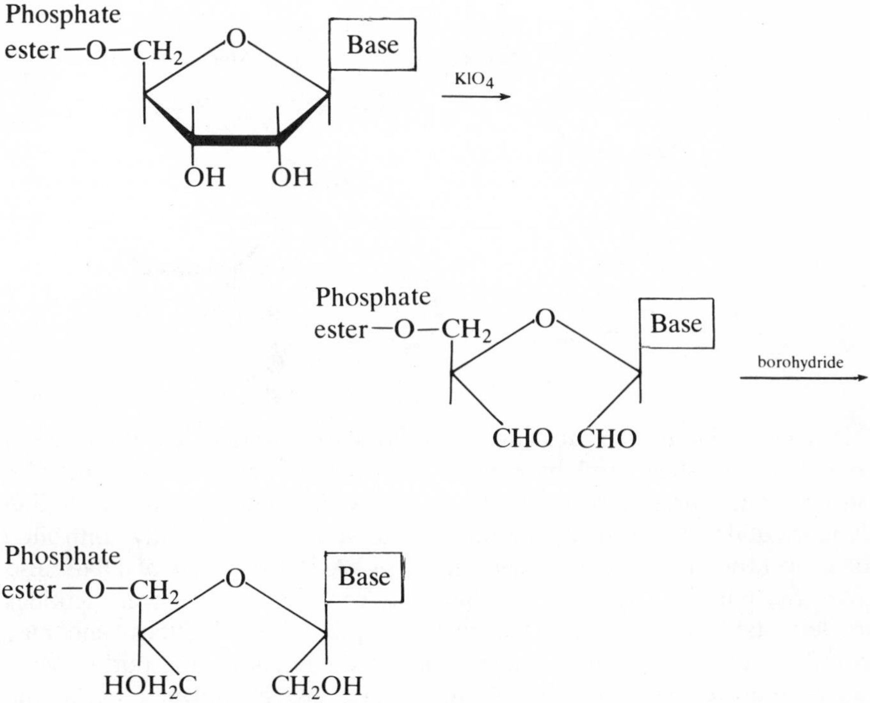

Thymine can undergo a photodimerization which is very important when deoxyribonucleic acids are damaged by UV light (see Section 5.5.2, p. 203):

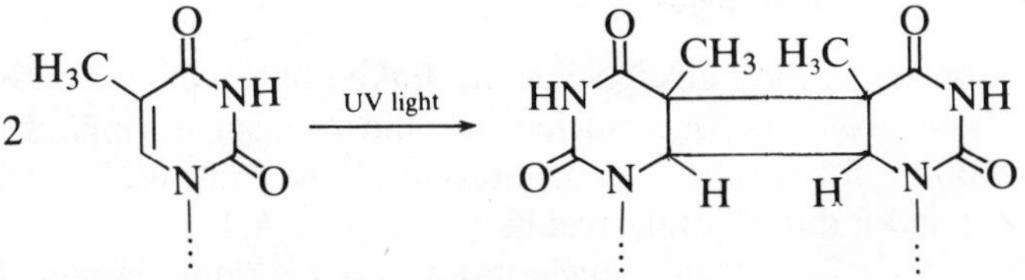

2 H_3C O NH N O $\xrightarrow{\text{UV light}}$ HN O CH_3 H_3C O NH O N H H N O

The nucleosides are fairly stable in alkali and, under the conditions normally used for the alkaline hydrolysis of ribonucleic acids, the mononucleotides are stable. DNA is resistant to alkaline hydrolysis, but RNA can be hydrolysed to a mixture of 2′- and 3′-mononucleotides. These mononucleotides are formed via a cyclic intermediate:

RNA $\xrightarrow{\text{alkali}}$ Cyclic 2′,3′-phosphate $\xrightarrow{\text{alkali}}$ 2′-Monophosphate and 3′-Monophosphate

In acid, purine nucleosides and nucleotides are unstable, and the glycosyl bond between sugar and base is readily hydrolysed to release the free purine. Thus, during acid hydrolysis of RNA, free purines and pyrimidine ribonucleotides are formed; the conditions have to be carefully controlled for concentrated acid deaminates the bases. Acid hydrolysis of DNA also gives free purines and pyrimidine nucleotides—these deoxyribonucleotides are less stable in acid than the corresponding ribonucleotides, and may hydrolyse to the free bases. Under controlled conditions, the purine bases can be removed from DNA without fragmenting the polymer backbone. The product is known as an apurinic acid. These properties are important during the hydrolysis of nucleic acids for the determination of base composition, or base sequence, of the polymers.

5.2.2. Intermonomer linkages

The monomers in the nucleic acids are linked by covalent bonds between the phosphate group of one nucleotide and the sugar ring of the next. During bond formation, pyrophosphate is released (from two nucleotides) and a dinucleotide results (see Figure 5.4a).

Condensation with further nucleotides gives a long polynucleotide, a *nucleic acid* (see Figure 5.4b). The phosphate groups act as bridges forming *phosphodiester links* between the C-3′ of the sugar of one nucleotide and the C-5′ of the next.

At one end of a polynucleotide chain, there is a group, not included in a polymer link, attached to the 5′-carbon of the sugar ring; this is referred to

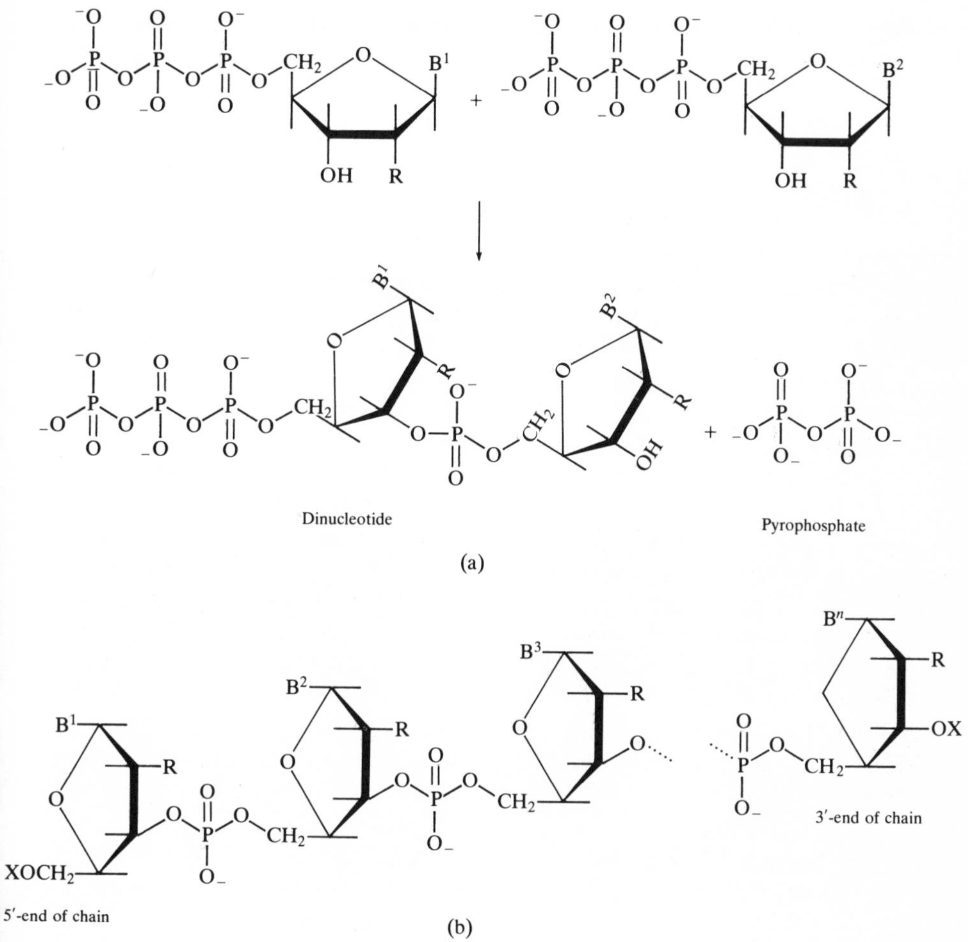

Dinucleotide

Pyrophosphate

(a)

5′-end of chain

3′-end of chain

(b)

$B^1p\ B^2p\ B^3p \ldots pB^n$

(c)

FIGURE 5.4 (a) Formation of a dinucleotide: R = H in DNA; R = OH in RNA; B^1, B^2 = A, G, T, or C in DNA; B^1, B^2 = A, G, U, or C in RNA. (b) The chain structure of a nucleic acid: X = H or a phosphate group. (c) Shorthand representation of the same nucleic acid with X = H

as *the 5′-end* of the chain, and normally consists of a mono- or tri-phosphate group on the sugar ring. At the other end of the chain, *the 3′-end*, a free hydroxyl group is often found on carbon 3′ of the ribose or deoxyribose.

Most DNA molecules consist of two very long, unbranched polynucleotide chains, while RNA molecules normally contain one chain only. The various nucleic acids within one organism differ in their base composition and sequence, but it is the base sequences of the deoxyribonucleic acids, coding for the inherited traits of all living things, which determine the nature of each individual organism.

5.2.3. Other bonds

Specific *hydrogen bonding* is thought to be very important in the three-dimensional structure of nucleic acids; it occurs between two bases on separate polynucleotide chains, or on different parts of the same chain. The pair of bases linked by hydrogen bonds always consists of one purine and one pyrimidine, and usually adenine is bonded to thymine or uracil, whilst guanine is bonded to cytosine (Figure 5.5). Under some circumstances three bases may be involved in hydrogen bonding. These triplets are discussed further in Section 5.4.2.2, p. 199).

Hydrophobic bonding is also extremely important in the structure of nucleic acids. The heterocyclic bases are hydrophobic and tend to associate with one another rather than with water. In double-stranded regions of a molecule, the bases lie parallel to one another and stack easily on top of each other in the interior of the molecule. The π-electrons of the parallel heterocyclic bases can then interact to strengthen the hydrophobic bonding.

As in the case of the proteins, heating and extremes of pH can disrupt

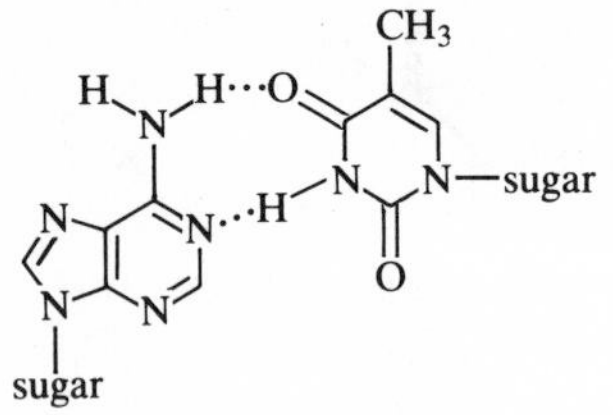

Adenine–thymine
DNA–DNA pairs

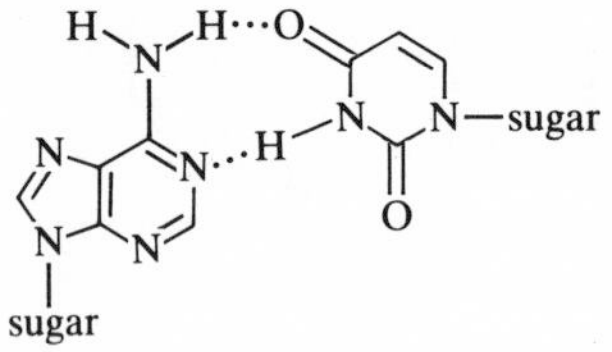

Adenine–uracil
DNA–RNA pairs
and RNA–RNA pairs

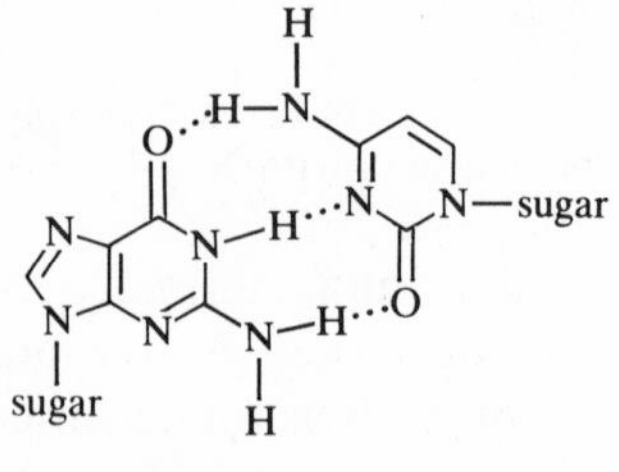

Guanine–cytosine
DNA–DNA pairs
and DNA–RNA pairs
and RNA–RNA pairs

FIGURE 5.5. Base-pairing in nucleic acids

the hydrogen bonds of nucleic acids. Changes of pH bring about the disruption by causing ionization of some of the groups involved in hydrogen bond formation. High concentrations of urea can also break the hydrophobic bonds, and rupture of hydrogen and hydrophobic bonds of nucleic acids bring about marked changes in the three-dimensional structures of the polymers. When this occurs, the nucleic acid is said to be *denatured*. As three hydrogen bonds exist between a G–C pair and only two between A–T and A–U pairs, structures with a high G–C content are slightly stronger and more resistant to heat denaturation.

5.3 SIZE AND COMPOSITION

5.3.1. Molecular size

The methods outlined in Chapter 3 may be used to measure the molecular weights of nucleic acids. Electron microscopy (see Section 3.6, p. 64) is used to measure the lengths of large double-stranded DNA molecules, and from this molecular weights can be estimated (10^6 Å ≅ molecular weight of 2×10^8). However, it is often difficult to isolate and purify many of the extremely large nucleic acids without degradation, so that, in some cases, the measured molecular weight may only be a lower limit.

Most DNA molecules contain two long polynucleotide chains, but some viral DNA consists of a single chain only. Conversely, most RNA molecules consist of one chain, while double-chain RNA molecules can be found in a few viruses. The size of a DNA depends on its source, but different deoxyribonucleic acids have been isolated with molecular weights from one to over one hundred million; in higher organisms, the molecular weight is often several hundred million. Thus the DNA of virus SV40 has a molecular weight of 3.2×10^6 whilst that of the bacterium *E. coli* has a weight of approximately 2.5×10^9. Sedimentation coefficients (see page 48) of over 100 S have been measured for DNA, and the molecules may be several million ångstroms long, and are usually 20 Å wide (in the double-chain form).

The three important groups of RNA found in living organisms—*messenger*, *ribosomal*, and *transfer RNA*—differ in function and molecular size, and will be discussed in greater detail later in this chapter. Messenger RNA usually forms 5–10% of the RNA of a cell, but may be the only nucleic acid present in a virus, where it may consist of two chains. Messenger RNA molecules from higher organisms are thought to contain one chain only. The molecular weight varies with the source, but can be as high as several million. Ribosomal RNA is found in a complex structure of nucleic acid and protein, known as a ribosome. Ribosomes contain 40–65% nucleic acid, and consist of two subunits. The smaller of the two subunits contains a nucleic acid with a sedimentation coefficient of

16 S in bacteria, and about 18 S in higher organisms; the corresponding molecular weights are 0.56×10^6 for the bacterial polymer, and ca. 0.65×10^6 for higher organisms. In the larger ribosomal subunit, two ribonucleic acids exist with sedimentation coefficients of 23 S and 5 S in bacteria, and 28 S and 5 S in higher organisms, corresponding to molecules of molecular weight 1.1×10^6 and about 4×10^4 for bacteria, and 1.7×10^6 and about 4×10^4 for higher forms of life. In addition, a nucleic acid with a sedimentation coefficient of 5.8 S and molecular weight of 5×10^4 exists in the ribosomes of higher animals. The third main type of RNA, transfer RNA, is a small molecule, containing approximately 80 nucleotides and with a molecular weight of 2.5×10^4. The molecular dimensions are believed to be ca. $25 \times 35 \times 85$ Å. Both ribosomal and transfer RNA molecules consist of one polynucleotide chain only.

Nucleic acids, like the proteins, are believed to have sharply defined structures, and so have a well-defined molecular weight with no molecular weight distribution.

5.3.2. Composition

The nature and amount of the nucleotides present in a nucleic acid is studied by hydrolysis of the polymer, followed by separation, identification, and estimation of the products.

DNA may be hydrolysed either by acid to give free purines and pyrimidine nucleotides (stronger acid gives free bases), or by enzymes, such as *deoxyribonucleases* or *phosphodiesterases*, to give nucleotides. The phosphate groups can be removed from the nucleotides using another enzyme, *phosphatase*, to yield nucleosides. The resultant hydrolysis products can be separated by ion-exchange chromatography and/or electrophoresis, and their amounts estimated by measuring UV light absorption. The base composition of DNA is characteristic of the organism from which it was extracted, but different cells of the same organism contain the same total DNA. In addition to the common bases, adenine, guanine, thymine, and cytosine, the nuclear DNA of higher organisms contains a little 5-methylcytosine (see Figure 5.6). Although the base composition of DNA can vary widely from organism to organism, it is always found that the amount of adenine equals the amount of thymine, and the amount of guanine equals the amount of cytosine plus 5-methylcytosine, in two-stranded DNA molecules.

RNA molecules may be hydrolysed by alkali to give ribonucleotides (concentrated bases used at high pressure give ribonucleosides), by acid to give free purines and pyrimidine nucleotides, or by enzymes, the ribonucleases and phosphodiesterases, to give nucleotides. Again *phosphatase* can be used to produce nucleosides from the nucleotides. The products are separated and estimated by methods similar to those used for DNA hydrolysis products. Messenger RNAs differ in base composition

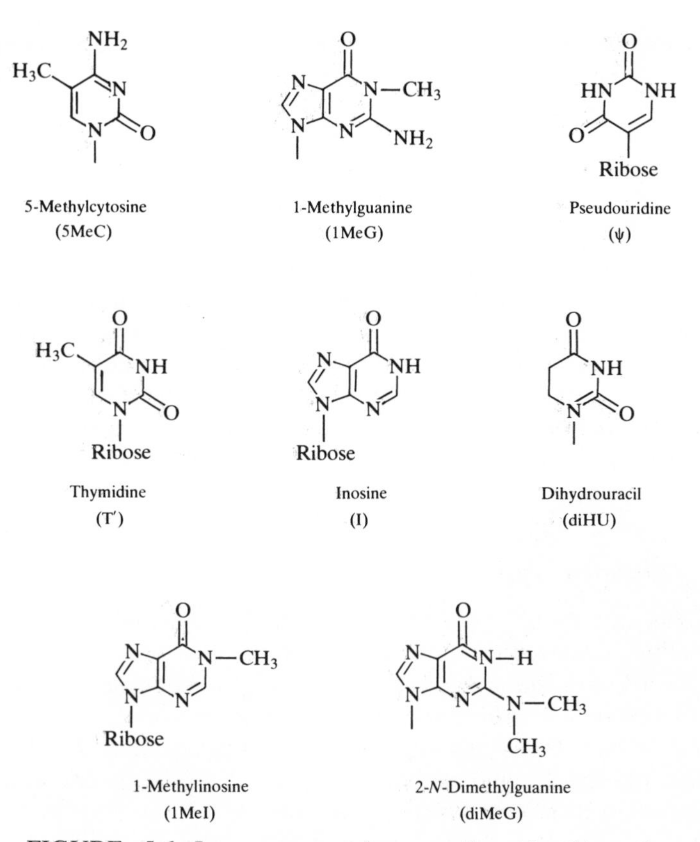

FIGURE 5.6. Less common bases and nucleosides of nucleic acids

within one organism, and there is no regularity in A/U or G/C ratios like that found for most DNA. The ribosomal RNAs of one organism have differing base composition, but in general any one type of ribosomal RNA, e.g. 5S-RNA, is fairly, but not completely, uniform in one organism. The two larger ribosomal RNAs both contain pseudo-uridine and some methylated bases, such as 1-methylguanine (see Figure 5.6), whilst the small 5S-ribosomal RNA appears to contain little, if any, of the less common bases. In all single-chain RNA molecules, the A/U and G/C ratios do not equal 1, and the base composition depends on the source of the polymer. The ribosomal RNA of higher organisms usually contain rather more (G + C) than (A + U). Transfer RNA molecules within one organism do not have identical base compositions, and these molecules differ from other types of RNA, for they contain relatively high proportions of the more unusual bases or nucleosides shown in Figure 5.6. In RNAs, some methyl groups are found as 2′-*O*-methylribosides, i.e. on the sugar ring rather than on the base.

As in the case of proteins, the sequence of monomers along a polynucleotide chain is much more important than the overall monomer composition.

5.4. STRUCTURE

5.4.1. Primary structure

The base sequence of a nucleic acid is believed to be directly related to its function, but until recently it was extremely difficult to determine the complete sequence for a large DNA or RNA molecule containing a thousand or more nucleotides. In addition to the difficulties caused by the size of the molecules, there were problems due to the fact that only four bases occur commonly in the large nucleic acids. Now, however, complete base sequences of DNA and RNA molecules over 3000 bases long have been determined.

5.4.1.1. Deoxyribonucleic acid

For DNA, the number of chains in the molecule should be known. This is usually discovered during purification procedures. Certain deoxyribonucleases, however, act only on single-chain DNA, whilst others act on double-chain molecules, and so these enzymes can be used to distinguish molecules containing one or two polynucleotide chains. The behaviour of single- and double-chain DNA is also different during density-gradient ultracentrifugation (see Section 2.3.3, p. 28).

The two chains of a double-stranded DNA may be separated by heat and density-gradient ultracentrifugation, so that the base sequence of a single chain can be studied. The nature of the bases at the ends of a DNA chain may be determined by specifically labelling the end-groups. For example, if there is a free hydroxyl group at the 3′-end of the polynucleotide chain (see Figure 5.4b), this may be acetylated, and the base linked to the acetylated sugar can be identified after careful hydrolysis.

The 3′-end can also be labelled with a radioactive phosphorus atom by elongating the polymer chain with a radioactive nucleoside triphosphate. The reaction is catalysed by a transferase enzyme, *DNA-polymerase*:

$$\cdots-B^1-p-B^2-p-B^3-OH + p-p-p^*-A \xrightarrow{\text{transferase}}$$

Existing *DNA* chain   Radioactive *ATP*

$$\cdots-B^1-p-B^2-p-B^3-\mathrm{p}^*-A-OH.$$

Hydrolytic enzymes which break the phosphodiester bonds between phosphorus and C-5′ of deoxyribose can then be used to give a mixture of nucleoside monophosphates, including a radioactive nucleotide from the original 3′-end of the chain. This unit can then be readily identified:

$$\cdots-B^1-p-B^2-p-B^3-p^*-A-OH \xrightarrow[\text{e.g. spleen phosphodiesterase}]{\text{nucleases}} B^1p + B^2p + B^{3*}p + \cdots$$

A free OH group on the 5′-end of the chain can be labelled using a phosphate group containing radioactive phosphorus, so that the end-group can be easily identified. The phosphorylation reaction is catalysed by an enzyme, *polynucleotide kinase*.

This method can be employed even when the 5′-end of a chain initially carries a phosphate group. Another enzyme, a phosphatase, or phosphomonoesterase, can remove the original unlabelled phosphate group, which can then be replaced by one containing radioactive phosphorus:

$$\underset{\text{Existing } DNA \text{ chain}}{p-B^1-p-B^2-p-B^3-\cdots} \xrightarrow[\text{monoesterase}]{\text{phospho-}} HO-B^1-p-B^2-p-B^3-\cdots$$

$$\Big\downarrow \; p^*-p-p-A \text{ (radioactive ATP)} + \text{polynucleotide kinase}$$

$$p^*B^1 + pB^2 + pB^3\cdots \xleftarrow[\text{e.g. venom phosphodiesterase}]{\text{enzymic hydrolysis}} p^*-B^1-p-B^2-p-B^3-\cdots$$

Initially various nuclease enzymes were used to hydrolyse a deoxyribonucleotide chain to oligonucleotide fragments. These fragments were separated by ion-exchange chromatography, often in the presence of 7 M urea, or by electrophoresis, and the base sequence of each studied, in a manner analogous to the investigation of oligopeptide sequences during the determination of the primary structure of a protein (see Section 4.4.1.3, p. 90). Most oligonucleotides produced by enzymic hydrolysis have a phosphate group at one end of the molecule and a free hydroxyl at the other; the size of the oligonucleotide could easily be determined by enzymic release of this phosphate end-group, and subsequent measurement of the ratio of free phosphate to total phosphate. The base sequence in the oligonucleotide was studied using hydrolytic enzymes with known specificities, e.g. *venom phosphodiesterase*, which hydrolyses from the 3′-OH-end to release nucleoside 5′-phosphates, or *spleen phosphodiesterase*, which hydrolyses from the 5′-OH-end to give nucleoside 3′-phosphates.

An example of their use in finding the sequence of a tetranucleotide triphosphate GpCpApT is shown below. (The base at the 5′-end of the chain is written at the left; p written to the right of a base indicates phosphate linked to C-3′ of that nucleotide, while p to the left of a base indicates phosphate linked to C-5′ of the nucleotide, cf. Figure 5.4c, p. 175.) The venom enzyme releases a nucleoside, in the case guanosine,

$$\text{GpCpApT} \xrightarrow[\text{phosphodiesterase}]{\text{venom}} \text{G} + \text{pC} + \text{pA} + \text{pT}$$

$$\text{GpCpApT} \xrightarrow[\text{phosphodiesterase}]{\text{spleen}} \text{Gp} + \text{Cp} + \text{Ap} + \text{T}$$

from the 5′-end of the oligonucleotide, whilst the spleen enzyme gives a nucleoside, deoxythymidine, from the 3′-end. Thus the sequence in the tetranucleotide must be *G*p(*C* or *A*)p(*A* or *C*)pT.

Another deoxyribonuclease, *exonuclease* 1 from the bacterium *E. coli*, can be used to resolve the ambiguity in the sequence, for it hydrolyses an oligonucleotide stepwise from the 3′-end to give 5′-phosphates, but leaves a dinucleotide at the 5′-end i.e.:

$$\text{GpCpApT} \xrightarrow{\text{exonuclease I}} \text{GpC} + \text{pA} + \text{pT}$$

Thus the complete sequence must be *GpCpApT*. (If the tetranucleotide had originally carried a phosphate group at either end, this could have been removed by phosphatase, prior to the sequence studies.) In this way, the base sequences of small oligonucleotides were determined, and in theory the order of the oligonucleotides in the original molecule was then investigated to give the sequence of the whole molecule (as is done for proteins, see Section 4.4.1.3, p. 94). In practice, this was difficult because there are only four bases present to any great extent in DNA, and so the same short oligonucleotide sequences may occur more than once in the complete polynucleotide chain.

The chromosomes of higher organisms contain a special type of DNA, known as *satellite DNA*, whose function is at present unknown. It can account for up to 10% of the DNA of some organisms, and appears to consist of short base sequences repeated, with minor changes, many thousands of times. It is much easier, therefore, to study the base sequence of satellite DNA than other DNA molecules. One method which has been employed involved the use of diphenylamine, which destroys the purine bases to leave strands containing solely pyrimidines. The pyrimidine sequence was determined, and as will be explained in Section 5.4.2.1, p. 194, it is possible to deduce the complete sequences of both chains of a double-stranded DNA, if the pyrimidine sequences of both chains are known. Thus it has been found that one guinea pig satellite DNA consists of the sequence CCCTAA on one chain and TTAGGG on the other (written with the 5′-end at the left) repeated, with some changes, about 10^7 times.

Two powerful methods are now used to give, directly, the sequences of up to 200 bases in a stretch of DNA. These are the *plus and minus method* of F. Sanger and A. R. Coulson, and the *chemical method* of A. Maxam and W. Gilbert.

For both methods, oligonucleotides of less than 200 bases are required, and these are usually prepared by enzymic hydrolysis using *restriction endonucleases*. These are bacterial enzymes of high specificity, which hydrolyse DNA at, or near, a few sites possessing particular unique base sequences. Each restriction enzyme is specific for a different sequence. Thus restriction enzymes are used in DNA sequence analysis in the same way that trypsin is used in protein sequence analysis (see Section 4.4.1.3, p. 90).

Many restriction enzymes hydrolyse double-stranded DNA only, so that the first stage in sequence analysis would be the production of double-helical fragments of the original DNA. The fragments are then separated and purified by electrophoresis.

In both sequencing procedures, effectively only single-stranded DNA is studied. Thus a double-stranded fragment obtained by enzymic hydrolysis should next be heated to separate the two polynucleotide chains, and each chain should be purified. Only one chain, however, need be studied, because, as explained in Section 5.4.2.1 (p. 194), the sequence of both strands can be deduced if only one is known.

In the *plus and minus method*, a smaller fragment is chosen as a 'primer' for the controlled enzymic synthesis of new DNA. The second strand of the double-stranded DNA is used as a pattern, or template, to direct the insertion of bases into the new DNA. (The details of DNA biosynthesis are discussed later, in Section 5.5.2.) To the primer and template are added a DNA-synthesizing enzyme, *DNA-polymerase*, and the four nucleoside triphosphates (cf. Figure 5.2a, p. 169), each radioactively labelled. Conditions are chosen so that successive nucleotides are added to produce extension products of every possible chain-length (see Figure 5.7a), up to a maximum determined by the template size (say 100 bases).

This mixture is then divided into eight, and any unused nucleoside triphosphates are removed. Synthesis with the same enzyme is continued in four of the eight subsamples, but to each is added only three of the four nucleotides necessary for DNA elongation. Each of the four subsamples lacks a different nucleotide. This constitutes the 'minus' treatment of the plus and minus method. Then in each subsample, chain-extension continues up to the position before the missing nucleotide (see Figure 5.7b).

The other four subsamples are used for the 'plus' treatment. Here chain hydrolysis is carried out by a special viral enzyme which hydrolyses nucleotides stepwise from the 3′-end of chains. If a free nucleotide, e.g. GTP, is added to the mixture, the enzyme stops hydrolysis when the DNA chain ends with the base of the added nucleotide (see Figure 5.7c). Thus, if a different nucleotide is added to each of the four subsamples, shorter chains ending with that nucleotide are obtained in each.

The products of the preliminary mixture and eight subsamples are separated according to size by gel electrophoresis, and are detected because each is radioactively labelled.

(a) Preliminary mixture containing all possible chain-lengths up to the maximum to be studied: in this case, an arbitrary maximum of 15 has been chosen:

Primer		*Chain-length*	*Positions of radio-active bands on gel*
pGCGTAG	TCGATAGCG	15	+
PGCGTAG	TCGATAGC	14	+
PGCGTAG	TCGATAG	13	+
PGCGTAG	TCGATA	12	+
PGCGTAG	TCGAT	11	+
PGCGTAG	TCGA	10	+
pGCGTAG	TCG	9	+
pGCGTAG	TC	8	+
pGCGTAG	T	7	+

Direction of movement on gel (arrow pointing downward)

(Note: The oligonucleotide of chain-length 15 contains the whole sequence to be investigated.)

(b) Final products of the minus-G subsample:

Original chain-length		*Final chain-length*	*Positions of radio-active bands on gel*
15	Primer—TCGATAGCG	15	+
14	Primer—TCGATAGC	14	+
13	Primer—TCGATAG**C**	14	
12	Primer—TCGATA	12	+
11	Primer—TCGAT**A**	12	
10	Primer—TCGA**TA**	12	
9	Primer—TCG**ATA**	12	
8	Primer—TC	8	+
7	Primer—TC	8	

(Note: The new bases added during extension are shown in bold type.
The same treatment is carried out with each of the other three nucleotides missing.)

(c) Final products of the plus-G subsample:

Original chain-length		*Final chain-length*	*Positions of radioactive bands on gel*
15	Primer—TCGATAGCG	15	+
14	Primer—TCGATAG	13	
13	Primer—TCGATAG	13	+
12	Primer—TCG	9	
11	Primer—TCG	9	
10	Primer—TCG	9	
9	Primer—TCG	9	+
8	Primer—	6	} not
7	Primer—	6	} radioactive

(Note: Again, the same treatment is carried out in the presence of the other three nucleotides.)

FIGURE 5.7. Plus and minus method of sequencing DNA: (a) preliminary mixture; (b) final products of the minus-G subsample; (c) final products of the plus-G subsample

Comparison of positions of bands appearing on the gel with the bands from the preliminary mixture allows the sequence to be read off directly (see Figure 5.8).

The sequence can be read directly from the plus system: i.e. oligonucleotide 7 ends in T, oligonucleotide 8 in C, 9 in G, etc., so that position 7 in the original sequence must be occupied by T, 8 by C, 9 by G, etc., and the initial part of the sequence must be TCG . . . Continuing this reasoning through all the oligonucleotides gives the whole sequence TCGATAGCG. This result can be confirmed by examination of the minus

Positions of radioactive bands from complete 'plus' and 'minus' treatment of oligonucleotide 15 of Figure 5.7

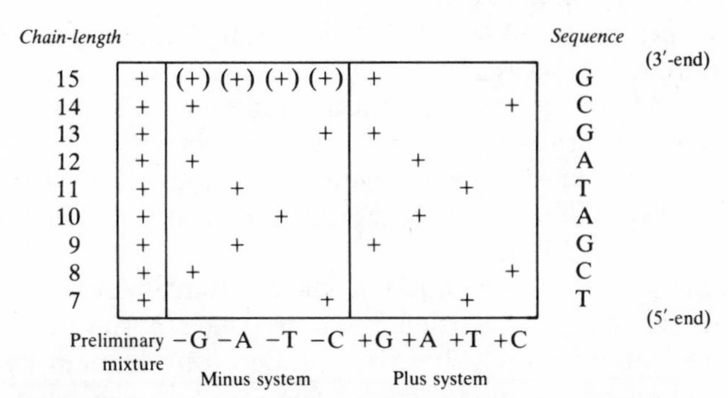

FIGURE 5.8. Reading off the sequence of DNA. [Oligonucleotides of chain length 15 occur in each minus subsample, because the template or pattern is 15 nucleotides long and can give no information on the sixteenth base.]

system, although the reading of the sequence is a little more indirect. Here oligonucleotide 7 ends *before* C, thus position 8 must be occupied by C. Oligonucleotide 8 ends *before* G, and so position 9 must be occupied by G. Then reading through all the oligonucleotides, the original sequence is XCGATAGCG. (Note that the minus system does not give information on the base at position 7.) A recent extension of this technique involves the use of deoxyribonucleotides.

By choosing different primers, different regions of a large DNA molecule can be sequenced. If the relative positions of the primers in the total DNA are known, then the order of investigated segments is also known, and the complete sequence can be obtained.

In the *chemical method of Maxam and Gilbert*, the sample of DNA studied is divided into four subsamples. The polynucleotide chains should be radioactively labelled, often with radioactive phosphorus at the 5′-end. Each of the subsamples is subjected to a chemical treatment which cleaves the chain specifically at one base (or at most two). Thus treatment with dimethylsulphate causes methylation of the guanine and adenine residues of the DNA. Subsequent treatment with acid cleaves the DNA chain on the 5′-side of an adenine residue e.g.

$$p^{*}\ldots.\ TA\ldots. \xrightarrow[\text{acid}]{\text{dimethyl sulphate}} p^{*}\ldots.T + \text{non-radioactive products.}$$

Alternatively if, after methylation, the modified polymer is heated first at neutral pH, and then with alkali, the DNA splits preferentially on the 5′-side of guanine residues, although some breakage also occurs at adenine residues.

Hydrazine and piperidine treatment cleaves polydeoxyribonucleotide chains equally at cytosine and thymine, but if sodium chloride is added to the hydrazine reaction, modification of thymine is suppressed, and chain-scission then occurs at cytosine only.

Conditions can be so arranged that, for example, after methylation and acid hydrolysis, a break occurs at only one adenine residue per chain, but each chain of the sample may be split at a different adenine residue. Thus cleavage products arise from every adenine in the chain, and each radioactive product consists of a fragment of the original chain stretching from the labelled 5′-end to a base which preceded an adenine in the sequence (see Figure 5.9a).

The products of the four different chemical treatments are separated according to size by gel electrophoresis, and their positions compared to those of fragments of all possible sizes produced by random hydrolysis of the original DNA section under study. The sequence can be read directly from the gel (see Figure 5.9b).

Those oligonucleotides which occur in the A- or G-lane were formed by cleavage at (on the 5′-side of) A or G residues, respectively. Oligonucleotides which occur in the (T + C)-lane, but not in the C-lane,

(a) Radioactive products separated from chemical treatment of p*TCGATAGCG

Treatment	*Products*	*Chain-lenth*
Adenine-specific cleavage	p*TCG	3
	p*TCGAT	5
Guanine-specific cleavage	p*TC	2
	p*TCGATA	6
	p*TCGATAGC	8
	+ traces of adenine-specific products	
(Thymine + cytosine)-specific cleavage	p*T	1
	p*TCGA	4
	p*TCGATAG	7
Cytosine-specific cleavage	p*T	1
	p*TCGATAG	7

(b) Gel electrophoresis of products

Chain-length	Random hydrolysis	A-lane	G-lane	(T + C)-lane	C-lane	*Sequence*
						(3′-end)
9	+					G
8	+		+			C
7	+			+	+	G
6	+		+			A
5	+	+				T
4	+			+		A
3	+	+				G
2	+		+			C
1	+			+	+	X (5′-end)

Direction of movement ↓

FIGURE 5.9. Chemical method of sequencing DNA: (a) radioactive products expected; (b) gel electrophoresis of products

arose by cleavage at T-residues, whilst those in both the (T + C)- and C-lanes arose by cleavage at C-residues.

Thus oligonucleotide 1 terminates before C, meaning that in the original sequence position 2 was occupied by C. (This is exactly analogous to the minus system of the plus and minus method.) Oligonucleotide 2 terminates before G, and so G occupied position 3 of the original sequence. Proceeding in this way, the sequence XCGATAGCG is obtained. No information is given about the base at position 1, but this may be determined by end-group analysis.

Using these techniques, the complete sequences of the DNA of three viruses—*ϕX*174, *G*4 and *SV*40—have been determined, each of these being over 5,000 nucleotides long. In addition, sequences are known for long stretches of the DNA of bacteria and higher organisms which code for different types of RNA, or which function as control regions. The significance of these results will be discussed later in the sections on nucleic acid function.

5.4.1.2. Ribonucleic acid

For RNA, end-groups may be determined by some of the methods outlined for DNA, i.e. incorporation of radioactive phosphorus on the 5′-hydroxyl groups. If a free 3′-hydroxyl group is present on the RNA, then periodate may be used to oxidize the 2′, 3′-diol system, and the resulting dialdehyde can be reduced with ^{3}H-labelled borohydride to give a radioactive end-group (see Section 5.2.1, p. 173). In addition, alkaline hydrolysis of ribonucleic acids can give useful information on end-groups, as shown in the following examples:

(*i*) From the chain pQpRpS . . . pT—OH, alkali releases pQp, Rp, Sp, . . . , and T—OH. Thus a diphosphate is released from the 5′-end, and a nucleoside from the 3′-end. Identification of these gives not only the nature of the bases at the chains-ends, but also shows that the 5′-end was phosphorylated, whilst the 3′-end carried a free hydroxyl group.

(*ii*) From the chain pQpRpS . . . pTp, alkali releases pQp, Rp, Sp, and Tp. In this case, only the base at the 5′-end can be identified from the diphosphate, but it can be deduced that both ends of the chain were phosphorylated.

(*iii*) From the chain QpRpSp . . . pTp, alkali releases Qp, Rp, Sp . . . Tp. Here the end-bases cannot be identified, but it can be deduced that the 5′-end of the chain carried a free hydroxyl group, whilst the 3′-end was phosphorylated.

Sequences near the ends of RNA molecules can be studied using chemical, or enzymic, stepwise degradations. For example, periodate oxidation, followed by base-catalysed elimination of the dialdehyde formed at the 3′-end of an RNA molecule, can remove one base at a time from the polymer. (If a phosphate group is present originally at the 3′-end, it must be removed enzymatically prior to oxidation.) The free base can be separated and identified, whilst the nucleic acid, shortened by one nucleotide, can be subjected to the same treatment again to release the next base.

Enzymes, such as venom phosphodiesterase, can also be used for stepwise hydrolysis of ribonucleic acids to give oligonucleotides of shorter and shorter chain-length, whose sequences can then be studied.

In general, however, the base sequence of an RNA molecule is usually investigated by partial hydrolysis, producing oligonucleotides which can be separated by ion-exchange chromatography or electrophoresis. If 7 M urea is used at neutral pH during ion-exchange chromatography, base-binding to the ion-exchange material is minimized, and only the charged phosphate groups bind to the resins. Thus separation takes place on the basis of the number of phosphate groups in an oligonucleotide fragment, i.e. on the basis of size of the fragments. With 7 M urea in acid or alkali, separation can be obtained according to base composition, for C, A and G can become protonated at low pH, whilst C, U, and G ionize partially at high pH.

The sequence of each oligonucleotide is then determined, and the order of oligonucleotides in the original polynucleotide is worked out to give the complete base sequence.

Two of the most useful enzymes for carrying out partial hydrolyses are *pancreatic ribonuclease* (discussed in Section 4.5.2.6, p. 131) and a fungal enzyme known as *T_1-ribonuclease*. The *pancreatic ribonuclease* hydrolyses linkages of the type . . . pyrimidine-p—X . . . to give pyrimidine 3′-monophosphates, or oligonucleotides with a pyrimidine residue at the 3′-end and a purine, or modified pyrimidine, at the 5′-end. The enzyme cannot, for example, hydrolyse links involving 3-methylpyrimidines or 2′-*O*-methylpyrimidines, both of which are present as minor constituents of some transfer ribonucleic acids. T_1-ribonuclease, on the other hand, hydrolyses bonds of the type . . . Gp—X . . ., producing guanosine 3′-monophosphate, or oligonucleotides ending with Gp.

Oligonucleotides produced by pancreatic or T_1-ribonuclease hydrolysis can be further hydrolysed by enzymes of known specificity, e.g. venom phosphodiesterase, to give the nucleotide sequence in each fragment.

Once the sequence of each fragment has been determined, the order of the fragments in the original nucleic acid molecule must be investigated. This sequence can be evaluated by allowing ribonucleases to act on RNA for a short time only, to give partial hydrolysis, resulting in large oligonucleotide fragments. In ribonucleic acids, some bonds are more susceptible to enzymic attack than others, and these bonds are split during partial hydrolysis. It is believed that the three-dimensional structure of the nucleic acid determines which are the susceptible bonds (see Section 5.4.2, p. 197). Other large oligonucleotides can be obtained by prior chemical modification of some bases in the nucleic acid so that the ribonucleases can no longer hydrolyse the bonds adjacent to the modified residues. Then by study of these large fragments, it is often possible to determine the order of the small oligonucleotides within the larger ones; if different large fragments contain readily identifiable overlap sequences, the base sequence of the complete nucleic acid may be found. This process is relatively easy to carry out for transfer RNAs, as they contain several unusual bases, which may be used as markers for determining overlap sequences. However, these unusual bases are easily destroyed by acid or alkali, so great care must be taken when hydrolysing transfer RNA and separating the hydrolysis products. In ribosomal RNA, the methyl bases can be used as markers, but even when these are present, it is extremely difficult to obtain the complete base sequence of a large nucleic acid. The situation for messenger RNA is even more complex, for few, if any, unusual bases are present.

Now that great progress has been made in sequencing DNA, an RNA sequence can be obtained by making a DNA transcript from the RNA, i.e. the RNA is used as a pattern, or template, to direct synthesis of a DNA molecule with base sequence complementary to that of the RNA. (For a discussion of complementary base sequences, see Section 5.4.2.1, p. 194; for RNA-directed DNA synthesis, see Section 5.5.6.2, p. 227.) The sequence of the new DNA molecule is studied by the methods outlined in the previous section, and from this sequence, the base sequence of the original RNA can be deduced.

In addition, methods analogous to the plus and minus method and the chemical method have now been developed for RNA molecules. In the modified plus and minus method, stretches of radioactive DNA are synthesized using a special DNA-synthesizing enzyme, *reverse transcriptase* (see Section 5.5.6.2, p. 227), and the RNA under study is used as a template or pattern. In the minus system, the reverse transcriptase lengthens DNA fragments until the position of the missing base is reached (compare the previous section), whilst in the plus system, a special nuclease cuts down the DNA fragments until the added base is reached. The sequence of the DNA can be deduced as shown in Figures 5.7 and 5.8, and because of the special relation between the DNA sequence and the RNA pattern on which it was made (compare Sections 5.4.2.1, p. 194 and 5.5.3, p. 204), the sequence of the RNA can be determined.

For RNA sequences, the method analogous to Maxam and Gilbert's chemical method involves specific enzymic, rather than chemical, chain scission. Enzymes are available which hydrolyse RNA chains specifically at the 3′-end of a guanine-containing nucleotide or an adenine-containing nucleotide. Other enzymes hydrolyse at pyrimidine residues (i.e. cytosine and uracil), or at all residues *except* cytosine. If these four enzymes are used, and the products are separated by gel electrophoresis, the sequence can be read directly off the gel (see Figure 5.10). (Again the original stretch of RNA should be radioactively labelled at the 5′-end.)

Oligonucleotides occurring in the A-lane or G-lane end in an A- or G-residue, respectively. Those oligonucleotides which occur in the C + U-lane, but not in the G + A + U-lane must end in C-residues. Those which occur in both the C + U- and G + A + U-lanes must end in U-residues. Thus nucleotide 1 terminates in A, 2 in G, 3 in C, etc. and the sequence must be AGCUAUCGC.

The complete base sequences of many transfer RNA molecules have been determined, mainly by partial enzymic hydrolysis and sequencing of small fragments, and it has been found that certain sequences are common to all transfer RNAs so far studied. The sequence of the first transfer RNA to be studied, a yeast alanine transfer RNA, is shown in Figure 5.11. The molecule is folded, so that a high percentage of its bases can pair by hydrogen bonding (as shown in Figure 5.5, p. 176). The first folding scheme, the so-called clover-leaf model, is hypothetical, but has been used for many years. The second, L-shaped scheme corresponds more closely to the folding as determined by X-ray crystallography.

The sequences of several small ribosomal ribonucleic acids are also known. The first to be studied were 5S-RNA from the bacterium, *E. coli*,

(a) Radioactive products expected from enzymic hydrolysis of p*AGCUAUCGC

Enzyme used	*Products*	*Chain-length*
Guanine-specific cleavage	p*AG	2
	p*AGCUAUCG	8
Adenine-specific cleavage	p*A	1
	p*AGCUA	5
(Cytosine + uracil)-specific cleavage	p*AGC	3
	p*AGCU	4
	p*AGCUAU	6
	p*AGCUAUC	7
	p*AGCUAUCGC	9
(Adenine + guanine + uracil)-specific cleavage	p*A	1
	p*AG	2
	p*AGCU	4
	p*AGCUA	5
	p*AGCUAU	6
	p*AGCUAUCG	8

(b) Gel electrophoresis of radioactive products

Chain-length	Random hydrolysis	G-lane	A-lane	(C + U)-lane	(G + A + U)-lane	
						(3′-end)
9	+			+		C
8	+	+			+	G
7	+			+		C
6	+			+	+	U
5	+		+		+	A
4	+			+	+	U
3	+			+		C
2	+	+			+	G
1	+		+		+	A
						(5′-end)

Direction of movement ↓

FIGURE 5.10. Direct enzymic sequencing of RNA: (a) radioactive products expected; (b) gel electrophoresis of products

and from a tissue-culture line of mammalian cells known as *KB*-cells. Both are relatively short nucleic acids, 120 nucleotides in length, and have similar, but not identical, sequences. The sequences of the first and second halves of the *E. coli* 5S-RNA molecules are very alike, but there is much less sequence duplication in the *KB*-RNA.

The complete sequence of the 16S-ribosomal RNA of *E. coli*, and partial sequences of the 23S-ribosomal RNA, have been obtained, and it has been found that many of the main sequences containing methylated bases occur twice in the 23S-RNA molecule, whilst a few sequences appear to be repeated in the 16S-RNA. The distribution of pseudo-uridine and methylated bases appears to be non-random in both types of large ribosomal RNA, but there is little similarity in sequence between the 16S- and 23S-RNA of one organism; the sequence of each type, e.g. 23S-RNA, varies from organism to organism.

The base sequences of messenger RNAs from higher organisms are now being studied (in fact several have been determined), and it has been found that most messenger RNA is transported from the nucleus of a cell, where it is synthesized, with a long stretch of consecutive adenosine phosphate residues (i.e. polyadenylic acid or poly-A) attached to the 3′-end of the molecule. Messenger RNA molecules coding for histones (see Section 4.5.6, p. 155) do not, however, contain a poly-A segment. Many *m*-RNA molecules of higher organisms and the viruses which infect them

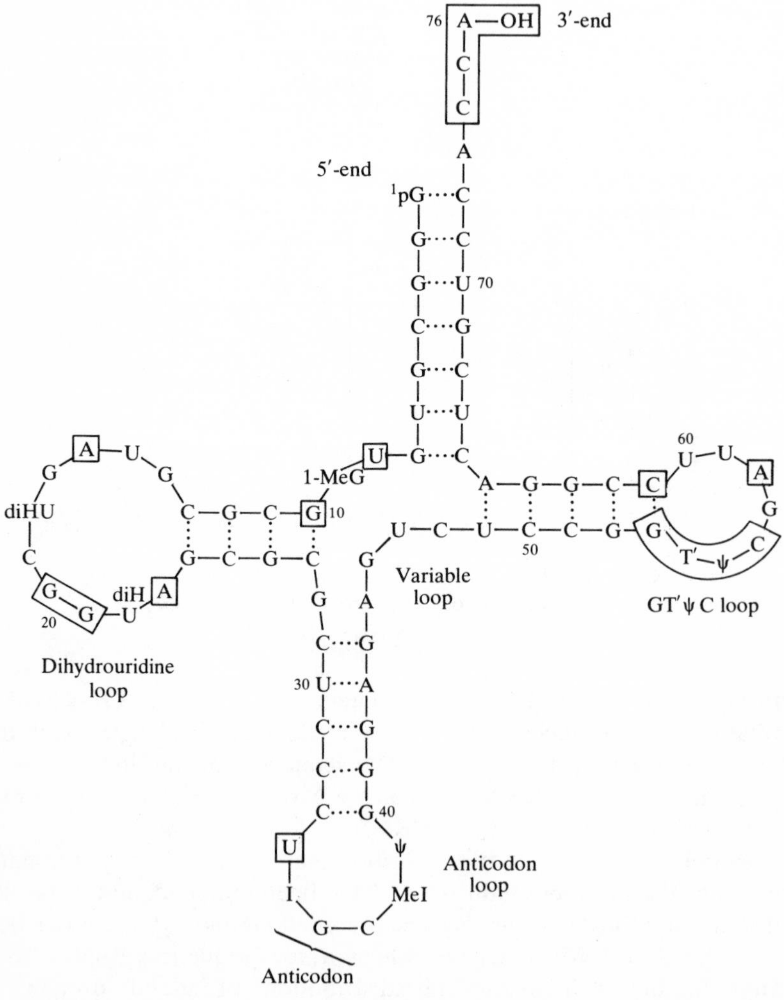

(a)

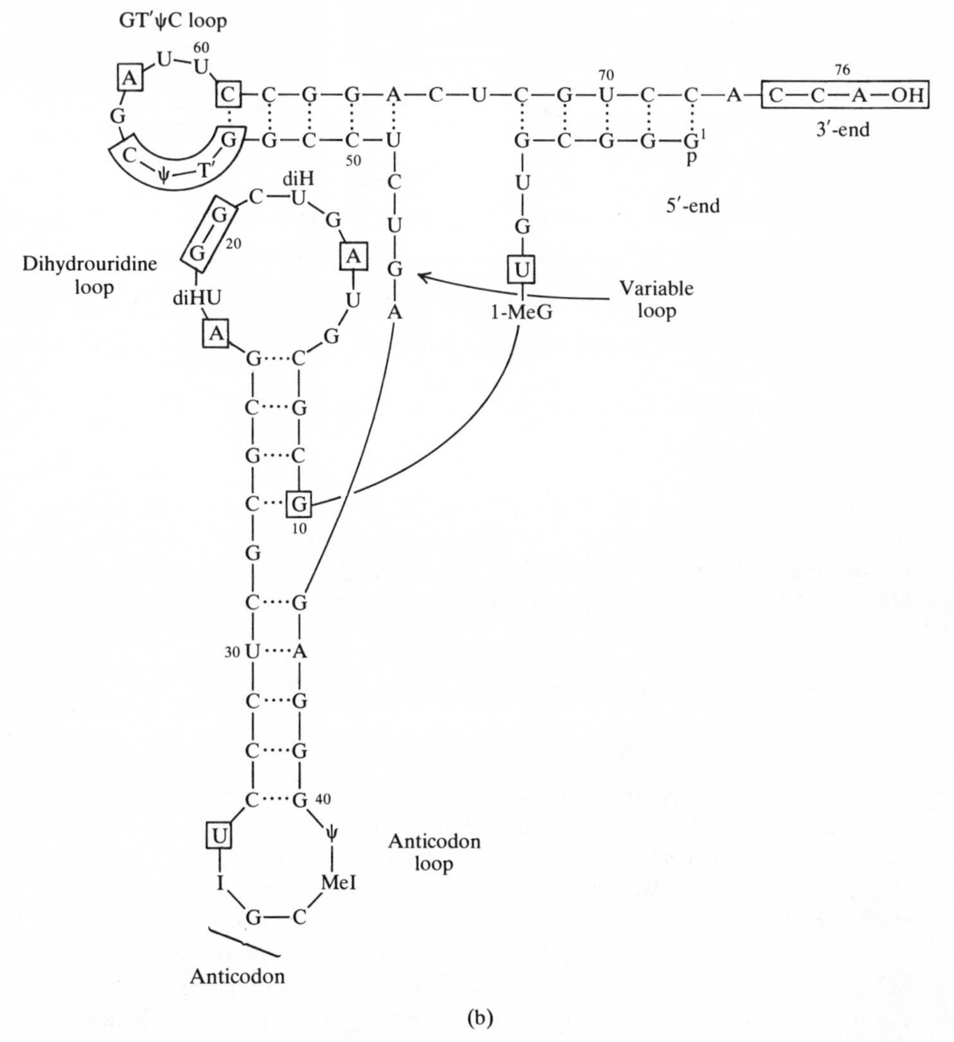

FIGURE 5.11. The base sequence of yeast alanine transfer RNA showing possible base-pairing: (a) clover-leaf model; (b) L-shaped model; . . . . hydrogen bonds; the bases outlined occur in all common transfer RNAs; unusual bases are shown in Figure 5.6, p. 179

carry a special 'cap' at the 5′-end, consisting of a methylated guanine nucleoside linked through a triphosphate group to the next nucleoside (see Figure 5.12).

Many viruses contain ribonucleic acid as their only genetic material, and this RNA is generally homogeneous within one type of virus. This polymer can be thought of as a special type of messenger RNA, and partial sequences of long viral RNAs, particularly those of *R*17- and *Qβ*-bacteriophages (viruses which attack bacteria), have been studied. The

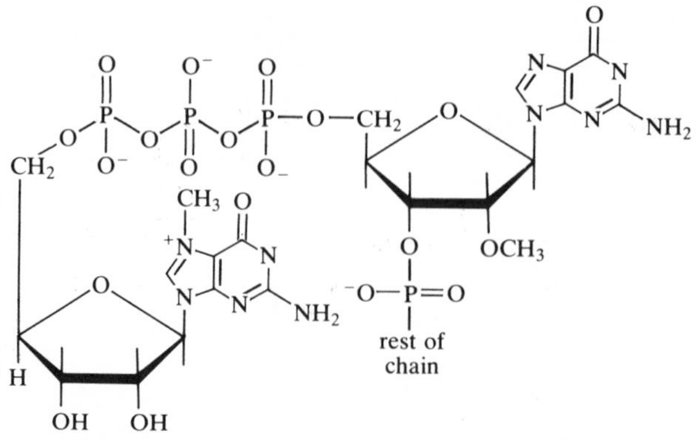

FIGURE 5.12. The 5′-cap of some *m*-RNA molecules

complete sequence of the RNA of virus *MS*2, which is 3300 nucleotides long, has now been determined. These sequences have been helpful in confirming the genetic code (and will be discussed later in this chapter), although base sequences of unknown function have also been found.

5.4.2. Secondary and tertiary structure

5.4.2.1. Deoxyribonucleic acid

The results of X-ray studies led by Watson and Crick to propose that DNA molecules consist of two polynucleotide chains in a double helical structure (see Figure 5.13). (For this work, these scientists were awarded the Nobel Prize in 1962.) The two chains run in opposite directions, and are held together by hydrogen bonds between pairs of bases, each chain contributing one base to the pair. Since the two chains of the molecule are always the same distance apart, a base pair must consist of a purine and a pyrimidine; hydrogen bonding between two pyrimidines would bring the chains closer together, whilst bonding between two purines would force the chains further apart.

In DNA, adenine on one chain is always bonded to thymine on the other, whilst cytosine on one chain is always bonded to guanine (as shown in Figure 5.5, p. 176). As a result, the bases of the two chains are not identical but are complementary to one another, and the base sequence of one chain determines the sequence of the other, e.g. where T appears in one chain, A must appear in the second chain. Thus the base sequence of a hypothetical section of DNA would be:

5′-end . . . GATCACGTA . . . 3′-end
3′-end . . . CTAGTGCAT . . . 5′-end

Thus in double-stranded DNA, the amount of adenine in a molecule must

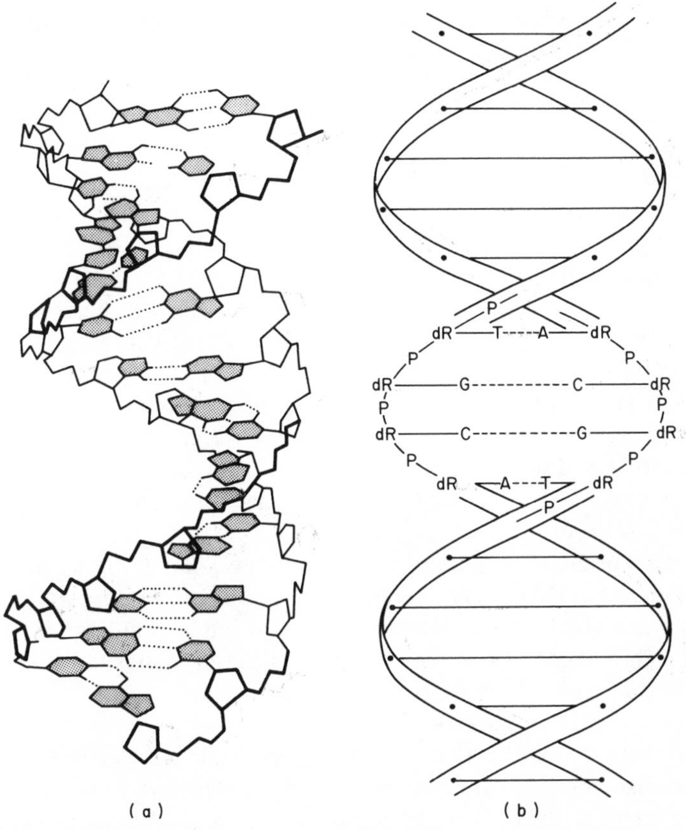

FIGURE 5.13. The structure of DNA: (a) helix showing bases and sugar units; (b) schematic double helix showing base-pairing. —P—dR—P—dR— etc. represents the deoxyribose phosphate backbone of a polynucleotide chain

equal the amount of thymine, and the amount of guanine must equal the amount of cytosine.

In the Watson–Crick model, the two chains are twisted round each other in a double helix with the bases on the interior and the sugar–phosphate backbone on the outside (see Figure 5.13). The bases lie at right-angles to the helix axis, and because they are flat, they stack easily on top of one another and any sequence of bases is possible. There are ten base-pairs in each complete turn of the double helix, which has a helical pitch of 34 Å. Thus each pair is 3.4 Å from the next one in the helix, and this distance is close enough for the π-electrons of the heterocyclic ring systems to interact, giving the bonding important for the maintenance of the three-dimensional structure of the molecule. The molecule is 20 Å in diameter, and has one shallow groove and one deep groove lying the whole of its length. Proteins such as histones (see Section 4.5.6, p. 155) fit into these grooves to give the nucleoprotein complexes found in *chromosomes*.

Other models for DNA structure have been proposed. In one, the two chains lie side by side, but do not twist round each other to form a double helix. Rather, the two strands are like sinusoidal curves lying lengthwise opposite each other, and so the molecules would have the form of a 'warped zipper'. For poly-CG sequences in DNA, it has been shown that a left-handed helix is possible (the Watson–Crick model is right-handed). In these alternative models, the base sequences of the two chains are complementary, as in the Watson–Crick model; thus the chains are held together by base-pairing.

Molecules of double-stranded DNA behave as slightly flexible rods if the molecular weight is less than one million, but very large double helices, with molecular weights *ca*. one hundred million, can behave more like random coils. Some DNA, from bacteria, viruses, the mitochondria (subcellular particles) of higher animals, and the chloroplasts of plants, can exist as circular double helices, i.e. each chain of the double helix is a covalently closed circle. If such loops are large enough, supercoiling within the loops can occur.

Since the original postulate by Watson and Crick of a double helix, single-stranded DNA has been isolated from some viruses. These molecules are much more flexible than the double-chain types, but where complementary base sequences exist in different parts of a single-chain molecule, the chain can fold-up on itself to give short double-helical regions linked by single-stranded loops.

Double-helical DNA, like the proteins, can be denatured, i.e. the three-dimensional structure can be altered, or destroyed, by rupture of the hydrogen and hydrophobic bonding. Denaturation can be studied by measuring the UV absorption, optical rotation, or viscosity of the polymer. When native DNA is denatured, the UV absorption increases and this is known as the *hyperchromic effect*. (This may be due to loss of π-bonding between the parallel bases.) The optical rotation is decreased as is the viscosity, when structural changes occur. Denaturation may be brought about by heat, low ionic strength, or extremes of pH, and is believed to involve the transition from a helical structure to a random coil. Heating causes 'melting' of regions of hydrogen-bonded base-pairs, and at any one ionic strength, there is a marked change in viscosity and UV absorption over a narrow temperature range. The temperature, at which half the total change has taken place, is known as the melting temperature (T_m). T_m depends on the total (guanine + cytosine)-content of the DNA. Three hydrogen bonds are involved in G–C pairing, as opposed to two in A–T pairing; thus DNA molecules with a high (G + C)-content are less easily denatured, and have a high T_m. On heating, the molecular weight usually remains unaltered, indicating that the two polynucleotide chains do not separate completely, and indeed partial renaturation can occur on slow cooling at an ionic strength greater than 0.4. However, the chains can be separated completely by heating at 100°C in a solution containing sodium chloride and sodium citrate. If the solution is cooled quickly, the strands remain separated. However, when the solution is held at a temperature about 25°C below T_m, the helical structure can reform. Extremes of pH can also cause denaturation, for the bases can become protonated or ionized, thus bringing about rupture of hydrogen bonds and breakdown of the double helix.

Much useful information concerning the organization of the genes in the chromosomes of higher organisms has been gained from quantitative studies of DNA renaturation. Mixed renaturation involving strands of DNA from different organisms has allowed measurements to be made of the degree of sequence homology between the different DNA molecules. In addition, if RNA complementary to one of the DNA strands is present, RNA–DNA hybrids can be formed. A study of this reaction has provided data on the distribution of specific genes within eukaryotic DNA.

5.4.2.2. Ribonucleic acid

X-ray work on the double-stranded, viral RNA molecules suggests that these can exist as double helices, again with the two chains running in opposite directions and held together by hydrogen and hydrophobic bonding. In this case, adenine in one chain bonds to uracil in the other, and guanine in one chain bonds to cytosine in the other, so that again the base sequences of the two chains are complementary. The bases lie parallel to one another in the interior of the helix, but are no longer at right angles to the helix axis. There are believed to be 11 base-pairs in each complete turn of the double helix, with helical pitch 28 Å. (The DNA double helix can, under certain circumstances, also exist in this form.)

Little is known of the three-dimensional structure of large single-stranded RNA molecules, particularly messenger and ribosomal RNA. Both types of RNA exist in living organisms as complexes with protein, and this protein may help stabilize the three-dimensional structure of the nucleic acid. Where complementary base sequences exist in different parts of a single polynucleotide chain, both base-pairing and helix formation can occur, so that the single chains probably consist of short stretches of double helix separated by non-helical regions. In addition, parts of the single strand may coil in such a way that the bases lie parallel to one another, i.e. *base-'stacking'* occurs, even if a double helix is not formed. This type of structure would be stabilized by hydrophobic bonding between the bases. If a solution of a large ribosomal RNA molecule is heated, or the ionic strength is decreased, a hyperchromic effect may be observed along with a change in viscosity, indicating that a helix–random-coil transition takes place, and showing that double-helical regions can be found in these single-stranded nucleic acids. Thus the presence of secondary structure, i.e. base-pairing to give helical regions, in some ribonucleic acids is confirmed. However, the details of tertiary structures, i.e. the mutual arrangement of helical regions within the molecules, are not yet known. (The position of bond scission during short exposure of an RNA to a ribonuclease enzyme is believed to be determined by the secondary and tertiary structure of the nucleic acid molecule. Thus single-stranded, non-helical parts of the polynucleotide chain lying on the surface of the molecule are usually the most susceptible to enzymic attack.)

Much work has been carried out on the three-dimensional structure of transfer nucleic acids. Some of these acids have been prepared in crystalline form, and their structure investigated by X-ray diffraction. The molecules possess secondary structure, i.e. the chains are folded to allow base-pairing and some helix formation. Different transfer RNA molecules all have very similar, complex, folding, as shown in Figure 5.14. The terminal stem (the part containing the ends of the molecule) and the loop, containing the sequence GT′ψC, form one almost continuous double helix. This is connected by a single-stranded region to another stretch of double helix, consisting of the dihydrouridine loop and the anticodon loop. These two principal double-helical stretches are folded up close to one another to give a tertiary structure. The molecule is essentially L-shaped with the chain-ends and anticodon more than 75 Å apart at opposite ends of the 'L'. The dihydrouridine and GT′ψC loops form the corner of the 'L'.

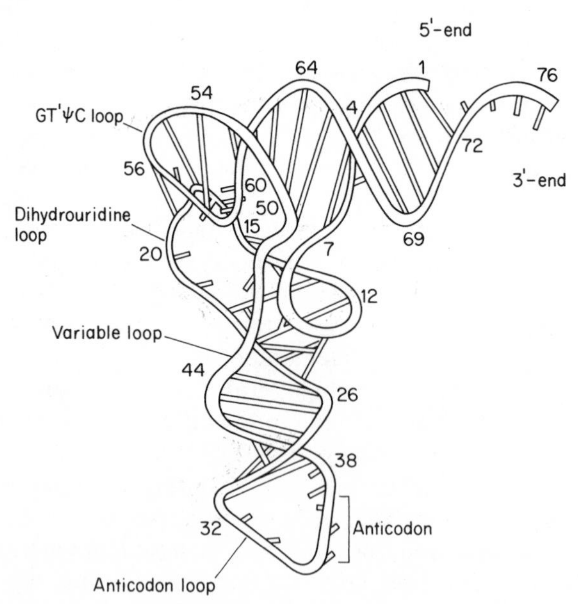

FIGURE 5.14. Schematic diagram of chain-folding and tertiary structure of yeast phenylalanine transfer RNA; the ribose phosphate backbone is represented by the continuous line; base interactions in the double-helical sections are shown by long lines, non-bonded bases by short lines. Reproduced with permission from A. Rich and S. H. King, Scientific American, January, 1978

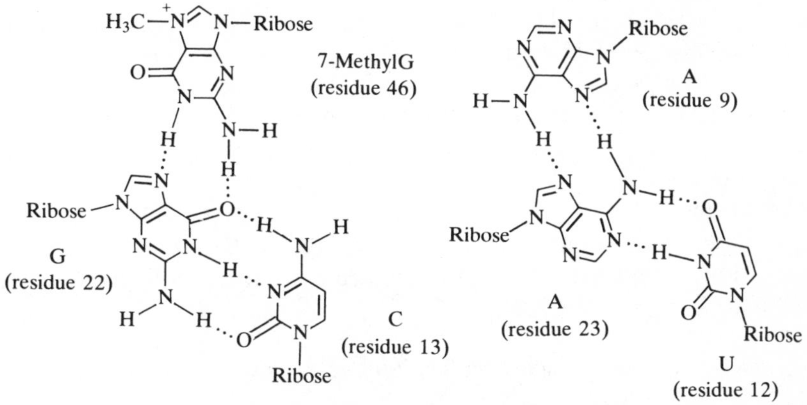

FIGURE 5.15. Base triplets in yeast phenylalanine *t*-RNA

Bases in variable regions are often unstacked on the surface of the molecule, whilst invariant bases are frequently involved in hydrogen bonding.

The tertiary structure brings bases together that are far apart in the sequence, and in two cases in the yeast phenylalanine transfer RNA, three bases come together and are held by hydrogen bonding (see Figure 5.15).

The three-dimensional structure of a yeast initiator *t*-RNA (see Section 5.5.5.2, p. 216) has also been determined by X-ray diffraction, and is very similar to that shown in Figure 5.14. The position of the anticodon loop is slightly different, but it is not known whether this position is fixed in space when a *t*-RNA functions in protein synthesis. Indeed, it is believed that conformational changes in the *t*-RNAs take place as these molecules carry out their *in vivo* function.

5.5. FUNCTION

5.5.1. Introduction

An organism's genetic information is believed to be located and '*coded*' in the base sequences of the deoxyribonucleic acid of the cell. In higher organisms, most of the DNA is found in the *chromosomes* of the cell nuclei, but a small proportion is in subcellular particles, i.e. mitochondria and chloroplasts. (Mitochondria produce much of the energy of a cell, whilst chloroplasts are the sites of photosynthesis.) The part of a chromosome specifying one trait of the organism is a *gene*, which consists of a length of DNA complexed with proteins and sometimes with RNA. Most genes determine the structure of a single, or a group of, proteins. However, because the growth of normal cells must be kept under strict control, much of the DNA of the chromosomes is probably involved in control mechanisms.

The material of chromosomes, *chromatin*, is believed to have a structure like beads on a string. The ‘beads’, called *nucleosomes*, consist of units of DNA approximately 150 base-pairs in length associated with an aggregate of two molecules each of histones *H2A*, *H2B*, *H3*, and *H4* (see Section 4.5.6, p. 155). The ‘string’ or linker DNA is usually 60–80 base-pairs long, and is associated with histone *H1*. The importance of this structure in DNA function is not yet understood.

The nucleic acids function either by passing genetic information from one generation to the next, or by involvement in the process whereby the information coded on the DNA molecules is used to specify protein structure.

5.5.2. Deoxyribonucleic acid; replication and repair

As an organism grows, the number of its cells increases by division, and each new ‘daughter’ cell contains the same genetic information as the original cells. The new DNA produced is identical to that of the parent cells, and the formation of this new DNA, with base sequences identical to the original, is known as *replication*.

The mechanism of replication is semiconservative, i.e. the two strands of the original double helix separate and act as patterns, or templates, for the formation of a new strand. Thus, each new double-helical molecule consists of a newly synthesized polynucleotide strand and one strand from the ‘parent’ molecule. The use of one strand of DNA as the template for biosynthesis of a new strand of correct base sequence is easily understood in terms of base-pairing. For example, whenever adenine is found in the parent strand, the enzymes catalysing DNA synthesis will join a thymine-containing nucleotide into the new strand; where cytosine is found in the old strand, a guanine nucleotide will be incorporated. Hence the new strand contains a base sequence complementary to that of the old strand, and identical to that of the strand in the original molecule which it has replaced, e.g.:

···—*P*—*dR*—*P*—*dR*—*P*—*dR*—*P*—*dR*—···
A ⋮ **T**   *C* ⋮ **G**   *G* ⋮ **C**   *A* ⋮ **T**
···—**P—dR—P—dR—P—dR—P—dR**—···

···—*P*—*dR*—*P*—*dR*—*P*—*dR*—*P*—*dR*—···
A ⋮ *T*   *C* ⋮ *G*   *G* ⋮ *C*   *A* ⋮ T
···—*P*—*dR*—*P*—*dR*—*P*—*dR*—*p*—*dR*—···

‘Parent molecule’

⟶

\+ **Newly synthesized strands**

···—**P—dR—P—dR—P—dR—P—dR**—···
A ⋮ *T*   **C** ⋮ *G*   **G** ⋮ *C*   **A** ⋮ *T*
···—*P*—*dR*—*P*—*dR*—*P*—*dR*—*P*—*dR*—···

In this way, new molecules of DNA are produced with base sequences identical to those of the original molecules.

Although this general scheme of DNA replication has been accepted for many years, the complete details of the process are still unknown.

DNA synthesis is generally believed to be catalysed by enzymes, *DNA-polymerases*, which require the presence of a primer molecule (an oligo- or poly-nucleotide to which the new strand becomes covalently linked), a region of single-stranded DNA to act as template, and the four deoxyribose nucleoside triphosphates (Figure 5.2a, p. 169). The primer must have a free 3′-hydroxyl group, and a short stretch of RNA (8–11 bases long) acts as the primer in most living organisms; this primer is quickly excised once the synthesis of the new DNA chain is under way. Another protein, a so-called '*swivelase*' may be involved in introducing a transient nick near the site of replication allowing the helix to rotate, whilst other proteins, *helix-destabilizing proteins*, bind to the single chains to prevent reformation of the helix. Replication is believed to occur at a 'replicative fork' (see Figure 5.16).

The polymerase binds to the primer and the single-stranded template at a specific initiation site; this site probably consists of an A–T-rich region between two G–C-rich stretches. Symmetrical sequences, such as

—TATT—AATA—
—ATAA—TTAT—

may be important as recognition sites for DNA-polymerase. In addition, secondary structures, such as hairpin loops on a DNA strand, may also be necessary to specify the initiation site. Other proteins are probably involved in initiation. If the base of an incoming nucleoside triphosphate can hydrogen bond to the first available base on the template, a phosphodiester bond is formed between the 3′-OH of the primer and the

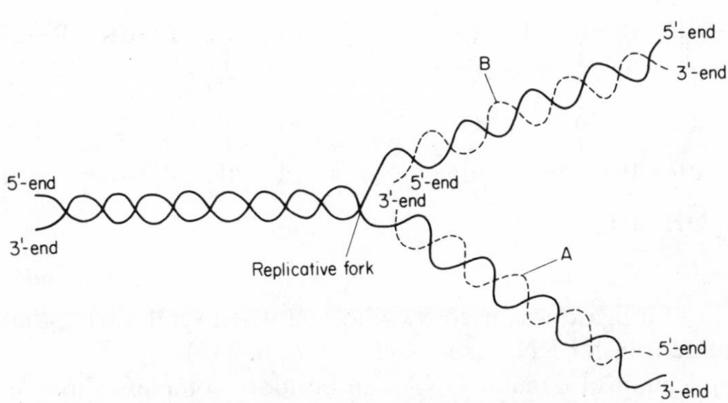

FIGURE 5.16. Replicative fork in DNA synthesis: —— 'parent' DNA; ---- newly-synthesized DNA

5′-phosphate of the new nucleotide, and pyrophosphate is released, e.g.:

```
      Template

··—P—dR—P—dR—P—dR—···                      ···—P—dR—P—dR—P—dR—···
     |       |      |                              |       |       |
     A       G      T     DNA-polymerase           A       G       T
     :                   ─────────────→            :       :            + P2O7^4−
     U                         +                   U       C
     |                        dCTP                 |       |
··—P—R—OH                                  ···—P—R—P — dR—OH

   Primer
```

The polymerase then 'checks' that the new base indeed pairs with the opposite base on the template. If it does not, it can immediately be excised and replaced by the correct base. This double action of the enzyme, i.e. polymerization and 'checking', ensures that the incorporation of mismatched bases is kept below 1 in 10^9. The polynucleotides then move with respect to the enzyme, so that the enzyme can link, to the growing chain, another nucleotide with base complementary to the next available base on the template. Magnesium ions, ATP, and other proteins are required for the reaction to proceed. The process is repeated, and the new chain grows in the 5′- to 3′-direction at the rate of up to 1000 bases added per minute. When the new chain reaches a certain (but as yet unknown) length, the RNA portion is removed enzymically from the 5′-end.

In the scheme shown in Figure 5.16, one new chain would grow in the 5′- to 3′-direction (*B*), while the other new chain would be synthesized in the 3′- to 5′-direction (*A*). No enzyme has yet been discovered which can replicate DNA in a 3′ → 5′-direction, and so various theories have been put forward to explain the synthesis of the second strand (*A*). It now seems likely that, for both *A*- and *B*-strands, short pieces of DNA are synthesized in the 5′ → 3′-direction, and are later joined together by an enzyme known as a *ligase*. The ligase can link 5′-phosphate on an oligonucleotide and 3′-hydroxyl on another, if the two pieces are hydrogen-bonded side-by-side to longer complementary strands:

```
···—P—dR—P—dR—P—dR—P—dR—···            ···—P—dR—P—dR—P—dR—p—dR
     |      |      |      |                    |       |       |      |
     A      G      T      C      ligase        A       G       T      C
     :      :      :      :     ──────→        :       :       :      :
     T      C      A      G                    T       C       A      G
     |      |      |      |                    |       |       |      |
··P—dR—P—dR  P—dR—P—dR—···            ···—P—dR—P—dR—P—dR—P—dR—
            /   \                                    ↑
          OH  HO                                     |
         ↙       ↘                                   |
    a 3′ end    a 5′ end                          new bond
```

The action of a ligase was used by Khorana when synthesizing the gene for yeast alanine transfer RNA (see Section 5.6, p. 233).

The replication of circular DNA can be more complex. Circular DNA is found in some viruses and bacteria, in the mitochondria of higher animals, and the chloroplasts of plants.

Firstly, the two strands of the circular DNA must separate slightly so that the strands may act as templates. For complete separation of the parent strands, so that each can form part of a new daughter molecule, a nick (or cut) must be introduced into one of the parent strands. This is later repaired by a ligase.

The first synthesis of circular DNA in the laboratory was carried out by Kornberg and coworkers using a DNA-polymerase isolated from a bacterium, *E. coli*. (For this work, Kornberg received the Nobel Prize in 1959.) A single-stranded circular DNA from a bacteriophage was used as template, and an oligonucleotide was used as primer. In the presence of a ligase, the DNA-polymerase synthesized a double-helical circular molecule consisting of the original and one new strand. A single-stranded break was introduced into the molecule by a nuclease, and the two chains were separated. Using the new strand as template, a fully synthetic duplex molecule was made which was indistinguishable from the original.

For some time, it was believed that the DNA-polymerase isolated by Kornberg was the main enzyme involved in the cell in DNA synthesis. However, studies of mutant bacteria, deficient in this polymerase, have shown that DNA replication can still take place normally. It is now thought that the Kornberg DNA-polymerase is important for the repair of damaged DNA in the living organism, but may also be one of three DNA-polymerases actually involved in DNA synthesis.

Single-stranded breaks can be introduced into DNA by radioactivity or exposure to X-rays, and UV radiation can bring about the formation of thymine–thymine dimers, where these bases are adjacent to one another in a polynucleotide strand (see Section 5.2.1, p. 173). Such changes in the molecules impair the correct replication of DNA, and thus endanger the life of the organisms in which they occur. The Kornberg DNA-polymerase can, however, excise the dimers and other mismatched bases; can insert the correct bases into the gaps; and, in the presence of a ligase, can fully repair the damaged DNA molecules. Other repair enzymes are known which can remove chemically altered bases, so that the original bases can be reinserted into the DNA.

5.5.3. Ribonucleic acid; transcription and replication

Although the deoxyribonucleic acids contain the information specifying protein amino acid sequences, DNA is not directly involved in protein synthesis. The different types of ribonucleic acid—messenger, ribosomal, and transfer RNA (*m*-, *r*-, and *t*-RNA)—are necessary, however, for the translation of base sequences on DNA into amino acid sequences in protein. In cells, ribonucleic acid is synthesized using DNA as template, so that the base sequences of RNA are complementary to sequences on the strands of DNA. This process of biosynthesis of RNA, giving molecules whose base sequences are determined by a DNA template, is known as *transcription*. In any one section of DNA, only one strand of the double helix is transcribed, i.e. only RNA base sequences complementary to one DNA strand are produced, but in another section of the same DNA molecule, the other polydeoxyribonucleotide chain may be transcribed.

Transcription can take place in the nuclei, mitochondria, and chloroplasts of higher cells. Each type of RNA (messenger, ribosomal, and transfer) is transcribed from a different section of an organism's DNA, and there may be multiple copies of the genes for some RNA. For example, in higher organisms there may be several hundred genes for each ribosomal RNA in a cell, and less than fifteen genes for each transfer RNA.

Details of the mechanism of transcription have been worked out for bacteria, particularly *E. coli*, and the mechanism for higher organisms is believed to be similar, but not necessarily identical. The enzyme catalysing ribonucleic acid biosynthesis is known as *RNA-polymerase*, and uses double-helical DNA as template, transcribing one strand of the DNA in the $3' \rightarrow 5'$-direction, i.e. the new RNA chain grows in the $5' \rightarrow 3'$-direction. Ribonucleoside triphosphates are linked together with the elimination of pyrophosphate, the order of bases in the RNA chain being determined by the base sequence of the template strand. Thus where guanine appears in the DNA, cytosine is inserted into RNA; where adenine appears in DNA, uracil is linked into RNA, etc (see Figure 5.17). (In DNA, thymine hydrogen bonds to adenine, but in RNA, thymine is replaced by uracil.)

The first stage in the initiation may involve non-specific reversible binding of RNA-polymerase to double-helical DNA. Another protein, the *σ-protein* (which can bind to the enzyme molecule), is thought to be important for ensuring that initiation takes place on the correct site on DNA. (It may also be important for ensuring that only one strand of DNA is transcribed.) Thus the enzyme, with the bound σ-protein, locates a site on the DNA, specific for the initiation of transcription. A specific binding site on DNA for RNA-polymerase is known as a *promoter*. In viruses and some bacteria, an AT-rich sequence of 6 or 7 bases is usually found on DNA shortly before the initiation site for transcription, and in many viruses another AT-rich stretch occurs about 35 bases before the initiation site. These sequences are probably important as promoters. The σ-unit probably helps bring about separation of the two DNA strands at initiation, so that one chain can be used as template.

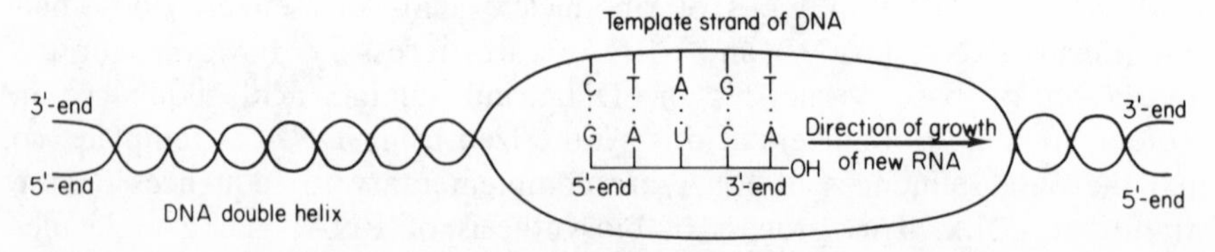

FIGURE 5.17. Biosynthesis of RNA by transcription

The first ribonucleoside triphosphate, usually with a purine base, binds to the enzyme at the 'initiating binding site'; this nucleotide will form the 5′-end of the new RNA molecule. A second nucleotide then binds to the enzyme, its base being hydrogen-bonded to the opposite base on the DNA template strand. Thus the nature of the incoming base is determined by the base sequence of the template. The enzyme catalyses the formation of a phosphodiester bond between the two nucleotides, (compare page 175), and pyrophosphate is released. The growing chain and the template are translocated one position, so that the second nucleotide is now in the 'initiating binding site' on the enzyme. A third nucleotide can be bound to the enzyme, and so the new RNA chain grows in the $5' \rightarrow 3'$-direction.

A transient, short stretch of double DNA–RNA helix may form near the growing end of the new RNA chain, but this would quickly dissociate leaving single-chain RNA, and double-helical DNA. (The template and other strand of the DNA are thought to reform the double helix soon after passage of the RNA-polymerase.) The σ-protein (usually called the σ-factor) is released soon after initiation, and can combine with another RNA-polymerase molecule to ensure that RNA chains are begun at the correct sites on DNA. Although the RNA-polymerases of bacteria and higher organisms consist of several subunits, this type of structure is not essential for activity, but may be important for the control of transcription.

In higher organisms (eukaryotes), three RNA-polymerases have been identified. One of these catalyses the transcription of *m*-RNA, another the transcription of *t*-RNA and the 5S-RNA of ribosomes, while the third is specific for the 5.8S- 18S-, and 28S-ribosomal RNAs.

In living organisms, RNA chains are synthesized at the rate of up to 50 nucleotides added per second. Additional protein factors are involved in the RNA biosynthesis. It is believed, for example, that in some cases a protein called the ρ-factor is important for ensuring that new RNA chains of the correct length are made, i.e. that the RNA-polymerase action stops at specific termination sites on the DNA. An AT-rich sequence, possibly coming at the end of a hairpin loop, seems likely to form a termination site.

Most RNA molecules are not transcribed from DNA in the form in which they play an active part in protein synthesis; the new RNA is generally longer than 'active' RNA, and contains no modified bases. These precursors are subsequently cleaved by specific enzymes, and bases are methylated, or modified, to give products such as shown on page 179. Thus the three larger ribosomal RNAs from higher organisms are transcribed together giving one ribonucleic acid molecule with a sedimentation coefficient of 45 S. Cleavages and base modification finally yield the ribosomal components with sedimentation coefficients 5.8 S, 18 S, and 28 S. Similarly the three *r*-RNAs of bacteria are transcribed as one unit, which is later cut and trimmed to give the biologically active ribosomal components.

Transfer RNAs may undergo a similar type of processing after transcription. Thus 'unwanted' segments are trimmed from both the 5′- and 3′-ends of the transcript, and bases are modified to give the unusual bases characteristic of *t*-RNAs. In addition, transcripts of some eukaryotic *t*-RNAs contain 'unwanted' base sequences in the interior, and these must be excised, and the two remaining parts joined together to form an active *t*-RNA.

Messenger RNA molecules are also formed by cleavage of larger precursors. In the processing of eukaryotic *m*-RNA, the 3′-end of the transcript is trimmed, and to most, but not all, *m*-RNAs a stretch of polyadenylic acid, i.e. 100–200 adjacent adenine nucleotides, is added at the 3′-end. Then the special 'cap' is added to the 5′-end (see Figure 5.12, p. 194).

The direct transcripts of many eukaryotic *m*-RNAs contain internal sequences which are not present in the final active messenger RNA. Such sequences are known as *introns*, and are excised as a last step in processing. The remaining sequences, called *exons*, must then be joined together in the correct order to produce an active *m*-RNA (see Figure 5.18). The enzymes, which catalyse the excision of introns and the linking of exons, are believed to recognize specific sequences on the RNA transcript at the intron–exon junctions; RNA-sequencing techniques have indicated that —CAGG— or similar sequences at junctions may fulfill this rôle.

Not all RNA is formed by transcription, the main exception being the RNA of certain viruses. (Some viruses contain RNA but no DNA, the ribonucleic acid in this case constituting the genetic material of the virus.) The RNA of many of these viruses is *replicated* by an RNA-polymerase, in a manner analagous to DNA replication by DNA-polymerase. Here, the existing RNA acts as the template for the formation of new RNA, otherwise the process is very similar to transcription.

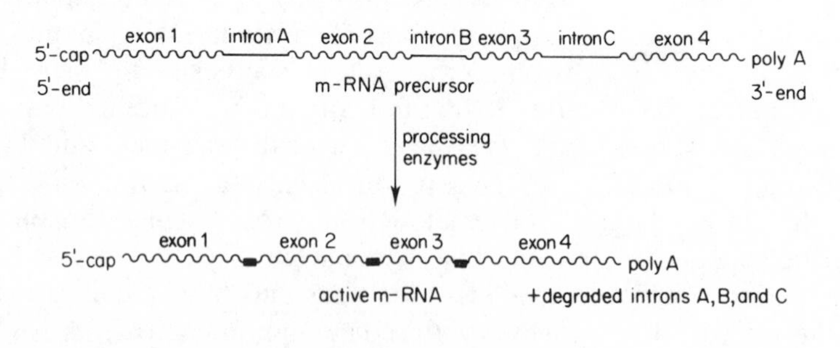

FIGURE 5.18. Final step in processing an eukaryotic *m*-RNA transcript

5.5.4. Messenger RNA and the genetic code

Whilst the DNA of mitochondria and plant chloroplasts codes for some proteins, *t*-RNA, and *r*-RNA, in higher organisms most of the genetic information of a cell resides in the DNA of the nucleus. But protein synthesis takes place in the cytoplasm, and therefore, the specifications for protein structure must be transmitted from the DNA through the nuclear membrane to the protein synthesis machinery. This transfer of information is carried out by messenger RNA(*m*-RNA). The base sequence of one strand of the nuclear DNA, coding for protein structure, is accurately transcribed into the complementary base sequence on a messenger RNA molecule. Each *m*-RNA codes for one protein (a monocistronic messenger), or a number of proteins taking part in closely related reactions (a polycistronic messenger). (The latter are found only in lower organisms.) Thus many different messengers can be found in one cell.

As mentioned earlier, most *m*-RNA species of higher organisms contain stretches of poly-A at the 3′-end. This stabilizes the messenger RNA, and helps to protect it from degradative attack by nucleases. The poly-A sections may also be involved in binding some of the proteins with which *m*-RNA is associated both in the nucleus and the cytoplasm. The function of these proteins is not yet known. However, the *m*-RNAs for histones (see Section 4.5.6, p. 155) and those of lower organisms are not thought to contain polyadenylic acid regions, and so poly-A cannot be vital for all *m*-RNA functions.

The sequence of amino acids in a protein is believed to be determined by the base sequence of the *m*-RNA coding for that protein, and hence by the base sequence of the DNA from which the *m*-RNA was transcribed. Because 20 different amino acids are incorporated into proteins, there must be at least 20 'code-words', or *codons*, in the genetic code. To form 20 or more codons from the four bases of RNA, each codon must contain at least three bases. (There can be only four different codons if each codon consists of one base only, and $4^2 = 16$ different codons if each codon consists of two bases only.) However, $4^3 = 64$ different triplet codons can be formed, and so each amino acid can be specified by more than one codon, if all possible codons are used in living organisms. It is now known that the 'code-words' do indeed consist of three bases.

The genetic code was first investigated using the protein synthesis machinery of the bacterium *E. coli*, but it is now believed that the code is more or less universal i.e. that one particular code-word of three bases is translated into the same amino acid in all living organisms. The code was deciphered by Nirenberg, Khorana, Ochoa, and their coworkers using chemically synthesized messengers. (Nirenberg and Khorana shared the Nobel Prize in 1968 for this work.) It was first discovered that poly-U (polyuridine phosphate) directs the incorporation of the amino acid, phenylalanine, into growing peptides; thus the base sequence U—U—U must be the codon for phenylalanine. Work with more complex messengers enabled the whole of the code to be deciphered (see Table 5.1). Three triplets (UAA, UAG, and UGA) do not code for amino acids, but are the signals to stop polypeptide chain synthesis. The codon AUG is the common start signal as well as the code-word for methionine, but GUG is also used occasionally as a start signal.

It has in fact been found that most amino acids are coded for by more than one triplet, e.g. six different triplets can code for leucine, and so the genetic code is said to be degenerate. The code on the messenger is read in a *non-overlapping* manner—thus the sequence AUGCAUGCG is read during protein synthesis as AUG.CAU.GCG and not as AUG.UGC.GCA.CAU . . . etc.

Amongst the messengers which have been studied most intensively are the RNA molecules of bacteriophages such as *M*2, *R*17, and *Qβ*. These RNA molecules constitute the genetic material of the phages (which contain no DNA), but may also be regarded as polycistronic messenger RNAs. The RNA of these phages codes for an assembly protein, a protein which forms the 'coat' of the phage, and an RNA-polymerase. Translation of the base sequence of the *m*-RNA into an amino acid sequence proceeds in the $5' \rightarrow 3'$-direction on the RNA, and begins at the *N*-terminal end of the protein.

The complete base sequence of the *M*2-RNA, 3300 nucleotides long, and sequences of long stretches of the *R*17- and *Qβ*-RNA molecules have been determined. It has been found that the code for the assembly protein, the first protein encoded in the messenger, does not begin right at the 5′-end of the *m*-RNA. Indeed, AUG, the 'start' signal for assembly protein synthesis, is believed to be situated at residues 102–104 from the 5′-end of *Qβ*-RNA. As far as is known, the sequence of the first 101 bases is not translated into protein, and the function of this region of the RNA molecule is not well understood. Untranslated portions with similar base sequences have been found at the 5′-end of *M*2, *R*17, and several other viral RNAs. Because the sequences are similar for a number of different viruses, these untranslated regions may be important for maintaining the RNA molecules in the corect conformation, i.e. these regions may be able to base-pair with other sequences in the molecules, and so determine the secondary and perhaps tertiary structures of the molecules. Untranslated sequences also occur at the 3′-end of the molecules, and part of these may be important as the binding site for the RNA-polymerase which replicates the viral RNA. Short untranslated regions also occur as 'spacers' between the messages coding for the three proteins.

The sequences of parts of these virus messengers coding for proteins have been determined. The amino acid sequences of some of the proteins are already known, and so it has been possible to confirm the genetic code. It has been found that different codons for the one amino acid may occur in one messenger—this variation in choice of codon may be dictated by base-pairing and the secondary structure of the RNA molecule.

TABLE 5.1. The genetic code[a]

First base of codon	Second base: U	Second base: C	Second base: A	Second base: G	Third base of codon
U	UUU, UUC: Phe UUA, UUG: Leu	UCU, UCC, UCA, UCG: Ser	UAU, UAC: Tyr UAA, UAG: Terminate	UGU, UGC: Cys UGA: Terminate UGG: Trp	U C A G
C	CUU, CUC, CUA, CUG: Leu	CCU, CCC, CCA, CCG: Pro	CAU, CAC: His CAA, CAG: Gln	CGU, CGC, CGA, CGG: Arg	U C A G
A	AUU, AUC, AUA: Ile AUG: Met, Start	ACU, ACC, ACA, ACG: Thr	AAU, AAC: Asn AAA, AAG: Lys	AGU, AGC: Ser AGA, AGG: Arg	U C A G
G	GUU, GUC, GUA, GUG: Val GUG: Start	GCU, GCC, GCA, GCG: Ala	GAU, GAC: Asp GAA, GAG: Glu	GGU, GGC, GGA, GGG: Gly	U C A G

[a]There are no codons for cystine, hydroxyproline, or hydroxylysine: it is believed that cysteine, proline, and lysine are incorporated into proteins and subsequently modified. (For abbreviations of amino acids see Table 4.1, p. 76.)

The binding of ribosomes (see the following Sections) to m-RNA at the initiation sites for the different proteins of $M2$, $R17$, and $Q\beta$ viruses has also been investigated. The base sequences of these regions of the m-RNAs have been studied in the hope that common base sequences would be found at the initiation sites, and indeed the ribosome-binding site is now believed to be a purine-rich region in the untranslated section of RNA which precedes the segment coding for protein. In higher organisms (eukaryotes), however, the AUG coding for the beginning of a protein is thought to suffice as a recognition site by the ribosomes. But it is not well understood how ribosomes can bind specifically near this AUG, and not near an AUG which specifies a methionine residue in the interior of a protein. However, the secondary structure of the m-RNA may well be important in determining the ribosome binding sites.

Recently, much work has been carried out on the messenger RNAs of higher organisms. In a mammalian cell, many different m-RNA molecules can exist, each of which may be present in an extremely small amount. Because of this fact, it was originally difficult to isolate and purify a mammalian messenger. However, these messengers can now be purified by selective binding to polythymidylic acid (the poly-A region on mammalian m-RNA interacts specifically with poly-T), and the sequences of several of these molecules, including the messengers for the chains of human haemoglobin, have been determined.

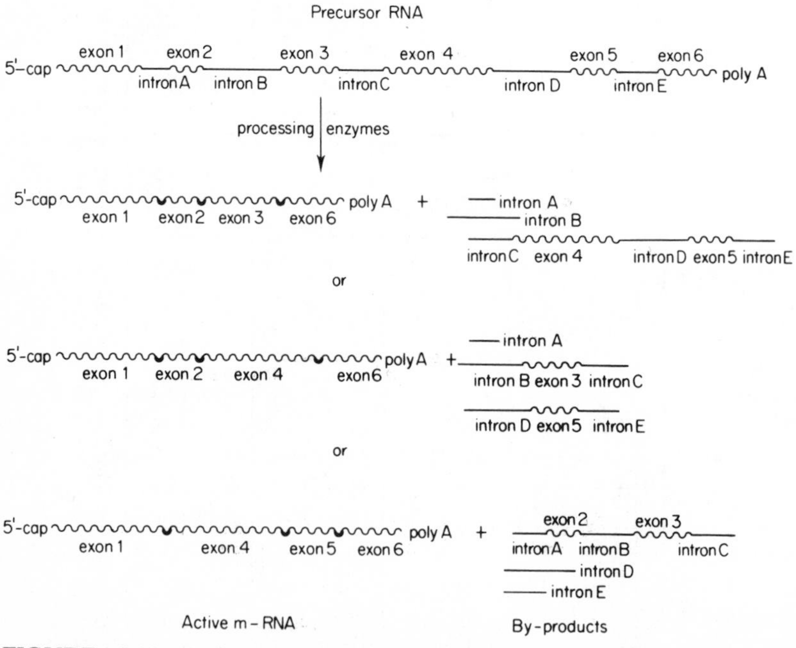

FIGURE 5.19. Possible production of different m-RNAs from one precursor

No specific ribosome binding site has been found, and so it has been concluded that the initiator AUG sequence functions as a recognition site. The 5′-cap (see Figure 5.12, p. 194) is believed to facilitate initiation of protein synthesis, but is not an essential requirement for the initiation. Eukaryotic *m*-RNA molecules usually code for one protein only, and have untranslated base sequences preceding and following the coding section.

The most remarkable characteristic of most eukaryote *m*-RNA molecules is that their active forms are produced by extensive modification after transcription, i.e. by excision of extra sequences from the interior of the transcript (see Figure 5.18). At first sight, this mechanism appears disadvantageous to the organism concerned—there seems to be a much greater risk of removing 'incorrect' sections from the transcript, and thus producing a non-functional messenger, than in the situation where an intact *m*-RNA is synthesized directly by transcription. The high specificity of the processing enzymes minimizes this risk, however, and the mechanism may instead be advantageous, allowing for greater flexibility in protein production. Thus it may be possible to produce *m*-RNA molecules coding for related, but different, proteins by removing and linking different sections of one transcript (see Figure 5.19). It has also been pointed out that this flexibility could speed up the the evolution of new proteins and hence new life-forms.

5.5.5. Mechanism of protein synthesis

5.5.5.1. The ribosome

Ribosomes are roughly spherical, nucleoprotein particles on which protein synthesis takes place. Both the size and composition of these particles depend on their source. Thus, most bacteria contain ribosomes with a sedimentation coefficient of 70 S, corresponding to a particle weight of about 2.6×10^6, whilst the cytoplasmic ribosomes of higher organisms have sedimentation coefficients around 80 S (particle weights varying from 3.5 to 5×10^6). The mitochondria of the cells of higher organisms also contain ribosomes quite distinct from, and usually smaller than, the ribosomes of the cell's cytoplasm. Bacterial ribosomes consist of 60–65% RNA and 35–40% protein, whilst animal and plant ribosomes contain only 40–50% RNA.

In the laboratory, ribosomes may be dissociated into two subunits by lowering the magnesium concentration (magnesium ions are necessary for ribosome activity), and it is believed that ribosomes are dissociated by a protein factor in the living organism after the synthesis of a polypeptide chain. The subunits from bacterial ribosomes have sedimentation coefficients of 50 S and 30 S; the larger animal ribosomes dissociate to give subunits with sedimentation coefficients 60 S and 40 S.

In bacteria, each 50 S-ribosomal subunit contains one molecule of

23 S-RNA and one molecule of 5 S-RNA (see Section 5.3.1, p. 178). These molecules are believed to possess definite secondary and tertiary structure, which at present is unknown. Comparison of the known sequences of 5 S-RNA and sections of 23 S-RNA indicate a 10-base sequence on the 5 S-RNA complementary to a segment of the 23 S-RNA. Thus the two RNA molecules may be hydrogen-bonded together within the ribosome at this sequence. In addition 34 different proteins are present, some bound directly to the RNA molecules.

The 30S-subunit of the ribosome consists of one molecule of 16S-RNA and 21 different proteins, at least six of which are bound directly to the RNA. Again, the RNA is believed to be folded-up. Thus, each subunit is thought to consist of a 'core' of RNA to which are bound a number of proteins. The remaining proteins are bound to the outside of the 'core' proteins. It should not be thought, however, that the RNA is buried within the ribosome: some stretches of RNA must be accessible on the surface in order to participate in protein synthesis. The conformations of the RNA molecules of the ribosome may not be fixed, and conformational changes may be required during protein synthesis.

Each protein occurs as a single copy per ribosomal subunit, and only one protein is common to both subunits, so that each ribosome contains over 50 different proteins. There is one exception to this rule—three to four molecules of a pair of proteins are found in the 50S-subunit, one of the pair occurring in an unmodified form and the other acetylated. However, it appears that other proteins may associate with the ribosomes at times during the cycle of polypeptide synthesis.

Attempts have been made to study the protein binding sites on the ribosomal RNA molecules, and it is known that sections of an RNA molecule, far apart along the chain, may be involved in binding one protein molecule. This fact suggests that the RNA molecule may be extensively folded within the ribosome. It is also known, however, that many of the protein molecules are highly extended within the ribosome, and work is currently in progress on the location of the proteins with respect to each other and the RNA molecules.

The amino acid sequences of 33 of the 54 ribosomal proteins of *E. coli* have been determined. Some of these proteins are highly basic, and are believed to interact directly with RNA.

During protein synthesis, ribosomes must be capable of binding both messenger and two transfer RNA molecules (see following sections), and of promoting the interaction between these two types of RNA. The 30S-subunit of bacterial ribosomes is believed to contain at least part of one of the *t*-RNA binding sites (the *A*-site), and the *m*-RNA binding site. A pyrimidine-rich stretch near the 3′-end of the 16S-RNA is complementary to the purine-rich segment of *m*-RNA molecules diagnosed as the ribosome binding site. The 50S-subunit contains at least part of the other *t*-RNA binding site (the *P*-site); indeed, a base sequence has been found in 5S-RNA, which is complementary to the GT′ψC sequence common to almost all *t*-RNA molecules (see Figure 5.11). The enzyme responsible for forming the new peptide bond in the protein being synthesized is also located in the 50S-subunit.

Although bacterial ribosomes have been most extensively studied, the cytoplasmic and mitochondrial ribosomes of higher organisms are believed to be similar, but not identical. The cytoplasmic ribosomes are larger than bacterial ribosomes, and contain larger RNA molecules and proteins. Also, each subunit contains more proteins than the corresponding bacterial subunit, giving in total ca. 80 proteins, in contrast to the bacterial total of 54. Both the RNA and proteins of the mitochondrial ribosomes of an organism differ from the corresponding molecules of the cytoplasmic ribosomes. The *r*-RNA of mitochondria is transcribed from mitochondrial DNA, but synthesis of the ribosomal proteins of the mitochondria is thought to take place in the cytoplasm on the cytoplasmic ribosomes.

5.5.5.2. Transfer RNA and aminoacylation

m-RNA contains, in code form, the information specifying the amino acid sequence of a protein, but during protein biosynthesis, the growing polypeptide chain cannot interact directly with the triplet codons on *m*-RNA. Adapter molecules, the transfer RNAs, are necessary to carry the amino acids to the ribosome, and to interact with the correct codons on the messenger RNA (see, for example, Figure 5.20, p. 217). Thus a *t*-RNA molecule, with its amino acid, binds to a ribosome at the codon on *m*-RNA (also bound to the ribosome) for that particular amino acid. Hence the amino acid is brought into protein synthesis in the correct position for linking into the growing polypeptide chain. In living organisms, there must be at least one transfer RNA specific for each amino acid, and in many cases, more than one *t*-RNA exists for each amino acid. Two sets of *t*-RNA molecules may occur in higher organisms, one set in the cell cytoplasm transcribed from nuclear DNA, and another set in the mitochondria transcribed from mitochondrial DNA.

The native secondary and tertiary structures of all transfer ribonucleic acids are believed to be similar (see Figure 5.14), and magnesium ions have been found necessary to maintain the correct conformation for activity. Each molecule must contain three recognition sites—one for the ribosome, one for the enzyme which links the correct amino acid to the *t*-RNA, and one for the codon on *m*-RNA corresponding to that amino acid. This last site is believed to be a triplet of three bases complementary to the bases of the codon on the *m*-RNA, and is known as the *anticodon* (see Figures 5.11 and 5.14). In many cases, the unusual nucleoside, inosine (I), (see Figure 5.6, p. 179) is found at the anticodon.

During protein synthesis, the transfer RNAs must become 'charged' with their corresponding amino acids, i.e. become aminoacylated. The enzymes catalysing these reactions are *aminoacyl-t-RNA-synthetases*, and there must be at least one such synthetase for each different amino acid. Recognition by an enzyme of the correct *t*-RNA molecule is extremely important for the living organism; if a *t*-RNA becomes charged with an incorrect amino acid, it can deliver that acid to the growing polypeptide chain. It is not yet known how, for example, an alanyl-*t*-RNA-synthetase recognizes a *t*-RNA specific for alanine. Not all synthetases may 'recognize' the same parts of a *t*-RNA molecule. For instance, some synthetases are believed to be able to distinguish between *t*-RNA molecules on the basis of base sequences in the dihydrouridine loop (see Figure 5.11, p. 192) and near the ends of the *t*-RNA, whilst other synthetases may 'recognize' bases of the anticodon.

When an amino acid is being linked to the corresponding *t*-RNA molecule, the synthetase first reacts with ATP (see Figure 5.2b, p. 170) and the amino acid, then with the *t*-RNA:

$$^{-}O{-}\overset{\overset{\large O}{\|}}{C}{-}\underset{\underset{\large R}{|}}{\overset{\overset{\large H}{|}}{C}}{-}NH_3^+ + ATP \longrightarrow$$

$$\text{Adenosine}{-}O{-}\underset{\underset{\large O}{\|}}{\overset{\overset{\large O^-}{|}}{P}}{-}O{-}\overset{\overset{\large O}{\|}}{C}{-}\underset{\underset{\large R}{|}}{\overset{\overset{\large H}{|}}{C}}{-}NH_3^+ + \text{Pyrophosphate}$$

Activated amino acid

$$\text{Adenosine}{-}O{-}\underset{\underset{\large O}{\|}}{\overset{\overset{\large O^-}{|}}{P}}{-}O{-}\overset{\overset{\large O}{\|}}{C}{-}\underset{\underset{\large R}{|}}{\overset{\overset{\large H}{|}}{C}}{-}NH_3^+ + t{-}\text{RNA}{-}OH \longrightarrow$$

(transfer RNA
with free OH on ribose
at 3′-end of molecule)

$$t\text{-RNA}{-}O{-}\overset{\overset{\large O}{\|}}{C}{-}\underset{\underset{\large R}{|}}{\overset{\overset{\large H}{|}}{C}}{-}NH_3^+ + \text{Adenosine}{-}O{-}\underset{\underset{\large O}{\|}}{\overset{\overset{\large O^-}{|}}{P}}{-}O^-$$

The amino acid becomes bound to the terminal adenosine residue at the 3′-end of the *t*-RNA molecule, i.e. the carboxyl group of the amino acid becomes attached to one of the free hydroxyl groups on the last ribose of the *t*-RNA. There is probably a conformational change in the *t*-RNA molecule on aminoacylation. The aminoacyl-*t*-RNA complex can then move to the ribosome to take part in protein synthesis.

During polypeptide growth, codons on the ribosome-bound *m*-RNA corresponding to the amino acids of the new protein become exposed successively at the aminoacyl-*t*-RNA binding site of the ribosome (the

A-site; see, for example, Figure 5.21, p. 218). When the codon GCA corresponding to alanine, for example, becomes exposed, then the complex of alanine and its transfer RNA (with anticodon complementary to GCA) can become bound to the ribosome at the *A*-site. The nature of the region of the *t*-RNA which binds to the ribosome is at present uncertain, although the base sequebce of GT′ψC is generally accepted as being important (see Figure 5.11). As mentioned earlier, the sequence GT′ψC on *t*-RNA is believed to interact with the 5S-RNA of the ribosome to promote binding.

If the sequence GT′ψC does indeed base-pair with a stretch of 5S-ribosomal RNA, then the *t*-RNA must undergo a conformational change on binding to the ribosome. In the conformation determined by X-ray analysis (Figure 5.14, p. 198), part of the GT′ψC sequence is involved in hydrogen bonding within the *t*-RNA molecule. Thus a change in molecular shape is necessary to free the GT′ψC for bonding with a ribosomal RNA.

The three bases of the anticodon of a *t*-RNA molecule may be locked in a conformation which allows a section of one strand of a double helix to be presented to the codon on *m*-RNA. In turn, the ribosome probably holds the codon in a similar conformation so that codon–anticodon recognition can involve base-pairing and the transitory formation of a short segment of RNA double helix.

It has been found for some *t*-RNAs that one anticodon can recognize more than one codon on *m*-RNA. For example, alanyl-*t*-RNA with anticodon IGC recognizes GCU, GCC, and GCA on *m*-RNA. (For base-pairing with the codon, the anticodon should be read from right to left.) To explain this effect, Crick put forward the 'wobble' hypothesis of base-pairing, suggesting that the third base of a messenger codon can pair with the first anticodon base in a way other than those shown on page 176. Thus inosine in the *t*-RNA anticodon can pair with either U, C, or A in the messenger codon, even although inosine, resembling guanine, might ordinarily be expected to pair with cytosine only. Table 5.2 shows possible base-pairs in the codon and anticodon according to the 'wobble' hypothesis. In many *t*-RNA molecules, a highly modified purine residue is found after the anticodon (see Figure 5.11); a hypermodified adenine residue always follows an anticodon ending in A or U. A pyrimidine

TABLE 5.2. Base-pairing according to the 'Wobble' hypothesis

Anticodon base	Can pair with codon base
U	A, G
C	G
A	U
G	U, C
I	U, C, A

residue followed by uracil always precedes the anticodon, and these bases are believed to be important for facilitating codon–anticodon base-pairing.

In all living organisms, there exists a special *t*-RNA specific for methionine which can initiate protein synthesis. This *t*-RNA recognizes the codon AUG (and possible GUG) when it signals the 'start' of a new polypeptide, but not when it codes for a methionine residue in the interior of a protein. The mechanism of this specificity is unknown. The base sequences of initiating *t*-RNAs have been studied, and have been found to differ little from that of the methionine-*t*-RNA, which inserts methionine into the interior of a polypeptide chain. The difference in base sequence occurs near the ends of the molecules, in the region of the GT′ψC sequence, and in the base after the anticodon. The differences in the GT′ψC region may allow the initiating methionine-*t*-RNA to be the first *t*-RNA molecule bound to the ribosome at the start of a new polypeptide chain. X-ray analysis of a yeast initiator *t*-RNA has shown that the three-dimensional structure in crystals is very similar to that found for other *t*-RNA molecules (Figure 5.14, p. 198). It is not yet known whether the detected differences in position of the anticondon loop along with sequence differences at GT′ψC are sufficient to permit the ribosome to distinguish between the initiator *t*-RNA and other *t*-RNAs.

5.5.5.3. Translation and protein synthesis

The first steps in protein synthesis can be considered as production of the messenger RNA by transcription from DNA and subsequent processing, and the 'charging' of transfer RNA molecules with their respective amino acids (as described in the preceding sections). As for the processes of DNA replication, RNA transcription, and *t*-RNA aminoacylation, the details of protein synthesis have been investigated using bacterial systems, but the mechanism is believed to be very similar in higher organisms.

Subsequent steps in protein synthesis can be summarized as follows:

(*a*) Messenger RNA binds to the 30S-subunit of the ribosome, followed by binding of the initiator *t*-RNA charged with its amino acid. Then the 50S-ribosomal subunit becomes bound to the complex. This stage completes the *initiation steps* (see Figure 5.20).

(*b*) *Polypeptide chain elongation* takes place in three stages, which are repeated over and over again until the protein molecule is completed. In the first of these stages, a second *t*-RNA charged with its amino acid becomes bound to the ribosome. The choice of which aminoacyl-*t*-RNA binds is dictated by the codon on the *m*-RNA adjacent to the initiator codon.

The second stage of chain elongation involves the formation of a peptide bond between the amino acid on the initiator *t*-RNA and the amino acid on the second *t*-RNA. The new dipeptide is left attached to the second *t*-RNA.

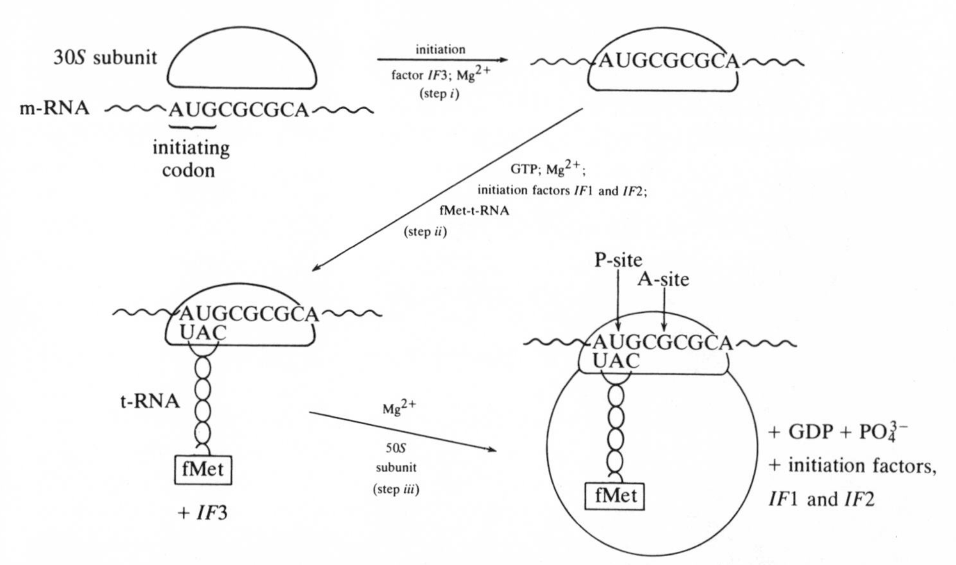

FIGURE 5.20. Schematic initiation of polypeptide chain synthesis in bacteria (steps *i–iii*); fMet = formyl methionyl residue = $-O-\overset{\overset{\large O}{\|}}{C}-\underset{\underset{\large (CH_2)_2-S-CH_3}{|}}{CH}-NH-CHO$

Lastly, the first (initiator) *t*-RNA leaves the ribosomes, the peptidyl *t*-RNA moves on the ribosome to the position previously occupied by the first *t*-RNA, and the *m*-RNA moves in step on the ribosome so that a third codon becomes exposed ready for recognition of a third aminoacyl-*t*-RNA. This process is known as *translocation* (see Figure 5.21).

(*c*) *Termination of protein synthesis* occurs when a termination codon i.e. UAA, UGA, or UAG becomes exposed on the *m*-RNA at the ribosome. The bond between the polypeptide and the last *t*-RNA is hydrolysed, the polypeptide chain and *t*-RNA are released from the ribosome, and the ribosome itself dissociates into subunits ready to participate in another round of protein synthesis (see Figure 5.22, p. 221).

These steps will now be described in greater detail.

In bacteria, *initiation* takes place on the 30S-subunit of the ribosome, and with the larger ribosomes of higher animals, the 40S-subunit is involved (see Figure 5.20). During the first stage of initiation, the 30S-subunit of the ribosome binds to the protein initiation site on the *m*-RNA i.e. near the AUG codon signalling the beginning of the code for the amino acid sequence (step *i*, Figure 5.20). As stated earlier, a pyrimidine-rich stretch of the 16S-ribosomal RNA probably 'recognizes' a purine-rich region of *m*-RNA preceding the initiator AUG codon. This stage is facilitated by a protein initiation factor, *IF*3. Then, in the presence of GTP (for the structure see page 170), the initiating aminoacyl-*t*-RNA binds to the

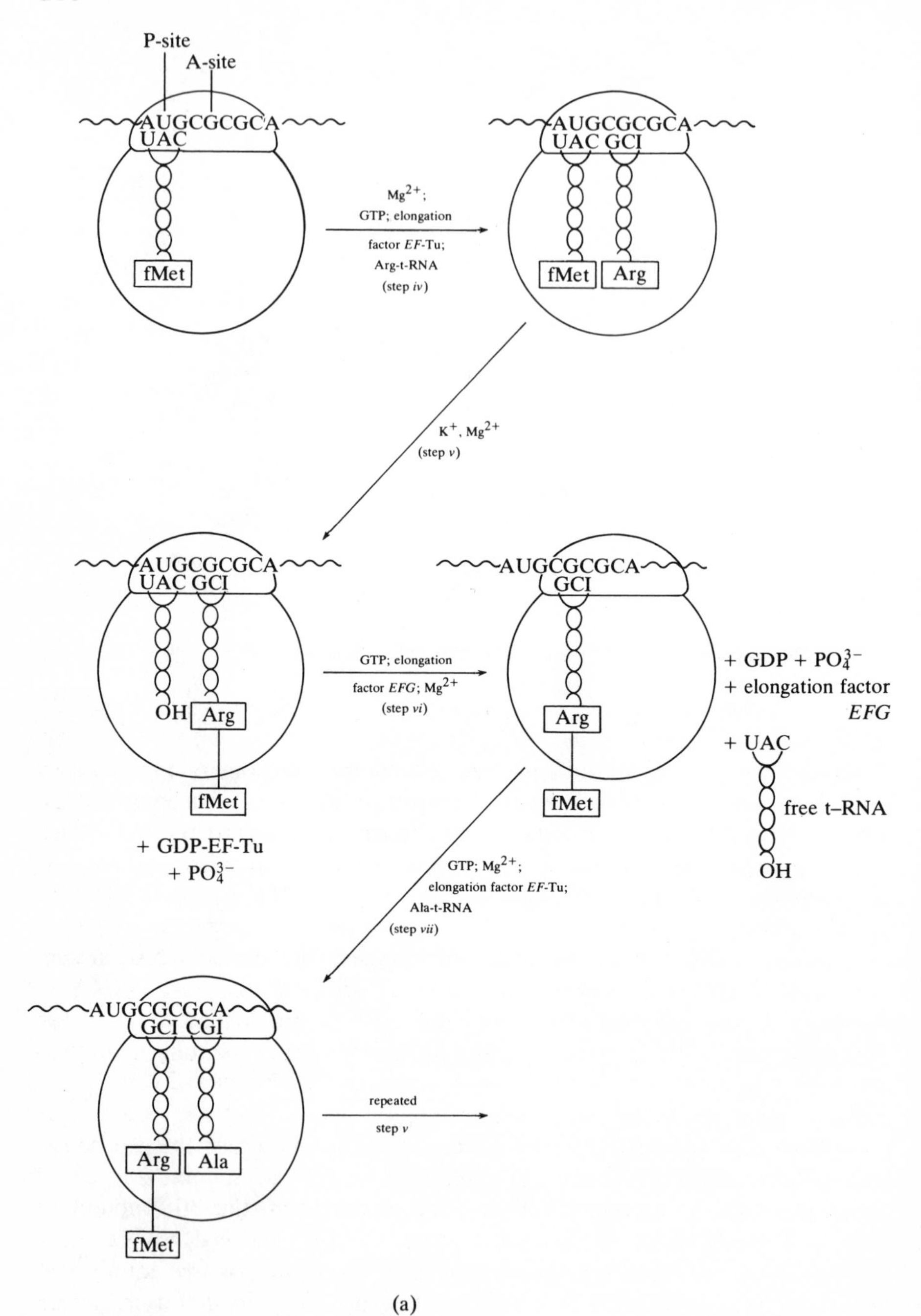
P-site
A-site
AUGCGCGCA
UAC
fMet
Mg^{2+};
GTP; elongation
factor *EF*-Tu;
Arg-t-RNA
(step *iv*)
AUGCGCGCA
UAC GCI
fMet
Arg
K^+, Mg^{2+}
(step *v*)
AUGCGCGCA
UAC GCI
OH
Arg
fMet
+ GDP-EF-Tu
+ PO_4^{3-}
GTP; elongation
factor *EFG*; Mg^{2+}
(step *vi*)
AUGCGCGCA
GCI
Arg
fMet
+ GDP + PO_4^{3-}
+ elongation factor
EFG
+ UAC
free t–RNA
OH
GTP; Mg^{2+};
elongation factor *EF*-Tu;
Ala-t-RNA
(step *vii*)
AUGCGCGCA
GCI CGI
Arg
Ala
fMet
repeated
step *v*

(a)

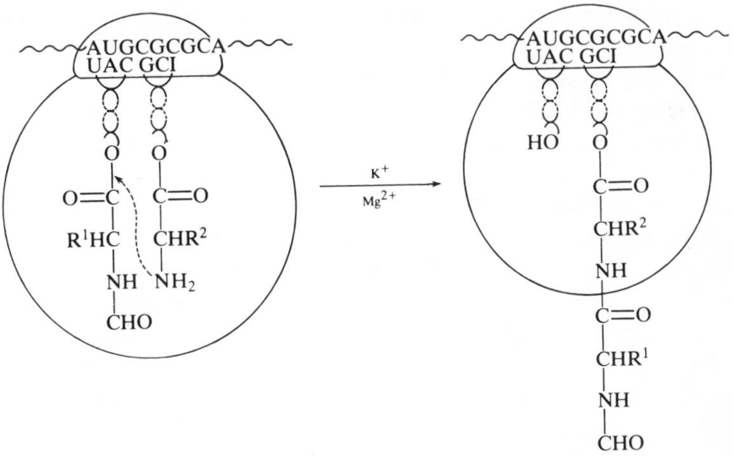

(b)

FIGURE 5.21. (a) Schematic representation of polypeptide chain elongation (steps *iv–vii*). (b) Schematic representation of peptide bond formation (step *v*); R^1 = side-chain of methionyl residue; R^2 = side-chain of arginine

m-RNA 30S-subunit complex (step *ii*, Figure 5.20). A further two protein initiation factors, *IF*1 and *IF*2, are required for binding. In bacteria, the initiating aminoacyl-*t*-RNA consists of an *N*-formylmethionine residue attached to the special methionine-*t*-RNA, which recognizes the initiator codon AUG. Thus all bacterial proteins are synthesized with an *N*-formylmethionine residue at the beginning (the *N*-terminal end) of the polypeptide chain. In higher organisms, the initiating methionyl-*t*-RNA lacks the *N*-formyl group of the bacterial system, but is in other ways similar. There is now some evidence that, at least in higher organisms, step (*ii*) may precede step (*i*).

The complex of GTP, *IF*1 and *IF*2, formylmethionyl-*t*-RNA (f-Met-*t*-RNA), *m*-RNA, and 30S-subunit is called the *initiation complex*. After its formation, the 50S-subunit of the ribosome becomes bound, the GTP is hydrolysed, the initiation factors are released, and the initiation is complete (step *iii*, Figure 5.20). (Hydrolysis of GTP provides energy for the processes taking place.) Interaction between the 16S-RNA of one subunit and the 23S-RNA of the other subunit probably helps to bring the two parts of the ribosome together.

During protein synthesis, a ribosome has two *t*-RNA binding sites. One, the *P*-site, holds the *t*-RNA attached to the growing peptide chain, whilst the other, the *A*-site, binds the incoming aminoacyl-*t*-RNA. At the end of the initiation stage just described, the fMet-*t*-RNA is believed to be bound to the ribosome at the *P*-site (see Figure 5.20).

In higher organisms, a similar mechanism is believed to operate, again with protein factors involved in initiation, but there may be seven initiation factors instead of the three found for bacteria.

The next steps bring about the *elongation of the polypeptide chain* (steps *iv–vii* of Figure 5.21). After initiation, a second aminoacyl-*t*-RNA becomes bound to the ribosome, this time at the *A*-site (step *iv*, Figure 5.21). The choice of which aminoacyl-*t*-RNA binds at the *A*-site is dictated by the codon next to the initiator AUG on the messenger RNA (see Figure 5.21). Some of the proteins of the 30S-ribosomal subunit are probably important in binding the correct aminoacyl-*t*-RNA at the codon; a protein elongation factor, *EF*-Tu, combines with GTP, and this complex is necessary for aminoacyl-*t*-RNA binding. *EF*-Tu does not interact with the initiator fMet-*t*-RNA, and so prevents it from attaching at the *A*-site of the ribosome if an *internal* AUG codon becomes exposed on *m*-RNA.

An enzyme, peptidyl transferase, which is part of the 50S-ribosomal subunit, then catalyses the formation of a peptide bond between the carboxyl group of the methionine and the amino group of the second amino acid (step *v*, Figure 5.21a; for details see Figure 5.21b). The dipeptide is attached to the *t*-RNA of the second acid, the whole complex remaining at the *A*-site of the ribosome. A GDP-EFTu complex is released after peptide bond formation. Translocation of the *m*-RNA and the dipeptide-*t*-RNA then occurs, so that the dipeptide-*t*-RNA now occupies the *P*-site (step *vi*, Figure 5.21). The *t*-RNA which had carried the formylmethionine residue is released, and the next codon on the *m*-RNA is aligned with the *A*-site. Another protein elongation factor, *EFG*, and further GTP hydrolysis are necessary for translocation.

A third aminoacyl-*t*-RNA can now bind at the *A*-site (step *vii*, Figure 5.21—note that step *vii* is really a repeat of step *iv*), a second peptide bond can be formed, and the process is repeated to give a growing polypeptide chain. The chain is lengthened at the rate of 1–20 amino acids added per second, the rate depending on the organism concerned.

Similar protein factors are involved in the mechanism of chain elongation in higher organisms.

The process, whereby the sequence of codons on messenger RNA becomes a sequence of amino acids in a new protein molecule, is known as *translation*. Translation of *m*-RNA takes place in the $5' \rightarrow 3'$-direction, and the proteins are synthesized from the *N*-terminal to the *C*-terminal end. As pointed out earlier, bacterial proteins are synthesized with *N*-formyl methionine at the *N*-terminus, whilst polypeptides synthesized in the cytoplasm of higher organisms begin with methionine. After initiation, these residues may be excised, so that not all proteins of living organisms have methionine as the *N*-terminal amino acid.

The last stage in protein biosynthesis is *chain termination*. When one of the terminator codons UAA, UAG, or UGA of the messenger RNA appears at the *A*-site of the ribosome, the bond between the completed polypeptide chain and its *t*-RNA is hydrolysed, and the protein is released (step *viii*, Figure 5.22). This hydrolysis may be catalysed by the same peptidyl transferase which forms the peptide bonds of the new protein. The last *t*-RNA molecule bound at the *P*-site then dissociates from the ribosome, the ribosome itself splits into its two subunits, and these can begin a new round of polypeptide biosynthesis (step *ix*, Figure 5.22).

Once again, protein factors are involved in the reactions. These factors seem to complex with the whole ribosome and the stop signal, and they may be able to alter the peptidyl transferase so that it can hydrolyse the polypeptide-*t*-RNA bond, instead of forming a new peptide bond. Two release factors are known, *RF*1, which recognizes the stop signals UAA or UAG, and *RF*2, which recognizes UAA or UGA. After release of the

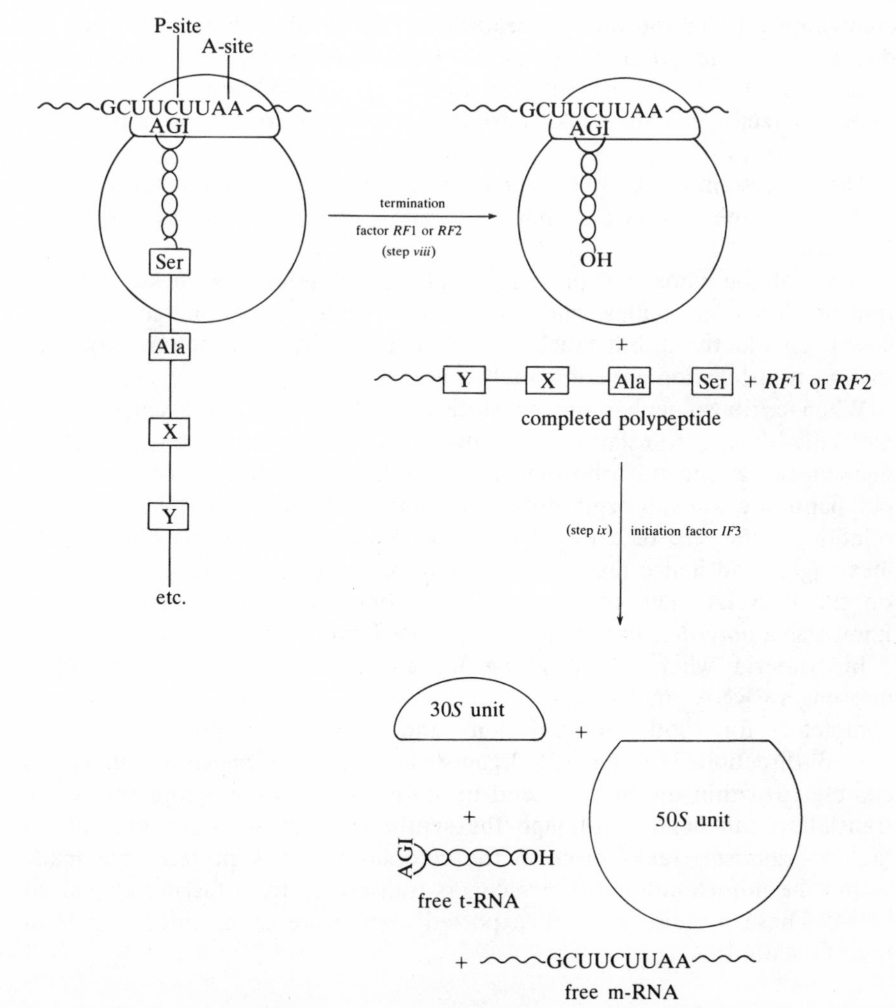

FIGURE 5.22. Schematic representation of the termination of polypeptide chain synthesis (step *viii* and step *ix*)

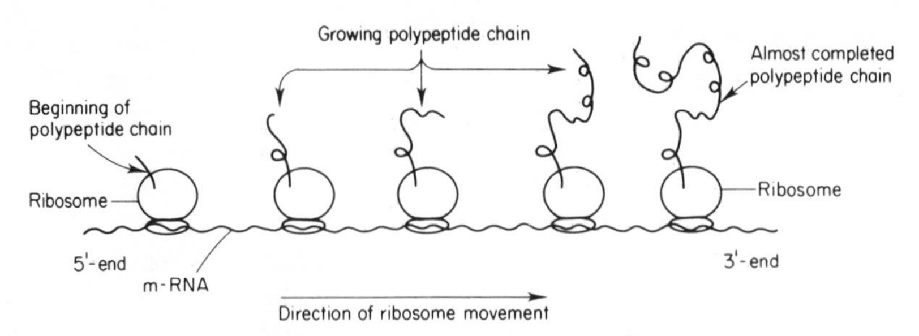

FIGURE 5.23. A polyribosome

completed protein and the last deacylated *t*-RNA, the whole ribosomes are dissociated by initiation factor *IF*3, which binds to the 30S-subunit, and causes the ribosomes to split in two. The 30S-subunit with attached initiation factor can then take part in the initiation of a new polypeptide chain.

The process is probably similar in mammalian systems, but there seems to be only one complex termination factor, instead of the two found in bacteria.

Some of the ribosomal proteins involved in binding the messenger and transfer RNA molecules, and the initiation and elongation factors, have now been identified, but much work remains to be done before processes such as translocation are completely understood.

When a ribosome has moved sufficiently far along a messenger RNA molecule during translation to allow some space at the 5′-end of the messenger, a second ribosome can bind, and synthesis of a second polypeptide chain can begin on the same *m*-RNA. If the *m*-RNA is long in relation to the size of a ribosome, then many ribosomes can bind to one messenger, and hence many protein molecules can be growing at one time on the *m*-RNA. The complex of many ribosomes with one *m*-RNA is known as a *polyribosome*, or *polysome* (see Figure 5.23).

In bacteria where there is no nuclear membrane, translation of a messenger RNA molecule can actually begin before transcription is complete, for both transcription and translation proceed in the $5' \rightarrow 3'$-direction. In higher organisms, however, *m*-RNAs must be completed within the nucleus, and be transported to the cytoplasm before translation can begin. Although the synthesis of most of the proteins of higher organisms takes place in the cytoplasm, a few proteins are made within the mitochondria, on messengers transcribed from the mitochondrial DNA. These proteins are not 'exported', but make up an integral part of the mitochondria.

5.5.6. Control of nucleic acid function

5.5.6.1. General

Because biosynthesis of functional proteins is so essential to all forms of life, nucleic acid replication, transcription, and translation must be kept under strict control. Complex controls are exerted over the three levels of nucleic acid function, but the complete details of these control systems are not yet known.

The replication of DNA is usually an 'all or nothing' process, i.e. when a cell divides, the 'daughter' cell receives a full complement of DNA. Thus there is little need to control which parts of a DNA molecule are replicated. However, the timing of DNA replication is extremely important, as the new DNA must be ready when the daughter cell separates. The details of how the synthesis of DNA is triggered at the right stage of cell division are not yet known, but regulation may take place at the level of protein factors necessary for the initiation of DNA replication, or for the production of RNA primer (compare Section 5.5.2, p. 200).

The methyl groups on some DNA can be important for controlling degradation of the molecules. Thus, some bacterial nucleases can degrade 'foreign' DNA (from an invading phage), but are inactive against the bacterial DNA, because the pattern of methylation is different for the two types of DNA.

Many of the base sequences of DNA molecules may not code for protein structures, but are important for control, particularly for the control of transcription. In most organisms, only a part of the DNA is being transcribed for protein biosynthesis at any one time. Indeed, in higher organisms where cells are differentiated (cells are not all the same in the adult organism), most cells contain the same DNA, but synthesize different proteins. Thus some parts of the DNA are transcribed only in particular cell types. Histones may be non-specific repressors, whilst the acidic proteins of the chromosomes may be involved in specific inhibition, or promotion, of transcription of certain sections of the DNA. It is thought that changes in DNA conformation could be important for transcriptional control, and histones may be involved in this process. The process of differentiation itself involves changes in control, but as yet the details are not understood.

Control of transcription must be particularly strict in some small viruses such as those for which the complete DNA base sequence has been determined (see Section 5.4.1.1, p. 187). It has been found that one stretch of DNA can be involved in coding for two or even three different proteins, the genetic code for each protein being translated in a different reading frame. For example, consider a base sequence on DNA, e.g.:

$$3'\text{-}A\ G\ T\ C\ C\ A\ A\ G\ C\ T\ A\ G\ C\ G\ U\text{-}5',$$

which become transcribed as part of messenger-RNA molecules with base sequence

$$5'\text{-}U\ C\ A\ G\ G\ U\ U\ C\ G\ A\ U\ C\ G\ C\ A\text{-}3'.$$

If this sequence is translated into amino acids, starting with the first codon UCA, we obtain

—serine—glycine—serine—isoleucine—alanine—.

If, however, the reading frame is shifted, so that the first base, U is disregarded, the first codon becomes CAG and the equivalent amino acid sequence is

—glutamine—valine—arginine—serine— [see the 'genetic code' Table 5.1, p. 209]

Thus two proteins of completely different amino acid sequence can be coded by one base sequence of DNA. It seems likely, therefore, that a complex control system exists to ensure that the correct amounts of each protein are produced. It is not known whether similar 'economies' are found in higher organisms, but currently it seems unlikely.

The control of bacterial transcription has been intensively studied, but at present it is uncertain whether similar types of mechanism operate in higher organisms. In bacteria, a transcription unit, or *operon*, is a length of DNA coordinating and controlling the synthesis of functionally related

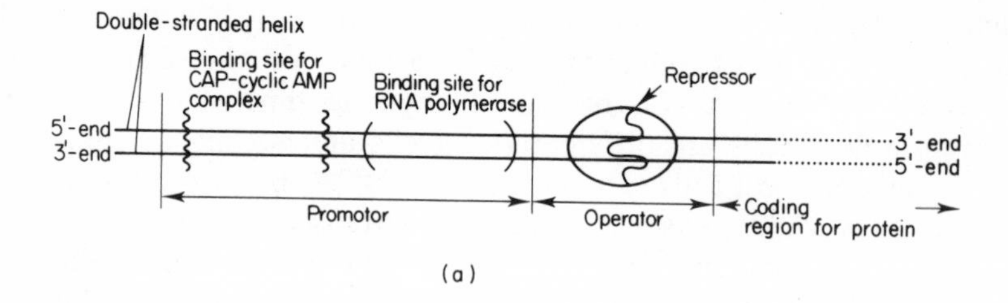

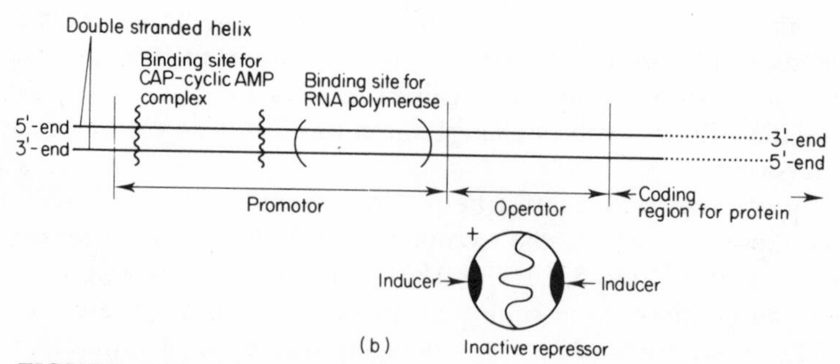

FIGURE 5.24. (a) Control region of one possible type of operon. Repressor is bound at the operator, and no transcription takes place. (b) Inducer binds to the repressor and alters the conformation, so that it can no longer interact with the operator. Transcription can now occur

proteins. The operon consists of a control element containing both the promoter region (where RNA-polymerase binds and where transcription begins) and a sequence known as the operator, as well as the section which is transcribed to messenger-RNA. The control element may be involved with positive or negative control, or both. The promoter region is not generally transcribed. This operon model for transcriptional control was suggested by Jacob and Monod, and for their work they received the Nobel Prize in 1965.

When negative control is exerted on transcription, a specific molecule known as a repressor binds to the DNA at the operator (see Figure 5.24). The repressor is usually an oligomeric protein capable of allosteric alteration (cf. Section 4.5.2.2, p. 119), and is believed to bind in the grooves of the DNA double helix. The operator site is at least 20 bases long.

The first operator, whose base sequence was determined, was the operator of the lactose operon of *E. coli*. (The lactose operon codes for enzymes involved in lactose metabolism.) A 27-base sequence, to which repressor could bind, was found to contain a region of local twofold symmetry (the sequence read 5′ to 3′ on one DNA strand is identical to a sequence read 5′ to 3′ on the other strand) e.g.:

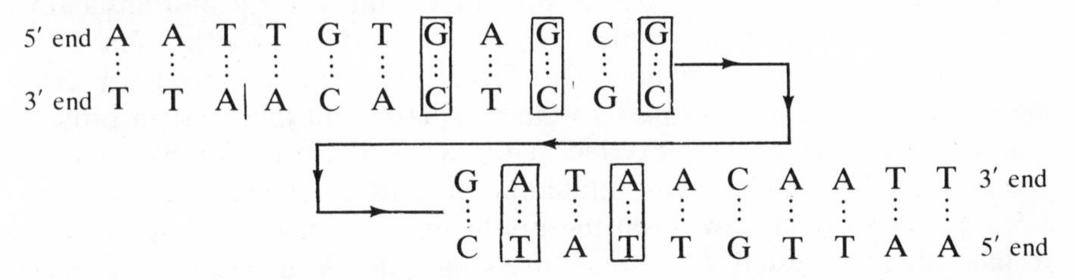

(Bases not conforming to this symmetry are boxed.) This symmetrical sequence is probably 'recognized' by a similar type of symmetry in the repressor protein.

Repressors can act by preventing RNA-polymerase binding to the DNA, or in some cases by blocking the initiation of transcription without interfering with enzyme binding; in the example shown in Figure 5.24 the latter mechanism probably operates. Repression of a transcription unit can be relieved by a small molecule, an *inducer*, which binds to the repressor and alters its conformation so that it can no longer bind to DNA (see Figure 5.24b). RNA-polymerase can then transcribe that section of the DNA, and derepression has been achieved.

In some cases, a repressor protein cannot by itself bind to DNA, but must first bind a small molecule, such as an amino acid, or a *t*-RNA. Such a small molecule is known as a *corepressor*, and only the repressor–corepressor complex is capable of binding to DNA to prevent transcription. In this way, the amino acid, tryptophan, can inhibit further synthesis of itself in *E. coli*, by acting as a corepressor of the tryptophan operon (the unit which controls the production of enzymes involved in tryptophan biosynthesis).

For positive control, a protein may bind to the DNA and enhance transcription, possibly by facilitating the binding of RNA-polymerase at the promoter. Some non-peptide hormones bind to a special receptor protein, and this hormone–protein complex binds to DNA to stimulate transcription. The action of several peptide hormones (see Section 4.5.3, p. 135) is brought about by an increase in an intermediate, cyclic AMP (see Figure 5.25). Cyclic AMP reacts with a specific protein, CAP (*catabolite activator protein*), and the CAP-cyclic AMP complex is thought to bind to DNA near the promoter site and increase the synthesis of RNA from that transcription unit (see Figure 5.24).

Another type of positive control has been identified. In some cases, an inducer not only causes release of repressor from DNA, but converts it to an activator which functions like CAP–cyclic AMP.

Control can also be exerted by changes in the RNA-polymerase. Thus, some viruses code for a subunit which attaches itself to the host RNA-polymerase. The modified enzyme now transcribes viral RNA in preference to host RNA. The mechanism involved is, at present, not understood.

Less information is available at present on the control of translation. Some *m*-RNA may exist in the nucleus in a form which cannot be 'exported' for use in translation. This form must then be activated before the *m*-RNA molecules can pass to the cytoplasm and take part in protein synthesis. Controls may be exerted at the level of the post-transcriptional processing of *m*-RNA. Thus inhibition of intron removal (see Section 5.5.3, p. 206) would slow down the production of active messenger RNA. It has been suggested that some messenger RNA molecules, or even *m*-RNA–ribosome complexes, may be repressed, so that translation can only take place after removal of the repressor from the messenger. The rate of translation could be controlled by the concentration of *t*-RNA molecules available, whilst the amount of one protein synthesized could be

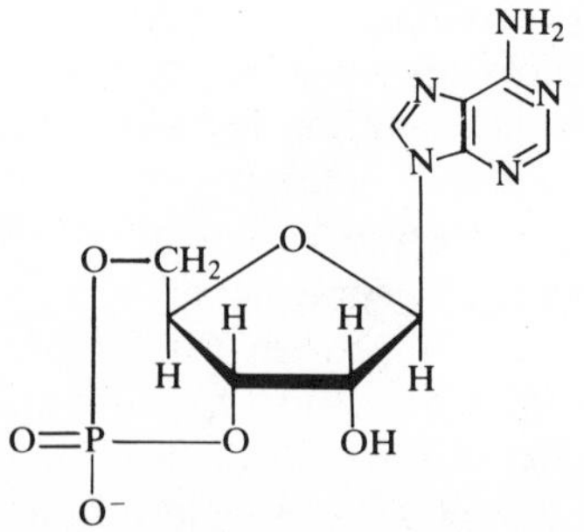

FIGURE 5.25. Cyclic adenosine monophosphate (cyclic AMP)

controlled, at least in part, by degradation of the messenger RNA. Alterations in ribosomal proteins, perhaps by phosphorylation mediated by cyclic AMP, may also be involved in translational control.

Initiation factors may be important in translational control. Thus a special initiation factor could promote preferential binding of ribosomes to one specific *m*-RNA, or to one translational initiation site of a polycistronic messenger. In viral infection, often virus *m*-RNA is translated in preference to host *m*-RNA, and this change could be mediated by a viral-specific initiation factor which causes ribosomes to bind preferentially to viral *m*-RNA. Also inactivation of initiation factors could be effective in the control of translation.

5.5.6.2. Cancer and reverse transcription

Although the causes of human cancer are not yet known, work with animals suggests that certain viruses may be involved in the development of the disease. Tumour-producing viruses of animals fall into two broad classes, DNA-viruses and RNA-viruses. It is believed that both types of virus can cause the incorporation of viral-specific DNA into the host cell DNA, thus producing host cells with altered genetic information. At first it was not understood how an RNA-containing virus could bring about incorporation of DNA into a host cell. An enzyme was discovered in tumour-producing RNA-viruses, however, which can catalyse the synthesis of DNA on an RNA template. This process is known as *reverse transcription*, and the enzyme is called *reverse transcriptase*. The enzyme requires an RNA primer, as well as an RNA template, and the new DNA strand is covalently linked to the primer. This enzyme alone can probably catalyse the synthesis of a double-helical DNA molecule, and such molecules are formed during infection by these viruses. This double-stranded DNA can then be incorporated into the genetic DNA of the host cell.

The ability of reverse transcriptase to produce DNA with a base sequence complementary to that of an RNA is now widely used in RNA sequencing. Radioactive DNA copies (often called *c*-DNA) are synthesized using the RNA under study as template, and the sequence of the DNA is determined by the methods of Sanger, or Maxam and Gilbert (see Section 5.4.1.1, p. 182).

Alterations in control mechanisms may be important in the development of cancer, for incorporation of viral DNA into the host genes may not always be sufficient to transform normal cells to cancerous cells. In some cases, addition of a chemical known to produce tumours (a carcinogen) to an untransformed cell after virus infection can bring about the transformation. Certainly it is known that chemical changes in DNA can transform a cell to the cancerous state; thus, several potent mutagenic chemicals (see next section) are also carcinogens. It has been suggested that the 'cancer genes' incorporated into the cells by the virus may be repressed and so the cells appear to grow normally, then the chemical carcinogen removes this repression and cells become cancerous. Indeed many control aspects of cellular metabolism seem to be altered in cancer, e.g. malignant cells are characterized by a loss of control of growth. In some cases, viruses implicated in the production of cancer may cause derepression of certain functions of the host cell, causing apparent loss of control.

Failure of DNA repair enzymes has been implicated in the development of cancer: individuals lacking the enzymes which repair damage caused by UV light (see Section 5.5.2, p. 203) are especially prone to skin cancer.

5.5.7. Mutations

Mutations are alterations in DNA base sequence, which can affect the RNA transcribed from that DNA or the control systems of the cell. Mutations of two main types are known: these are '*base substitution*' and '*frame-shift*' mutations. In the first, one base is changed to another, whilst in the second, a small number of bases have been added to, or deleted from, a section of DNA. Both types can occur naturally, and mutations can also be induced by irradiation with X- or γ-rays, or by treatment with certain chemicals. For example, nitrous acid causes base substitutions by deamination of cytosine to uracil (see Section 5.2.1, p. 171) and adenine to hypoxanthine, the base of inosine, which acts more like guanine in base-pairing.

Alkylating agents are extremely powerful mutagens and also, in many cases, potent carcinogens. They attack guanine residues preferentially, and seem to cause an increase in mismatching of bases when the altered DNA is replicated or transcribed.

Some chemical mutagens contain aromatic ring-systems which can slip between adjacent base-pairs in a DNA molecule and lie parallel to the bases. This causes local unwinding of the double helix, an effect which interferes with normal replication and transcription of that section of the DNA, often producing phase-shift mutations.

If mutations occur in control regions of the DNA, then the characteristics of the whole cell may be altered. Mutations in base sequences of DNA transcribed to *t*-RNA or ribosomal RNA could cause changes in many proteins. For example, altered *t*-RNA or ribosomes could misread the code on messenger RNAs to give several mutant proteins. In extreme cases, non-functional *t*-RNA or ribosomes could result from DNA mutations, and the cells containing these would die.

When mutations occur in a messenger RNA region of DNA, then only one protein may be altered. If a base substitution mutation is involved, then only one amino acid of a protein may be changed. Even this may have serious effects for the organism concerned. Thus, in human sickle-cell anaemia, a glutamic acid residue of a normal chain of haemoglobin (see Section 4.5.1.2, p. 106) is changed to valine. This change can be brought about by a *single* base substitution in DNA, so that the β-chain messenger RNA now carries a GUA or GUG codon (for valine) in place of a GAA or GAG codon (for glutamic acid), i.e. the central adenine residue of the original triplet is replaced by uracil in the mutant.

Of course not all base substitutions involving *m*-RNA result in mutant proteins. Base changes may occur, producing codons for the same amino acid. Thus mutations of GAA to GAG on *m*-RNA will give an unaltered polypeptide, for both triplets code for the amino acid, glutamic acid. At the other extreme, extensive changes in protein structure can result from a single base substitution. If a base is altered so that the codon, in which it lies, becomes a terminator codon, then a much shortened polypeptide chain may be synthesized. If this short chain cannot fulfill its normal function, the effect on the organism can be very damaging.

Frame-shift mutations affecting messenger RNAs usually produce much more extensive changes in protein structure (see Figure 5.26). Deletion of one base (a frame-shift mutation), and translation of the new messenger RNA would give a very different amino acid sequence. The complete amino acid sequence after the mutation would be altered, unless a second

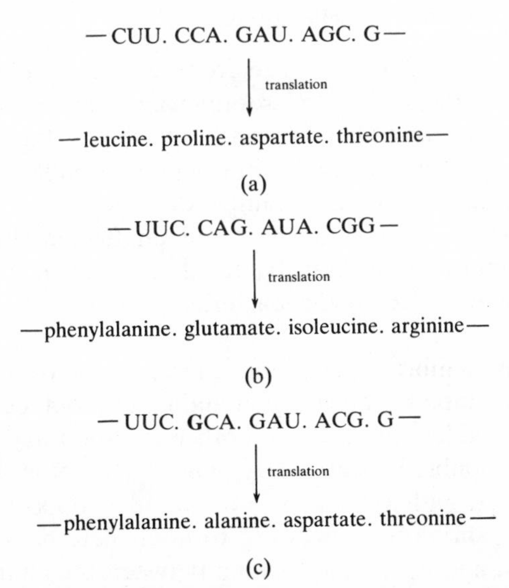

FIGURE 5.26. Results of frame-shift mutations. (a) Original base sequence of messenger RNA and corresponding amino acid sequence coded by the *m*-RNA. (b) Mutant *m*-RNA produced by deleting first base (C) of original, with resulting altered amino acid sequence. (c) Second mutant *m*-RNA produced from first mutant by adding a base, **G**, between the third and fourth bases of the first mutant. Only the first two amino acids are altered; the remaining amino acid sequence would be identical to the original

frame-shift mutation occurred to correct the first one. This correction could be brought about by adding a base into, or subtracting two bases from, the sequence at some point after deletion. Then only those amino acids whose codons fall between the first and second mutations are altered, whilst the remainder of the amino acid sequence remains unchanged.

Although mutations can have deleterious effects for the organism concerned, they may also be advantageous. Indeed, for evolution to proceed, mutations must occur and some of these must have selective advantage. Thus these organisms possessing advantageous mutations are more likely to survive and pass on the new characteristics in the genetic information to their offspring. Because the natural mutation rate is very low, evolution is an extremely slow process.

Artificial induction of bacterial mutations can on occasions be useful to man. Thus radiation-induced mutations which render bacterial DNA incapable of correct replication can cause the death of bacteria, and so radiation can be used for the sterilization of, for example, foodstuffs. Also much of our knowledge of the mechanisms and control of nucleic acid replication, transcription, and translation has been gained from work with bacteria carrying mutations in different parts of the protein synthesis or control machinery.

5.5.8. Antibiotics and nucleic acid function

Antibiotics are compounds obtained from certain microorganisms, which can inhibit the growth of other microorganisms such as bacteria or fungi. Many antibiotics are now known, and several of these interfere with nucleic acid function; here a small representative sample will be discussed. Ideally, the antibiotic should cause the death of the unwanted microorganism, whilst leaving the host cell unaffected. In practice, many antibiotics cause inhibition in both bacterial and human cells, but are used because the relative effect on the bacterial cells is much greater than on the human cells.

Antibiotics may inhibit replication, transcription, or translation. Thus *actinomycin D*, a complex molecule containing an aromatic ring system and two cyclic pentapeptides, slows down both replication and transcription. It binds tightly to double-helical DNA where the base sequence G—C appears on one strand (giving C—G on the opposite strand). The ring-system of the antibiotic is believed to intercalate between the guanine rings of the DNA, and hydrogen bonding between the amino groups of the guanine and oxygen atoms of the cyclic peptides is thought to be important. The peptides lie in the minor grooves of the DNA, and probably interfere with the action of DNA- and RNA-polymerases. Actinomycin D affects all types of living cells, but has a more disastrous action on bacteria, because they are normally growing and dividing (and hence making DNA and RNA) faster than animal cells.

Mytomycin C is another inhibitor of replication, this time by forming covalent bonds with DNA, which cross-link the two strands of a double-helical DNA and so interfere with DNA-polymerase action.

A useful inhibitor of transcription is *rifampicin*, which prevents the initiation of RNA synthesis in bacteria, and in the mitochondria of higher animals. It has no effect on the nuclear RNA-polymerase of higher animals, and so once again causes greater damage to bacteria than to the host cells. Rifampicin binds to one of the subunits of bacterial RNA-polymerase (see Section 5.5.3, p. 204) allowing the enzyme to bind to DNA, but blocks the incorporation of the first purine nucleotide at the beginning of the new RNA chain. If the enzyme is already bound to DNA before addition of the antibiotic, then rifampicin has no action, and so cannot inhibit RNA chain elongation. It is believed, therefore, that the RNA-polymerase undergoes a conformational change after binding to DNA, so that rifampicin can no longer interact with the enzyme.

Many antibiotics are known which interfere with translation. *Streptomycin* binds to one protein of the 30S-ribosomal subunit (see Section 5.5.5.1, p. 211), and may cause distortion of the ribosome surface. This effect can cause misreading of the code on the messenger, particularly at the first base of each codon, and can interfere with aminoacyl-*t*-RNA binding, hence inhibiting elongation of the new polypeptide chain.

Tetracyclines also bind to the 30S-ribosomal subunit and may inhibit *t*-RNA binding and, to a slight extent, the peptidyl transferase which catalyses peptide bond formation. Streptomycin and tetracyclines also inhibit peptide release by interfering with recognition of the termination codon.

Chloramphenicol, on the other hand, binds to the 50S-subunit, and may cause a change in the three-dimensional structure of the ribosome. This antibiotic seems to inhibit peptide bond formation, and the final peptidyl-*t*-RNA hydrolysis, and may interfere with the functional attachment to the ribosome of the aminoacyl end of the incoming aminoacyl-*t*-RNA. *Sparsomycin* may act in a similar way, although this is not yet certain.

Puromycin resembles the aminoacyl–adenyl end of an aminoacyl-*t*-RNA (see Figure 5.27), and so can interact with part of the *A*-site on the 50S-subunit of a ribosome. A 'peptide' bond can then be formed between the puromycin and the peptide at the *P*-site so that puromycin becomes the *C*-terminal end of the polypeptide (compare Figure 5.21). This peptidyl puromycin cannot bind at the *P*-site, and so an incomplete polypeptide chain is released from the ribosome. The incomplete chain may form part of a vital enzyme; thus the effects on the organism can be very serious.

Antibiotics have proved very useful, in addition to their medical applications, for slowing down or stopping certain stages of protein biosynthesis. Hence many details of transcription and translation have been gained by the use of antibiotics on bacteria.

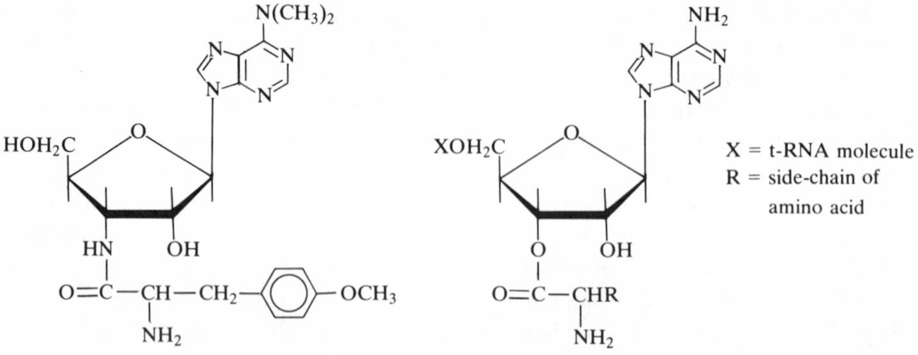

X = t-RNA molecule
R = side-chain of amino acid

Puromycin

Amino-acyl adenyl end of amino-acyl t-RNA

FIGURE 5.27. Structure of puromycin and the 3′-end of an aminoacyl-*t*-RNA

5.5.9. Recombinant DNA

Many viruses are now known which are capable of inserting viral-specific DNA into host cell DNA. In addition, bacteria often contain small circular DNA molecules, known as *plasmids*, which replicate independently of the bacterial chromosome, but can sometimes be integrated into the main bacterial DNA. The new DNA produced, by incorporation of viral or plasmid DNA, is known as *recombinant DNA*.

It is now possible to introduce DNA fragments from higher organisms into viral or plasmid DNA which, in turn, becomes incorporated into bacterial DNA. If the new DNA codes for a eukaroytic protein it is then possible to 'harvest' this protein from the bacteria. For example, human insulin might be produced in this manner on the industrial scale from *E. coli*; this protein has already been prepared in the laboratory by this method.

Because *E. coli* is a normal inhabitant of the human digestive tract, many fears have been expressed that new, extremely dangerous forms of *E. coli* could result from such 'genetic engineering'. Thus attempts have been made in many countries to restrict the types of recombinant DNA being synthesized. In particular it is hoped to avoid the introduction into *E. coli* of genes for antibiotic resistance, toxin production, or tumour production.

5.6 *IN VITRO* SYNTHESIS

Synthetic polynucleotides can be prepared by enzymic methods alone, or by using a mixture of chemical and enzymic reactions. Thus DNA may be synthesized from nucleoside triphosphates using *DNA-polymerase* as the catalyst if magnesium ions and DNA 'template' are present. The base sequence of the newly synthesized DNA is then complementary to that of the DNA template. Synthetic RNA may be prepared from nucleoside diphosphates by the action of another enzyme, *polynucleotide pyrophosphorylase*, in the presence of magnesium ions. If the diphosphates are added one at a time, and the product purified after each step, then an RNA of known sequence can be synthesized. A primer with a free hydroxyl group on the 3′-end must be present for the reaction to proceed:

$$\underset{\text{Primer}}{\text{XpYp}\cdots\text{pZ—OH}} + \underset{\text{Adenosine diphosphate}}{\text{ppA}} \xrightarrow[\text{phosphorylase}]{\text{polynucleotide}} \text{XpYp}\cdots\text{pZpA—OH} + PO_4^{3-}.$$

The 2′- and 3′-hydroxyl groups of the incoming diphosphate are usually protected, often by acetyl groups, to prevent unwanted side-reactions. After reaction, the blocking groups may be removed with alkali.

The first chemical synthesis was carried out by Khorana and coworkers, who prepared the double-stranded DNA coding for a yeast alanine transfer RNA. The method consists of condensing a mono- or oligo-nucleotide carrying a free 3′-hydroxyl group with a mono- or oligo-nucleotide carrying a 5′-phosphate; all reactive groups not participating in the condensation are first protected by blocking groups.

Firstly, the amino group of adenine, cytosine, or guanine must be blocked—a thymine nucleoside, or nucleotide, need not be protected in this way. An acid chloride, such as benzoyl, or anisoyl, or isobutyryl, chloride, is usually used, e.g.:

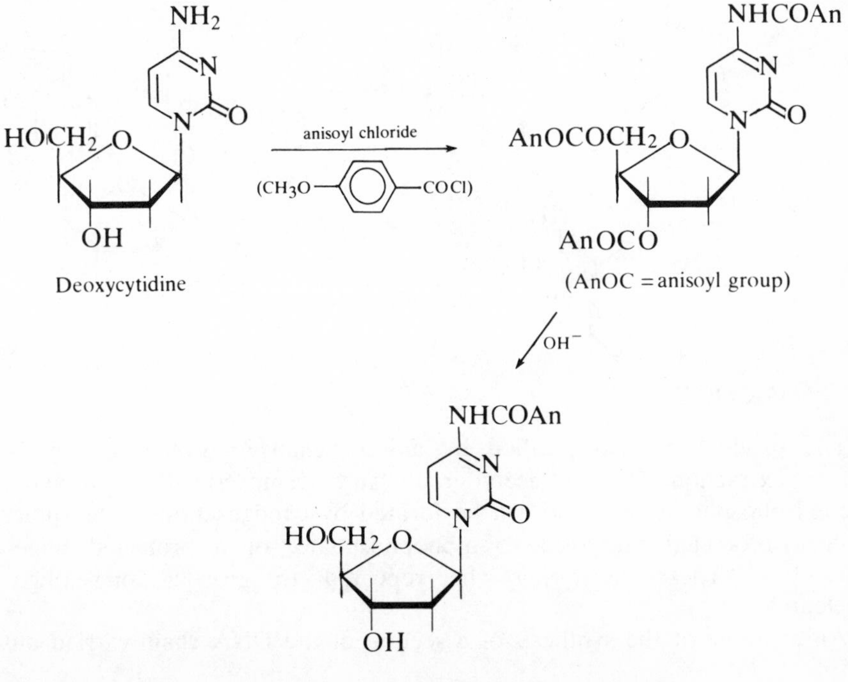

NH$_2$ … HOCH$_2$ … OH

Deoxycytidine

anisoyl chloride
(CH_3O—⟨benzene⟩—COCl)

NHCOAn … AnOCOCH$_2$ … AnOCO

(AnOC = anisoyl group)

OH^-

NHCOAn … HOCH$_2$ … OH

Synthesis of the polynucleotide is carried out from the 5′-end of the chain, and so the 5′-hydroxyl group of the first nucleoside must be protected. The blocking group for this purpose is usually a monomethoxytrityl group (MMTr), i.e. a monomethoxytriphenylmethyl group, $CH_3O—C_6H_4—C(C_6H_5)_2—$, when the reaction is:

NHCOAn ... HOCH₂ ... OH $\xrightarrow{\text{MMTrCl}}$ NHCOAn ... MMTrOCH₂ ... OH

The second nucleotide of the chain must have a phosphate group at the 5′-position and a protecting group, such as an acetyl group, on the 3′-hydroxyl position. The first two monomers can then be coupled using dicyclohexylcarbodiimide (DCC), or an aromatic sulphonyl chloride, e.g.:

NHCOAn ... MMTrOCH₂ ... OH + $^-O—P(=O)(HO)—O—CH_2$ ... H_3C ... NH ... H_3COCO $\xrightarrow{\text{DCC}}$ NHCOAn ... MMTrOCH₂ ... O ... $^-O—P{=}O$ ... $O—CH_2$ ... H_3C ... NH ... H_3COCO

The product is then purified by anion-exchange chromatography or solvent extraction. The 3′-*O*-acetyl group can be removed with alkali, and a second phosphodiester bond can be formed by condensation of the dimer with a protected nucleoside 5′-monophosphate, or a protected oligonucleotide. These steps may be repeated to give a long oligonucleotide.

An example of the synthesis of a section of the DNA chain carried out by Khorana and coworkers is shown below:

$$\text{MMTr—G}^{\text{iBu}}\text{—OH} \xrightarrow{\text{pA}^{\text{Bz}}\text{OAc}} \text{MMTr—G}^{\text{iBu}}\text{pA}^{\text{Bz}}\text{—OH} \xrightarrow{\text{pA}^{\text{Bz}}\text{OAc}}$$

$$\text{MMTr—G}^{\text{iBu}}\text{pA}^{\text{Bz}}\text{pA}^{\text{Bz}}\text{—OH}$$

$$\Big\downarrow \text{pC}^{\text{An}}\text{pC}^{\text{An}}\text{OAc}$$

$$\text{MMTr—G}^{\text{iBu}}\text{pA}^{\text{Bz}}\text{pA}^{\text{Bz}}\text{pC}^{\text{An}}\text{pC}^{\text{An}}\text{—OH}$$

$$\Big\downarrow \text{pG}^{\text{iBu}}\text{pG}^{\text{iBu}}\text{pA}^{\text{Bz}}\text{—OH}$$

$$\text{MMTr—G}^{\text{iBu}}\text{pA}^{\text{Bz}}\text{pA}^{\text{Bz}}\text{pC}^{\text{An}}\text{pC}^{\text{An}}\text{pG}^{\text{iBu}}\text{pG}^{\text{iBu}}\text{pA}^{\text{Bz}}\text{—OH}$$

$$\Big\downarrow \text{pG}^{\text{iBu}}\text{pA}^{\text{Bz}}\text{pC}^{\text{An}}\text{pT—OAc}$$

$$\text{MMTr—G}^{\text{iBu}}\text{pA}^{\text{Bz}}\text{pA}^{\text{Bz}}\text{pC}^{\text{An}}\text{pC}^{\text{An}}\text{pG}^{\text{iBu}}\text{pG}^{\text{iBu}}\text{pA}^{\text{Bz}}\text{pG}^{\text{iBu}}\text{pA}^{\text{Bz}}\text{pC}^{\text{An}}\text{pT—OH}$$

$$\Big\downarrow \begin{array}{l}\text{1. H}^+\\ \text{2. NH}_4\text{.OH}\end{array}$$

$$\text{GpApApCpCpGpGpApGpApCpT}$$

(MMTr = monomethoxytrityl on a 5′-OH group; OAc = 3′-*O*-acetyl group; superscript An, Bz, and *i*Bu = anisoyl, benzoyl, and isobutyryl groups on $—NH_2$ of bases.) Usually the amino group of cytosine is protected by an anisoyl grouping, that of adenine by a benzoyl grouping, and that of guanine by an isobutyryl grouping, $—\overset{\displaystyle O}{\overset{\|}{C}}—CH(CH_3)_2$

At each step, the product was purified by ion-exchange chromatography. When a fairly large oligonucleotide was obtained, the blocking groups were removed with acid and alkali, and the product further purified by ion-exchange chromatography in 7 M urea (on DEAE–cellulose).

The final stages of synthesis of the complete double-stranded DNA molecule were carried out enzymatically. The large oligonucleotide, now carrying free hydroxyl groups at both the 5′- and 3′-ends of the chain, was phosphorylated at the 5′-OH group, using an enzyme *T_4-polynucleotide kinase* and adenosine triphosphate, e.g.:

$$\text{HO—GpApA..pT—OH} \xrightarrow[\text{kinase + ATP}]{\text{polynucleotide}} {}^{-}\text{O—}\overset{\overset{\text{O}}{\|}}{\underset{\underset{\text{O}^{-}}{|}}{\text{P}}}\text{—O—GpApA..pT—OH} + \text{ADP.}$$

The phosphorylated oligonucleotide was mixed with the oligonucleotide which would be adjacent to it and its 5′-end in one strand of the complete DNA molecule, and to the mixture was added two or three oligonucleotides of the second strand, with base sequences complementary to those of the two oligomers of the first strand. Each oligonucleotide of the second strand was synthesized in such a way that the base sequence overlapped the sequences of *two* segments of the opposite strand. Then an enzyme, a ligase, was used in the presence of magnesium ions to link together the oligonucleotides of each strand. This process is shown below. (The sequences shown are not actual sequences from the DNA molecule synthesized, but are chosen to illustrated the method. The nucleotides joined by the new bonds are shown in heavy type.)

```
                                       O
                                       ||
                                   ⁻O—P—O⁻
                                       |
        5′-end                    HO   O                        3′-end
                                   |   |
        HO—G—A—C—T—T—A—G—A—C   A—A—C—T—G—C—C—OH
            :   :   :   :   :   :   :   :   :   :   :   :
HO—T—G—C—T—G—A—A   T—C—T—G—T—T—G—OH
                              |    |
3′-end                        O   OH                   5′-end
                              |
                         ⁻O—P—O⁻
                              ||
                              O

                    ligase | Mg²⁺
                           ↓

        HO—G—A—C—T—T—A—G—A—C—A—A—C—T—G—C—C—OH
            :   :   :   :   :   :   :   :   :   :   :   :
HO—T—G—C—T—G—A—A—T—C—T—G—T—T—G—OH
```

The processes of phosphorylation and condensation using the kinase and the ligase were repeated, joining together larger and larger fragments until the complete DNA molecule was synthesized.

Similar methods were used later to synthesize an almost complete gene, i.e. a stretch of DNA, with promoter, coding for a special type of *E. coli* tyrosine *t*-RNA. The synthetic gene was found to be biologically active.

Variations of these methods may be used to synthesize polyribonucleotides, but in this case the 2′-hydroxyl of ribose must also be protected by formation of an acetyl or benzoyl ester. For polyribonucleotides, it is more satisfactory to form the phosphodiester bond between a 3′-phosphate and a 5′-hydroxyl group, e.g.:

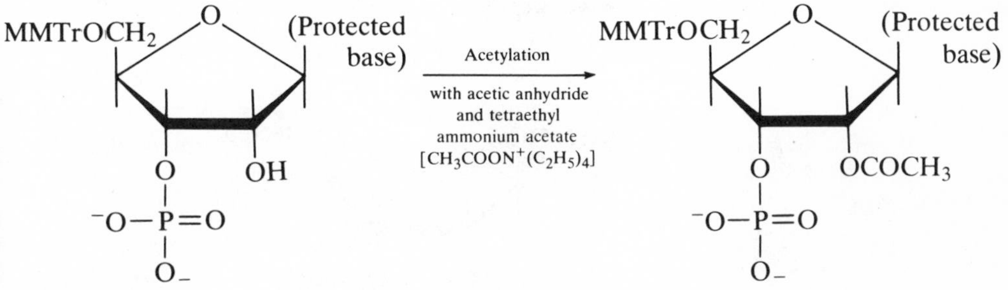

MMTrOCH$_2$ (Protected base); O; OH; $^-$O—P=O; O$_-$

Acetylation with acetic anhydride and tetraethyl ammonium acetate $[CH_3COON^+(C_2H_5)_4]$ →

MMTrOCH$_2$ (Protected base); O; OCOCH$_3$; $^-$O—P=O; O$_-$

Then

MMTrOCH$_2$ (Protected base)[2]; O; OCOCH$_3$; $^-$O—P=O; O$_-$

\+

HOCH$_2$ (Protected base)[1]; H$_3$COCO; OCOCH$_3$

DCC →

MMTrOCH$_2$ (Protected base)[2]; O; OCOCH$_3$; $^-$O—P=O; O—CH$_2$; *A* (Protected base)[1]; H$_3$COCO; OCOCH$_3$

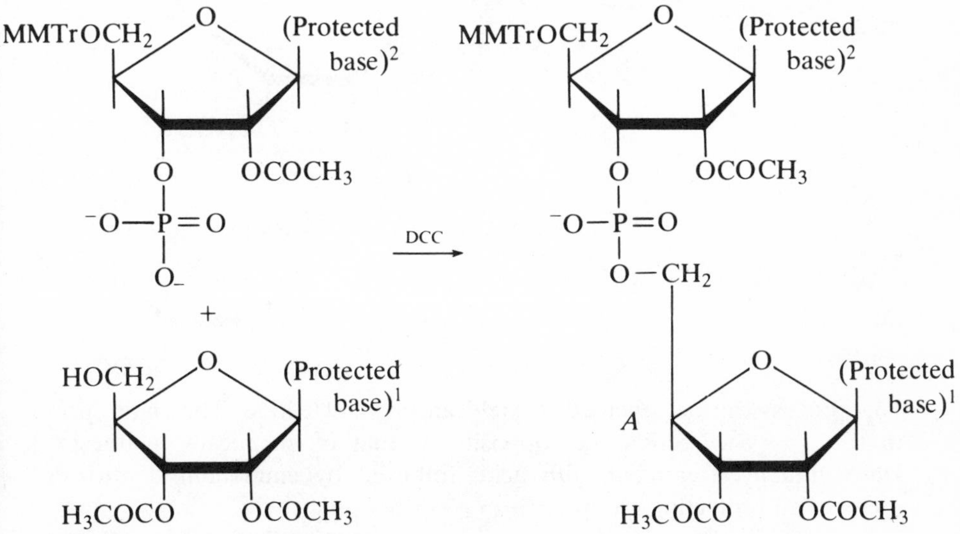

Acid treatment frees the 5′-OH on the upper ribose ring; the dinucleotide can then be condensed with a protected nucleoside 3′-phosphate to give a trinucleotide, e.g.:

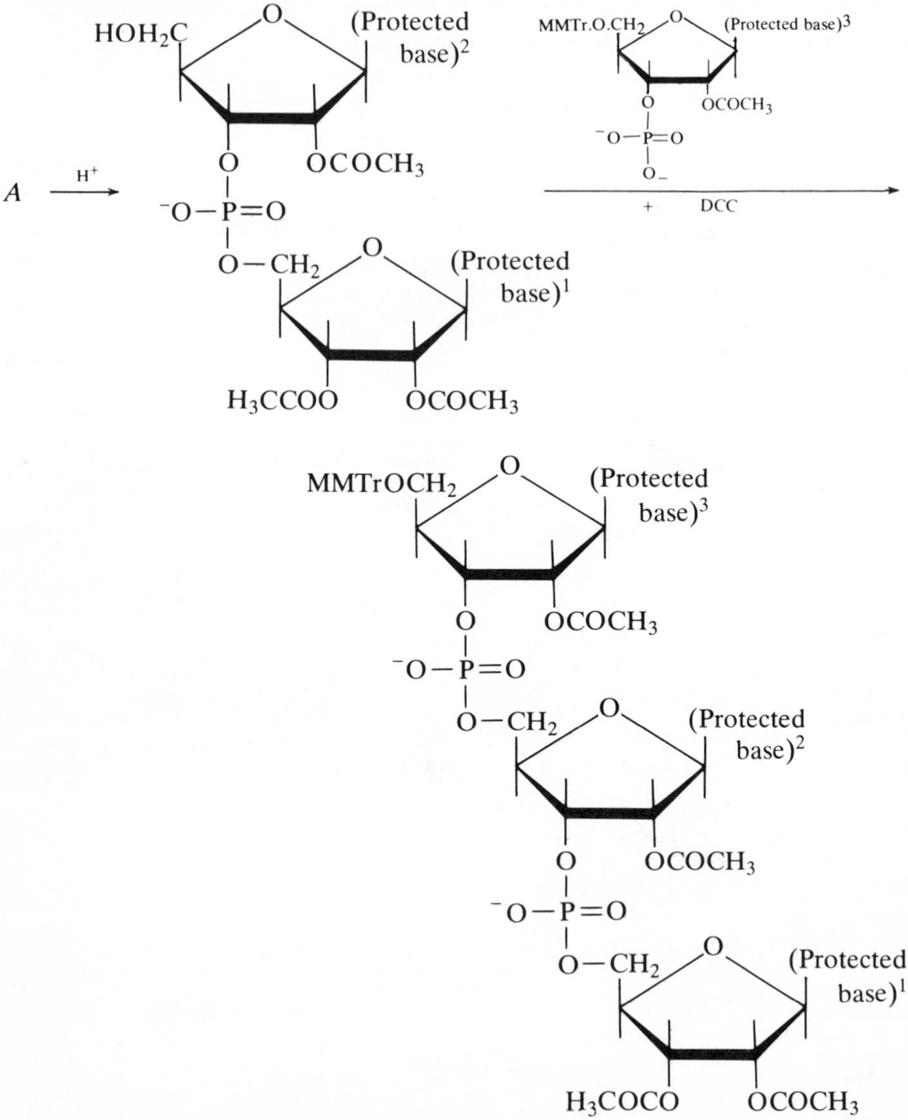

$$A \xrightarrow{H^+}$$ [structure: 3'-O-phosphorylated, 2'-acetylated nucleoside (Protected base)[2] with free HOH_2C, linked via $^-O-P=O$ to $O-CH_2$ of nucleoside (Protected base)[1] bearing H_3CCOO and $OCOCH_3$] $\xrightarrow[+\ \text{DCC}]{MMTr.O.CH_2\text{-nucleoside (Protected base)}^3\text{, }OCOCH_3\text{, }3'\text{-}O\text{-}P(=O)(O^-)O^-}$

[structure: $MMTrOCH_2$-nucleoside (Protected base)[3], $OCOCH_3$; O–$^-O-P=O$–$O-CH_2$-nucleoside (Protected base)[2], $OCOCH_3$; O–$^-O-P=O$–$O-CH_2$-nucleoside (Protected base)[1], H_3COCO, $OCOCH_3$]

This process can be repeated to yield an oligonucleotide. The chain grows in the $3' \rightarrow 5'$-direction, i.e. opposite to that of chemically synthesized DNA. Finally, treatment with acid, followed by ammonium hydroxide, removes all protecting groups.

5.7. ADDITIONAL READING

The Chemistry of Nucleosides and Nucleotides A. M. Michelson, Academic Press, London and New York, 1963.

The Chemistry of the Nucleic Acids D. O. Jordan, Butterworths, London, 1960.

Annual Reviews of Biochemistry (Ed. E. E. Snell), Annual Reviews Inc., California, 1968 onwards.

Nature, 1978 onwards.

Molecular Biology of the Gene J. D. Watson, W. A. Benjamin, New York, 1970.

Biochemistry—The Chemical Reactions of Living Cells D. E. Metzler, Academic Press, London and New York, 1977.

Progress in Nucleic Acid Research and Molecular Biology (Ed. W. E. Cohn), Vol. 1, Academic Press, London and New York, 1963, and subsequent volumes.

Protein Synthesis (Ed. E. H. McConkey), Marcel Dekker Inc., New York; Vol. 1 (1971), Vol. 2 (1976).

The Enzymes (Ed. P. D. Boyer), Vol. X, 3rd ed., Academic Press, London and New York, 1974.

DNA Synthesis A. Kornberg, W. H. Freeman & Co, San Francisco, 1974.

Studies on nucleic acids: total synthesis of a biologically functional gene H. G. Khorana, in *Bioorganic Chemistry*, Vol. 7, Academic Press, London and New York, 1978, p. 351.

Chapter 6

Polysaccharides

6.1. INTRODUCTION

Polysaccharides are found in great variety, and perform a number of different functions in plants, animals, and microorganisms. Most importantly, perhaps, some polysaccharides provide the essential structural elements in plants, whilst others serve as energy reserves in many organisms. Quantitatively, cellulose and starch—both plant polysaccharides—are amongst the most abundant biopolymers known. Special polysaccharides can also perform other functions, such as lubrication of bone-joints in animals, and protection of the bark of trees after damage.

Several polysaccharides have great commercial importance, and large industries are based on the utilization of these polymers. For example, starch is exploited not only by the food industry, but also in paper-making, and in the manufacture of adhesives, whilst cellulose forms the basis of the cotton textile as well as the paper industry.

A great diversity in structure is found amongst polysaccharides. Some may consist simply of linear molecules containing only one type of monomer. The most complex polysaccharides have branched molecules made up of four to six different monomers. Within one polysaccharide, the details of molecular architecture may vary from one molecule to the next. Thus when dealing with a complex polysaccharide, we are concerned with a collection of similar, but not identical, molecules.

Until recently, research on polysaccharides lagged far behind that on proteins and nucleic acids. But now greater attention is being focused on this type of natural polymer, and more knowledge is being gained, although much work yet remains. It is, for example, difficult to obtain single crystals of a polysaccharide large enough for the detailed examination of three-dimensional structure by X-ray crystallography, and so less precise information on structure is available for polysaccharides than for proteins.

Classifications of polysaccharides *in terms of structure* have been made, and so these biopolymers will be discussed in groups according to the composition of the 'backbone' of the main molecular chain.

6.2. THE MONOMERS

Polysaccharides are condensation polymers of *monosaccharides*, or *sugars* in common nomenclature. The simplest of these substances contain only carbon, hydrogen, and oxygen, and have the structural formula $(CH_2O)_n$. Hence they were first thought of as 'hydrates of carbon', and were given the name of *carbohydrates*. Monosaccharides are either polyhydroxy aldehydes known as *aldoses*, or polyhydroxy ketones, known as *ketoses*.

The majority of monosaccharides making up polysaccharides are aldoses. Most monosaccharides are optically active and can exist in two mirror-image forms. These two forms are commonly designated D- and L-, and Figure 6.1 shows the two common sugars D-glucose (an aldose) and D-fructose (a

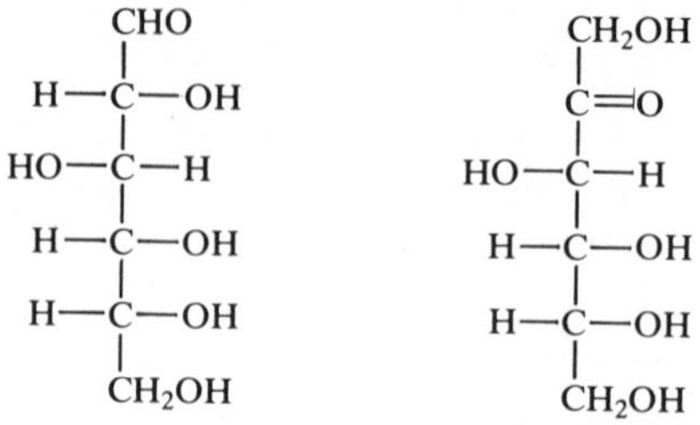

$$\begin{array}{ccc} & CHO & \\ & | & \\ H- & C & -OH \\ & | & \\ HO- & C & -H \\ & | & \\ H- & C & -OH \\ & | & \\ H- & C & -OH \\ & | & \\ & CH_2OH & \end{array} \qquad \begin{array}{ccc} & CH_2OH & \\ & | & \\ & C=O & \\ & | & \\ HO- & C & -H \\ & | & \\ H- & C & -OH \\ & | & \\ H- & C & -OH \\ & | & \\ & CH_2OH & \end{array}$$

D-Glucose D-Fructose

FIGURE 6.1. The structure of D-glucose and D-fructose

ketose). In most polysaccharides, only D-sugars are found but a few L-monosaccharides are incorporated into complex polymers. This situation contrasts with the amino acids in proteins, where only L-amino acids are found. [The designation D- or L- depends on the configuration of groups on the asymmetric carbon atom furthest from the aldehyde or ketone group of the monosaccharide. L-glucose and L-fructose have the following structures:

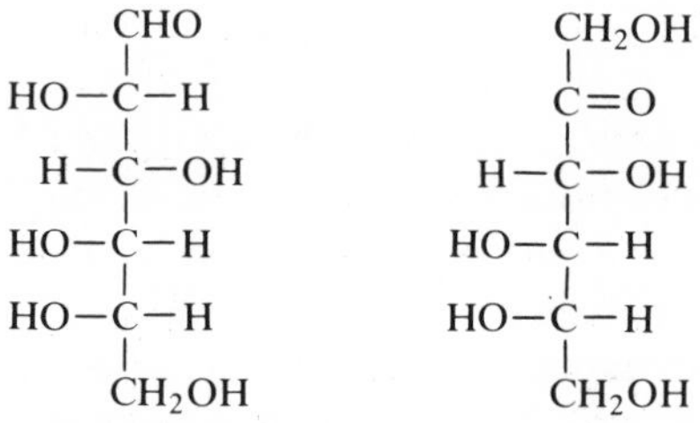

$$\begin{array}{ccc} & CHO & \\ & | & \\ HO- & C & -H \\ & | & \\ H- & C & -OH \\ & | & \\ HO- & C & -H \\ & | & \\ HO- & C & -H \\ & | & \\ & CH_2OH & \end{array} \qquad \begin{array}{ccc} & CH_2OH & \\ & | & \\ & C=O & \\ & | & \\ H- & C & -OH \\ & | & \\ HO- & C & -H \\ & | & \\ HO- & C & -H \\ & | & \\ & CH_2OH & \end{array}$$

L-Glucose L-Fructose

The horizontal bonds should be considered to project out of the plane of the paper (from the acrbon atom) towards the reader.]

In polysaccharides, the commonest monomers contain either five carbon atoms and are known as *pentoses*, or six carbon atoms and are called *hexoses*; thus glucose and fructose are both hexoses.

It has been found that few monosaccharide molecules exist in solution in the straight chain form shown in Figure 6.1, but rather they occur in the ring forms, as shown in Figure 6.2(a), (b), and (c). Indeed, for a free monosaccharide the straight chain form in solution is in equilibrium with several possible ring forms. These rings are formed by reaction of the aldehyde or ketone group of the monosaccharide with an hydroxyl group of the same molecule, i.e. a *hemiacetal* or *hemiketal* bond is formed:

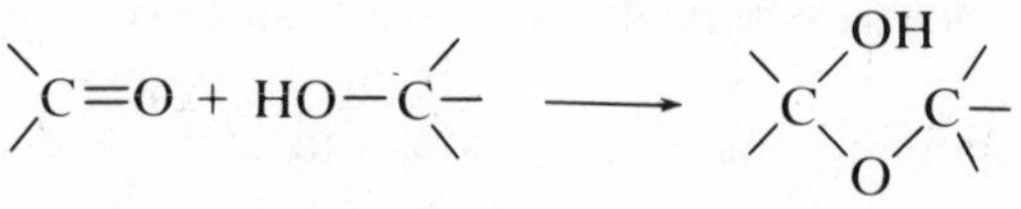

$$\rangle C=O + HO-C\langle \longrightarrow \rangle C(OH)-O-C\langle$$

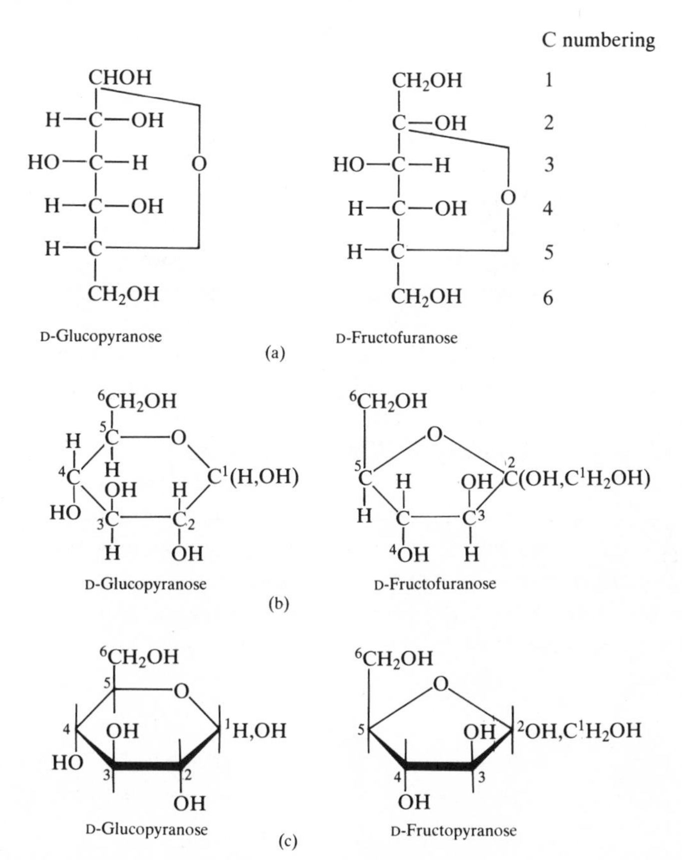

FIGURE 6.2. (a) The common ring forms of D-glucose and D-fructose. (b) Haworth projection formulae for common ring forms of D-glucose and D-fructose. (c) Simplified Haworth projection formulae for D-glucose and D-fructose. The numbering of carbon atoms in a monosaccharide is given. Note that the conventional way of representing ring-form monosaccharides is given in (c). The formulae in (a) and (b) are given for purposes of comparison with Figure 6.1, and for understanding the development of those given in (c). To convert from (a) to (b), the sugar is rotated clockwise through 90°, the ring oxygen is placed at the back, and the CH_2OH group on C-6 placed above the ring. Then groups on the right of the carbon chain in (a) are written below the ring in (b); those on the left of the carbon chain in (a) appear above the ring in (b). To convert from (b) to (c), the ring carbon atoms and hydrogen atoms attached to them (except at C-1) have been eliminated, for speed and ease of representation. In (b) and (c) the lower edge of the ring should be considered to be nearest to the reader

The rings may be five- or six-membered: a five-membered ring is known as the *furanose form* of a sugar (cf. fructose in Figure 6.2) and a six-membered ring is the *pyranose form* (cf. glucose in Figure 6.2). (The names come from the substance furan O with a five-membered ring, and pyran O or O with a six-membered ring. Pyranose and furanose are included in the name of the sugar to indicate the ring size of the molecule under discussion.)

Glucose can exist in the furanose form and fructose in the pyranose form, but for glucose, the form shown in Figure 6.2 is more stable and predominates in aqueous solution. For fructose derivatives one or the other ring size may predominate.

In any monosaccharide, e.g. D-*glucopyranose*, the ring structure may form in two ways giving the products shown in Figure 6.3(a). In the same way there are two possible structures for D-*fructofuranose* (Figure 6.3(b)). These forms of a sugar, differing only in the configuration at the carbon atom which would carry the carbonyl group in the straight-chain form, are known as *anomers*. The carbon atom at which the variation occurs is the *anomeric carbon atom*.

Haworth projection formulae (Figure 6.2(c)) are useful for indicating the relative positions of groups, i.e. whether above or below the ring, but should not be considered to represent the absolute stereochemistry of the monosaccharides. Neither the furanose nor pyranose ring forms of sugars are planar. Furanose rings are usually almost planar (with four atoms coplanar and one outside the plane) with the degree of puckering varying from one sugar derivative to another. Pyranose sugars may adopt several

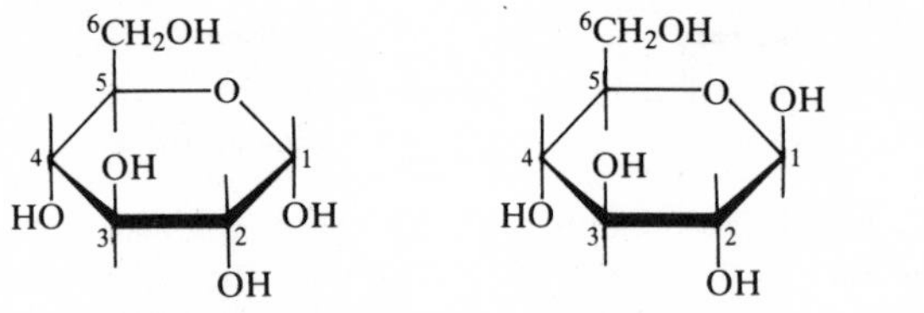

α-D-Glucopyranose

β-D-Glucopyranose

(a)

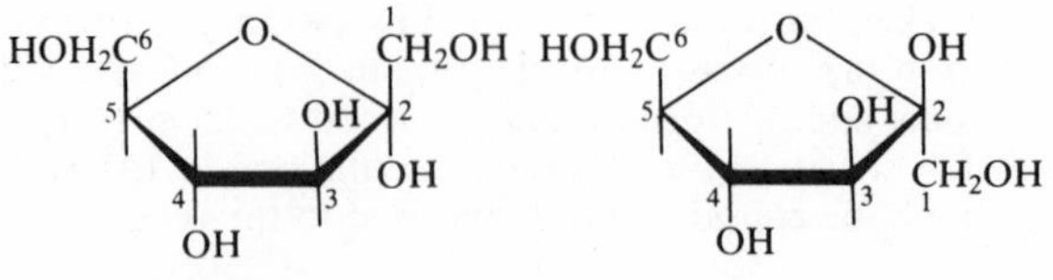

α-D-Fructofuranose

β-D-Fructofuranose

(b)

FIGURE 6.3. (a) Anomeric forms of D-glucopyranose. (b) Anomeric forms of D-fructofuranose. (Numbering of carbon atoms is shown in all cases)

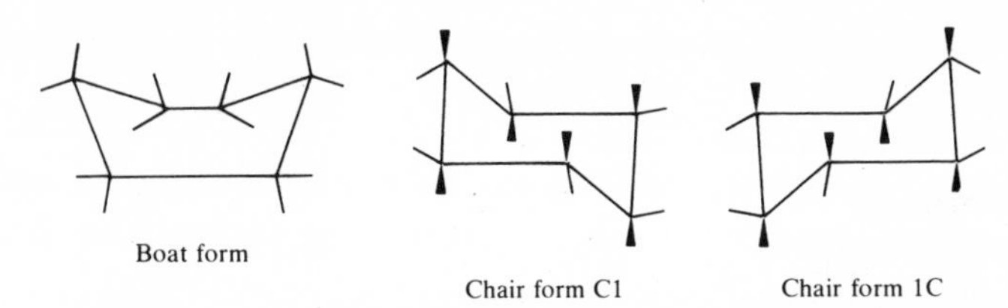

FIGURE 6.4. Schematic possible conformations of 6-membered rings. For the chair forms, axial bonds to substituent groups are shown in heavy type, the other bonds are equatorial

conformations (stereochemical arrangements achieved by rotation about single bonds), amongst which are *chair* and *boat* forms (see Figure 6.4). In general, chair forms are more stable and are thought to predominate in polysaccharides.

The substituent groups on a monosaccharide in a chair form are not all geometrically and chemically equivalent. Two types of substituent may be distinguished—*axial* and *equatorial* (see Figure 6.4), and usually, it is more difficult to insert a bulky grouping into an axial position. Also, monosaccharides tend to adopt the chair form which gives the highest number of large substituents in equatorial positions (see Figure 6.5).

Stereoisomers of sugars occur. For example, because of asymmetric carbon centres, the straight chain form of glucose shown in Figure 6.1 has four asymmetric carbon atoms (carbons 2, 3, 4, and 5). Inversion of the —H and —OH groups at one or more of these carbons gives a new sugar, i.e. $2^4 = 16$ stereoisomeric aldohexoses can exist, which correspond, in fact, to the D- and L-forms of eight different sugars. In polysaccharides, however, only a few of these are commonly found. (See Figure 6.6(a) where here the commonly occurring ring form is given.)

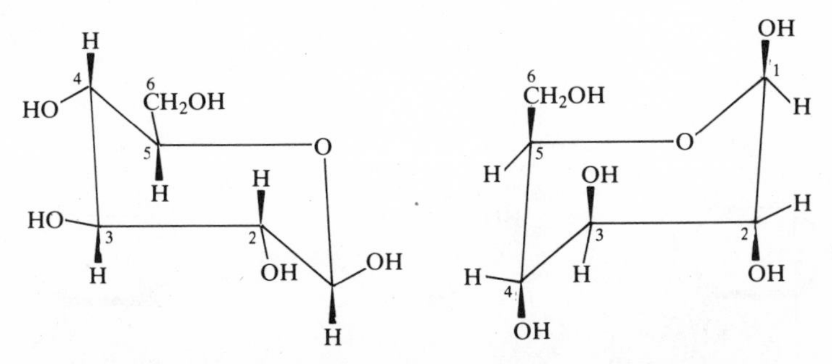

FIGURE 6.5. Schematic possible chair conformations of β-D-glucopyranose. (Numbering of the carbon atoms is shown.) The form shown on the left is the stable conformation of β-D-glucopyranose; all —OH groups and the $—CH_2OH$ group are equatorial. In the right-hand form, these bulky groups are axial, and steric hindrance results, decreasing the stability of this conformation

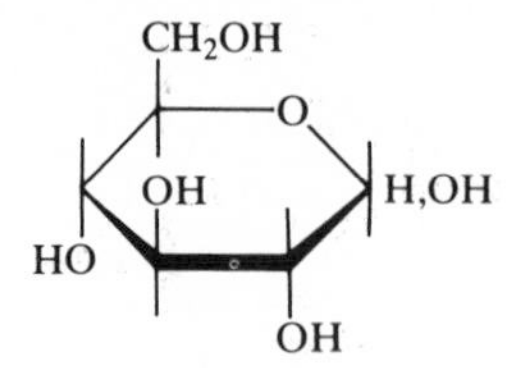

D-Glucose
D-Glucopyranose
Glc*p*

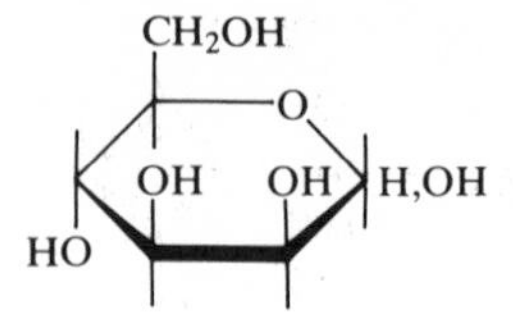

D-Mannose
D-Mannopyranose
Man*p*

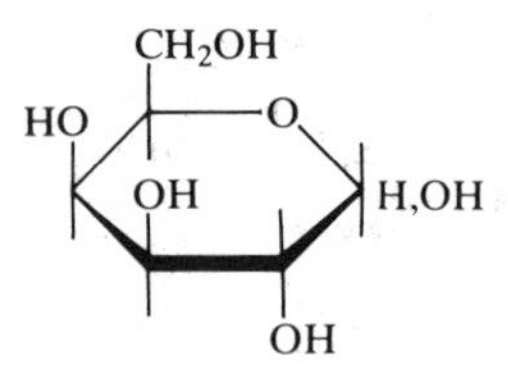

D-Galactose
D-Galactopyranose
Gal*p*

(a)

L-Arabinose
L-Arabinofuranose
Ara*f*

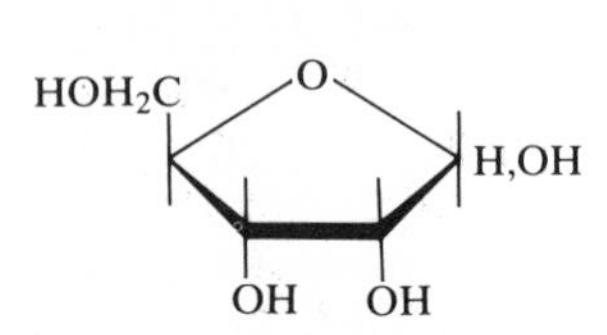

D-Ribose
D-Ribofuranose
Rib*f*

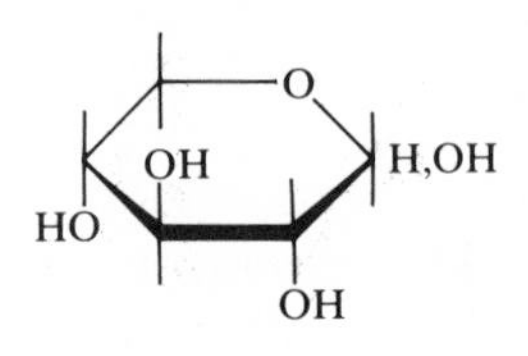

D-Xylose
D-Xylopyranose
Xyl*p*

(b)

2-Deoxy-D-ribose
2-Deoxy-D-ribofuranose
deRib*f*

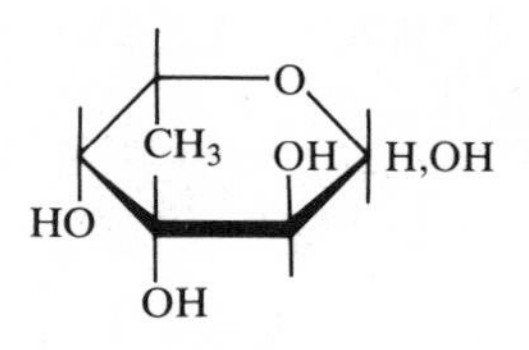

6-Deoxy-L-galactose
L-Fucose
Fuc *p*

6-Deoxy-L-mannose
L-Rhamnose
Rha*p*

(c)

D-Glucosamine
2-Amino-2-deoxy-D-glucopyranose
Glc N*p*

D-Galactosamine
2-Amino-2-deoxy-D-galactopyranose
Gal N*p*

N-Acetylglucosamine
Glc NAc *p*

(d)

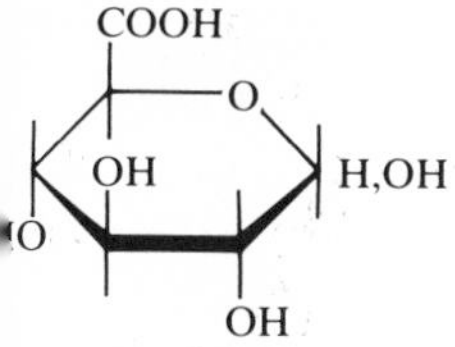

D-Glucuronic acid
Glc*p* A

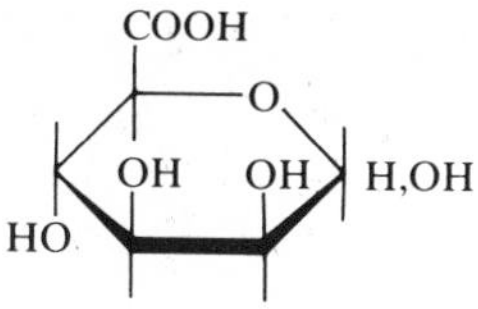

D-Mannuronic acid
Man*p* A

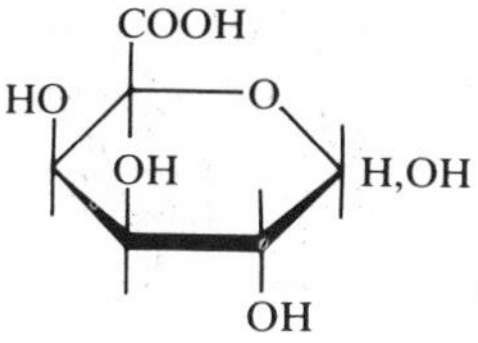

D-Galacturonic acid
Gal*p* A

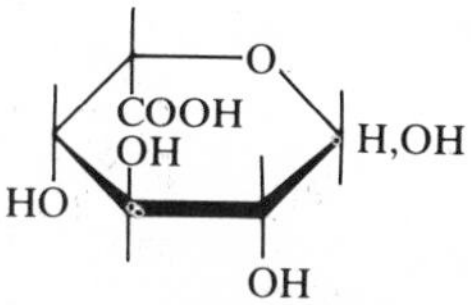

L-Iduronic acid
Ido*p* A

(e)

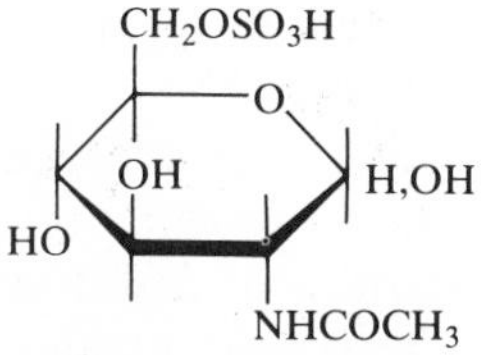

N-Acetylglucosamine-6-sulphate

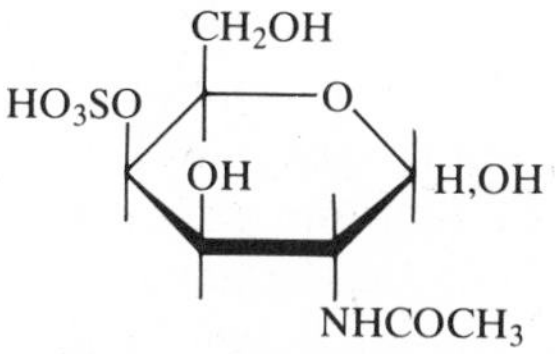

N-Acetylgalactosamine-4-sulphate

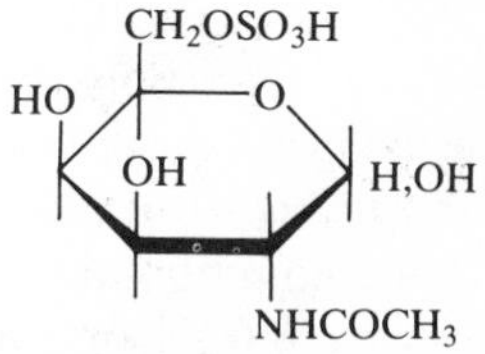

N-Acetylgalactosamine-6-sulphate

(f)

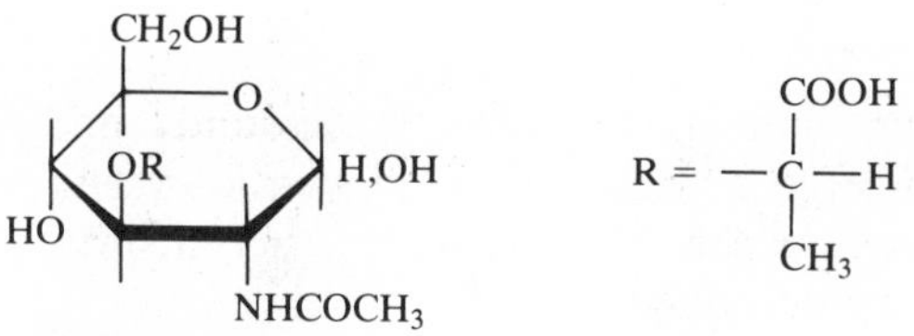

N-Acetylmuramic acid
2-Acetamido-3-*O*-(2,1-carboxyethyl)-2-deoxy-D-glucose
Mur NAc

5-*N*-Acetyl-D-neuraminic acid
5-Acetamido-3,5-dideoxy-D-glycero-D-galacto-2-nonulopyranosialonic acid

(g)

FIGURE 6.6. Monosaccharides commonly occurring in polysaccharides: (a) aldohexoses; (b) aldopentoses; (c) deoxy sugars; (d) amino sugars and derivatives; (e) uronic acids; (f) sulphated sugars; (g) muramic acid and neuraminic acid derivatives. The designation H, OH at C-1 of a ring denotes either an α- or β-sugar. Note: D-ribose and 2-deoxy-D-ribose occur in nucleic acids (see Chapter 5)

Similarly, the straight chain form of an aldopentose has three asymmetric carbon atoms, and so $2^3 = 8$ aldopentoses exist, i.e. the D- and L-forms of four different sugars. In practice, three of the four occur commonly in polysaccharides and are shown in Figure 6.6(b). Aldopentoses are often found in the furanose ring form.

Many derivatives of these simple aldo-hexoses and -pentoses are components of polysaccharides. Among the most frequent are *deoxy sugars*, where an OH group is replaced by H (see Figure 6.6(c)); *amino sugars*, where an OH group is replaced by $—NH_2$ and *N*-acetylated derivatives of the amino sugars (see Figure 6.6(d)); and *uronic acids* where the CH_2OH group on the carbon furthest from the anomeric carbon has been oxidized to COOH (see Figure 6.6(e)). Furthermore, *sulphated sugars* are found in many animal polysaccharides (see Figure 6.6(f)), whilst *complex acids* consisting of a six-carbon amino sugar linked to a three-carbon acid are found in bacterial polysaccharides and glycoproteins (see Figure 6.6(g)).

Some common abbreviations for the names of monosaccharides are given in Figure 6.6—*p* or *f* after the abbreviated name indicates the pyranose or furanose ring form, respectively.

6.2.1. Reactions

Monosaccharides, being both polyhydroxy compounds and potential aldehydes or ketones, undergo a wide variety of reactions. Here only the most important of those which are used in the determination of polysaccharide structure are described.

If a free —OH group is available at the anomeric carbon of a monosaccharide so that the ring form can open, the potential carbonyl group can be reduced using sodium borohydride. The product is a sugar alcohol, or alditol e.g.:

CH_2OH, O, OH, HO, OH, H,OH ⇌ CHO / H—C—OH / HO—C—H / H—C—OH / H—C—OH / CH_2OH $\xrightarrow{NaBH_4}$ CH_2OH / H—C—OH / HO—C—H / H—C—OH / H—C—OH / CH_2OH

D-Glucopyranose D-Glucose D-Glucitol

The carboxyl group of a uronic acid can be reduced to a primary alcohol, but the aldehyde or ketone group of the sugar must first be protected if it is to remain unaffected.

$$\underset{\text{R}}{\overset{\text{COOH}}{|}} \xrightarrow{\text{LiAlH}_4} \underset{\text{R}}{\overset{\text{CH}_2\text{OH}}{|}}$$

Again, whenever the ring form of a monosaccharide can open to give a carbonyl group, the sugar can be oxidized, i.e. can itself act as a reducing agent. Such monosaccharides are known as reducing sugars. These substances can be oxidized by alkaline Cu^{2+}, Ag^{+}, or ferricyanide solutions, and many techniques for the qualitative detection, and quantitative estimation, of reducing sugars are based on such reactions.

Aldoses and ketoses react with an alcohol in acid to give a product known as a *glycoside*, e.g.:

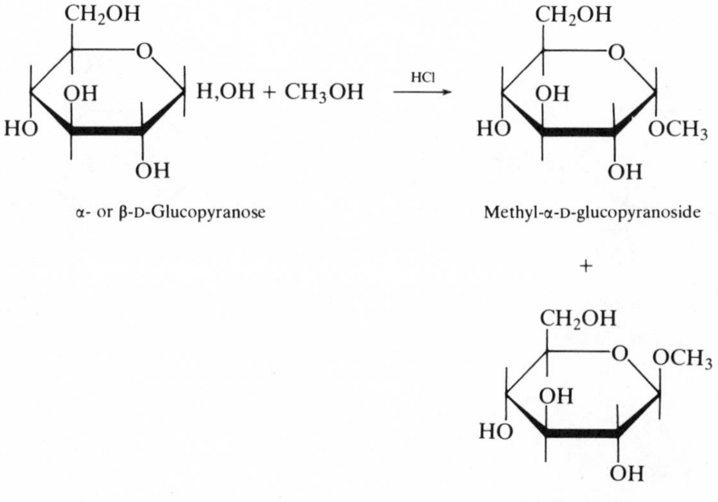

CH₂OH ring structure: α- or β-D-Glucopyranose (H,OH) + CH_3OH $\xrightarrow{\text{HCl}}$ Methyl-α-D-glucopyranoside + Methyl-β-D-glucopyranoside

α- or β-D-Glucopyranose

Methyl-α-D-glucopyranoside

Methyl-β-D-glucopyranoside

The anomeric carbon atom is asymmetric, and so both α- and β-glycosides can exist. The sugar ring cannot now open, and glycosides are non-reducing. They are alkali-stable, but are hydrolysed by acid to the original sugar and free alcohol.

A sugar aldehyde or ketone group can react with hydroxylamine to give an oxime which can exist in ring or straight-chain forms. This derivative may be dehydrated to give a straight-chain nitrile:

$$\underset{\text{R}}{\overset{\text{CHO}}{|}} \xrightarrow{\text{NH}_2\text{OH}} \underset{\text{R}}{\overset{\text{C=NOH}}{|}} \xrightarrow{-\text{H}_2\text{O}} \underset{\text{R}}{\overset{\text{C}\equiv\text{N}}{|}}$$

Monosaccharides form weakly anionic complexes with borate, and these can be used for the separation and identification of sugars by ion-exchange chromatography.

Reducing sugars are unstable in strong alkali, which causes rearrangements and breakdown of the molecules. When heated in strong acids, hexoses give derivatives of furfural (O CHO), which can react with phenols to give coloured compounds. Many quantitative estimations of hexoses are based on these reactions. Pentoses also give coloured compounds when heated in strong acid in the presence of a phenol.

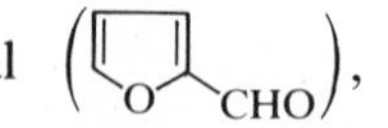

The hydroxyl groups of monosaccharides may be *esterified*. For example, a mixture of penta-*O*-acetyl-α-*D*-glucopyranose and penta-*O*-acetyl-β-*D*-glucopyranose can be obtained by the action of acetic anhydride with a catalyst on *D*-glucose:

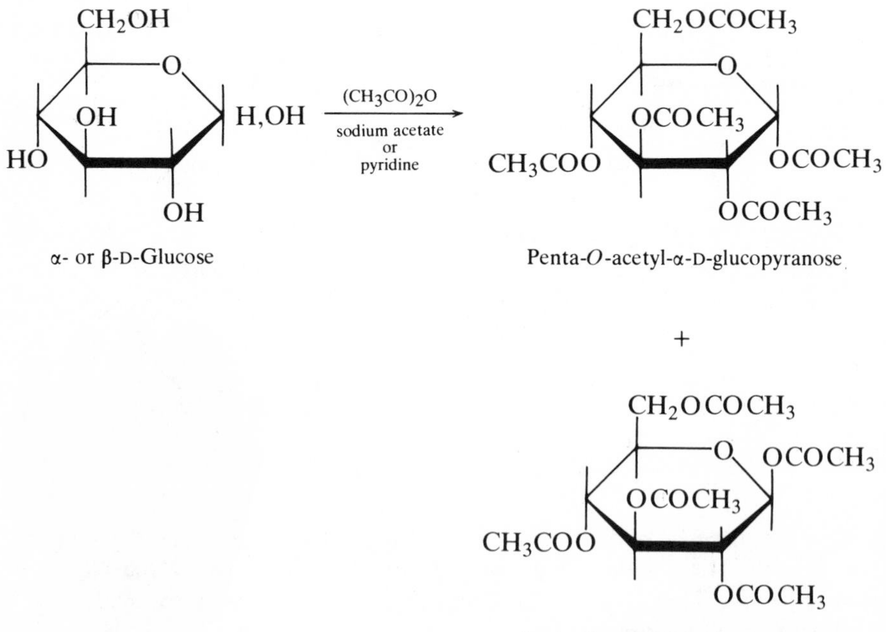

CH_2OH O OH H,OH HO OH

α- or β-D-Glucose

$\xrightarrow[\text{sodium acetate or pyridine}]{(CH_3CO)_2O}$

CH_2OCOCH_3 O $OCOCH_3$ CH_3COO $OCOCH_3$ $OCOCH_3$

Penta-*O*-acetyl-α-D-glucopyranose

+

CH_2OCOCH_3 O $OCOCH_3$ $OCOCH_3$ CH_3COO $OCOCH_3$

Penta-*O*-acetyl-β-D-glucopyranose

If the reducing group is first protected, say, by glycoside formation, then acetylation takes place at carbons 2, 3, 4, and 6, only. Alditol acetates or acetate oximes are often used in studies on sugar mixtures by gas chromatography.

Sugar hydroxyl groups may also be converted to *ethers*, and these derivatives are extremely important in structural determinations of polysaccharides. Often a monosaccharide is first converted to a glycoside before being, for example, methylated, i.e.:

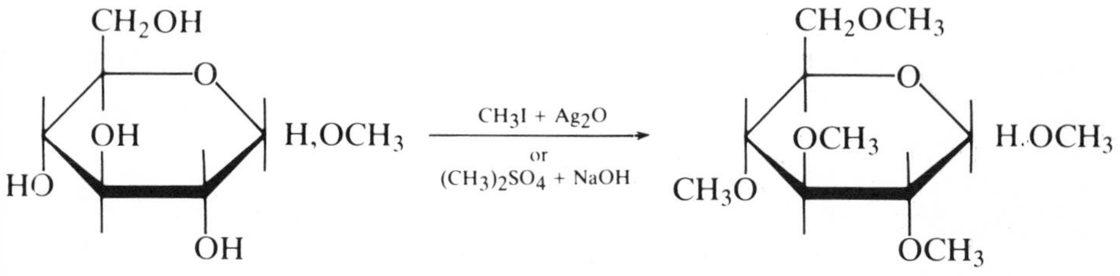

CH_2OH, O, OH, H,OCH_3, HO, OH

$\xrightarrow[\text{or } (CH_3)_2SO_4 + NaOH]{CH_3I + Ag_2O}$

CH_2OCH_3, O, OCH_3, H.OCH_3, CH_3O, OCH_3

Methyl-(α or β)-D-glucopyranoside

Methyl-2.3.4.6-tetra-*O*-methyl (α or β)-D-glucopyranoside

Diazomethane can also be used for methylation, but now very efficient methylation of polysaccharides is carried out in dimethyl sulphoxide solution with methyl iodide plus the dimethyl sulphinyl anion ($(CH_3)_2S{=}O$) produced by the action of sodium hydride on dimethyl sulphoxide). The methyl ether groups on the monosaccharide residues are stable to acid hydrolysis.

Identification of the methylated sugars obtained by hydrolysis of a methylated polysaccharide is usually an important step in the determination of polysaccharide structure.

With hexamethyl disilazane, silyl derivatives of monosaccharides are formed:

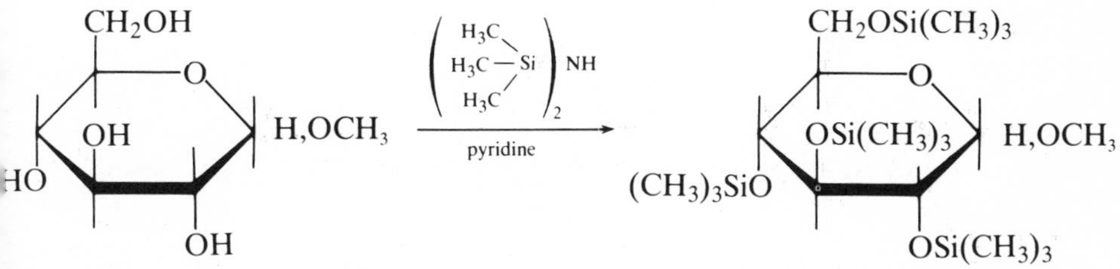

CH_2OH, O, OH, H,OCH_3, HO, OH

$\xrightarrow[\text{pyridine}]{((H_3C)_3Si)_2NH}$

$CH_2OSi(CH_3)_3$, O, $OSi(CH_3)_3$, H,OCH_3, $(CH_3)_3SiO$, $OSi(CH_3)_3$

Methyl-(α or β)-D-glucopyranoside

Methyl-2.3.4.6-tetra-*O*-tri-methyl silyl-(α or β)-D-glucopyranoside

This type of derivative is now widely used in the gas chromatographic examination of sugar mixtures.

Periodate can be used to oxidize polyhydroxy compounds such as carbohydrates whenever hydroxyl groups occur on two adjacent carbon atoms. One mole of periodate is consumed for each 'mole' of carbon–carbon bonds oxidized, and a dialdehyde derivative of the sugar is formed. Where —OH groups occur on three contiguous carbon atoms, a molecule of formic

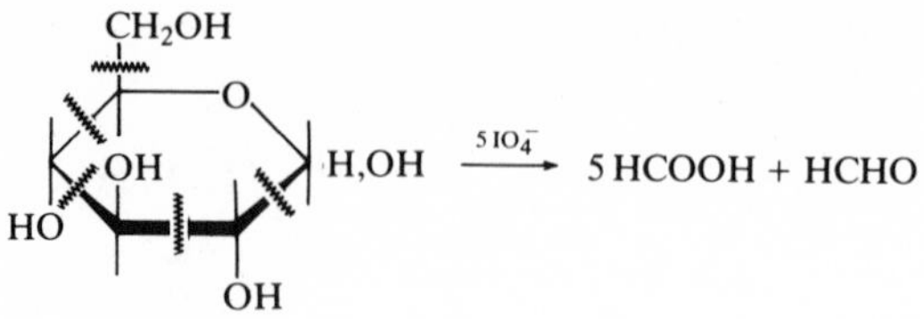

CH_2OH ... H,OH $\xrightarrow{5\,IO_4^-}$ 5 HCOOH + HCHO

HO OH OH

(a)

CH_2OH ... H,OH $\xrightarrow{3\,IO_4^-}$ 2 HCOOH + HCHO + OHC—C(OX)—CHO (*J*)

Initial reaction

HO OH OX

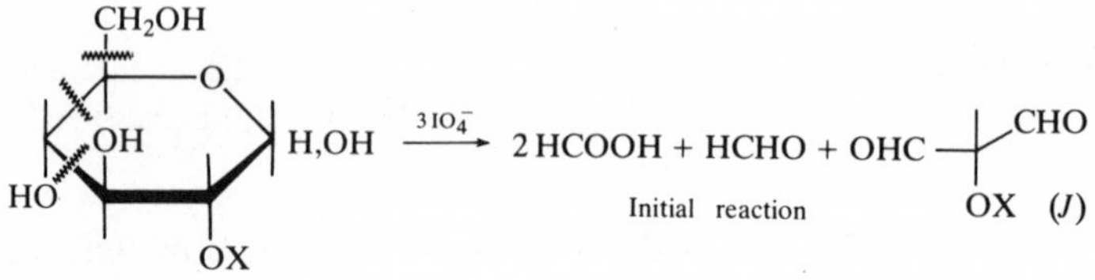

2-Substituted glucose (*A*)

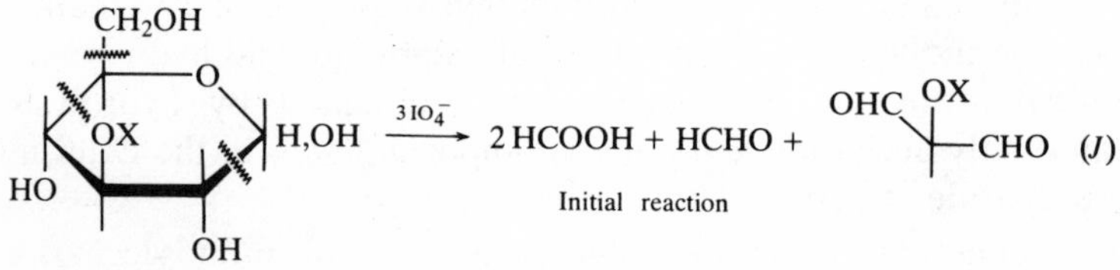

CH_2OH ... H,OH $\xrightarrow{3\,IO_4^-}$ 2 HCOOH + HCHO + OHC—C(OX)—CHO (*J*)

Initial reaction

HO OX OH

3-Substituted glucose (*B*)

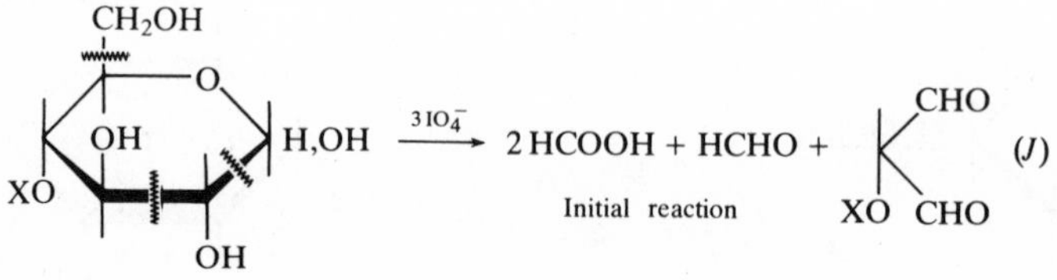

CH_2OH ... H,OH $\xrightarrow{3\,IO_4^-}$ 2 HCOOH + HCHO + CHO, XO CHO (*J*)

Initial reaction

XO OH OH

4-Substituted glucose (*C*)

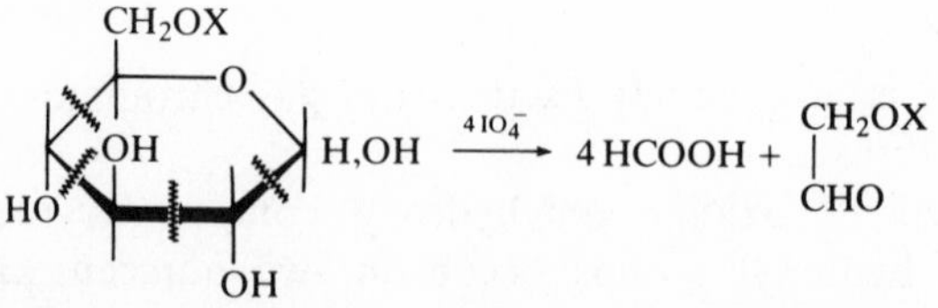

CH_2OX ... H,OH $\xrightarrow{4\,IO_4^-}$ 4 HCOOH + CH_2OX—CHO

HO OH OH

6-Substituted glucose (*D*)

(b)

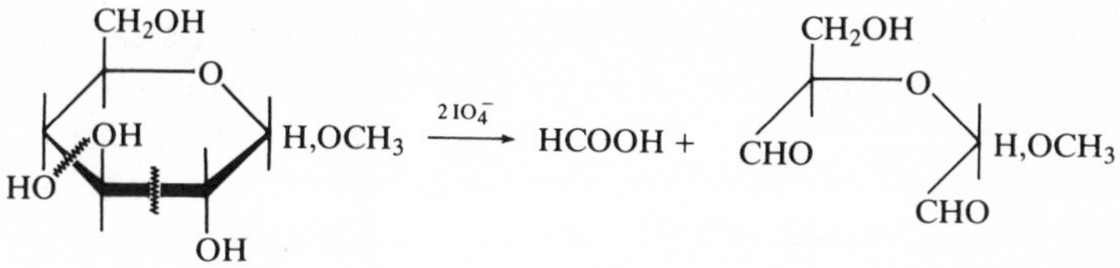

CH_2OH … OH, HO, OH, H,OCH_3 $\xrightarrow{2IO_4^-}$ HCOOH + CH_2OH … CHO, CHO, H,OCH_3

(c)

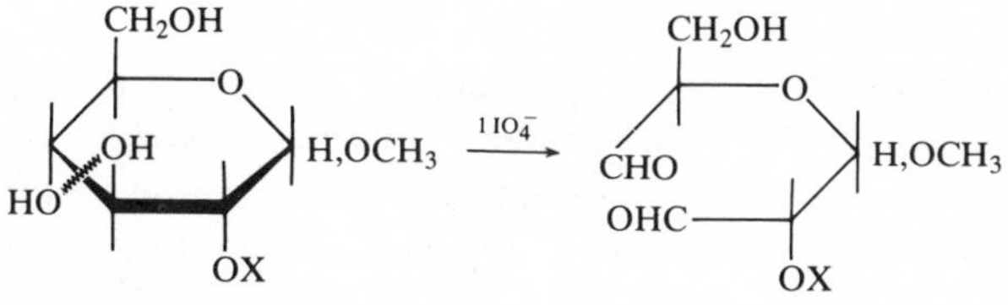

CH_2OH … OH, HO, OX, H,OCH_3 $\xrightarrow{1IO_4^-}$ CH_2OH … CHO, OHC, OX, H,OCH_3

Methyl-2-substituted glucoside

CH_2OH … OX, HO, OH, H,OCH_3 $\xrightarrow{IO_4^-}$ no reaction

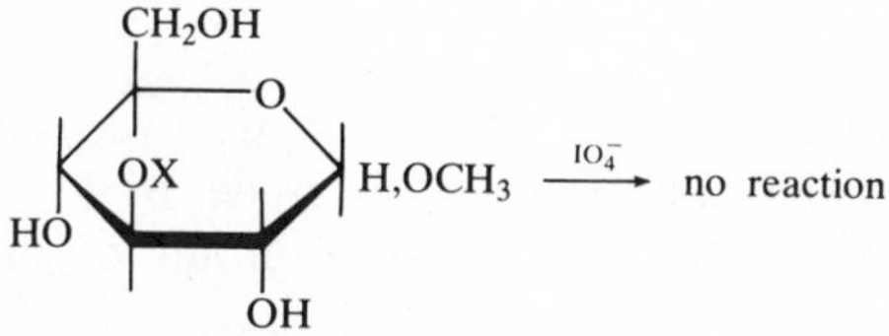

Methyl-3-substituted glucoside

CH_2OH … OH, XO, OH, H,OCH_3 $\xrightarrow{1IO_4^-}$ CH_2OH … XO, CHO, CHO, H,OCH_3

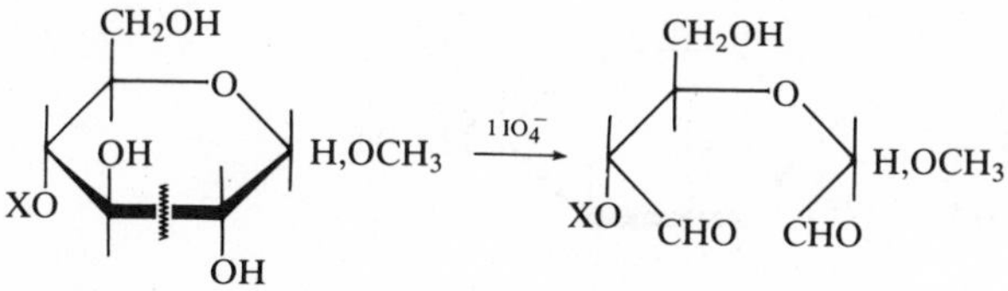

Methyl-4-substituted glucoside

CH_2OX … OH, HO, OH, H,OCH_3 $\xrightarrow{2IO_4^-}$ HCOOH + CH_2OX … CHO, CHO, H,OCH_3

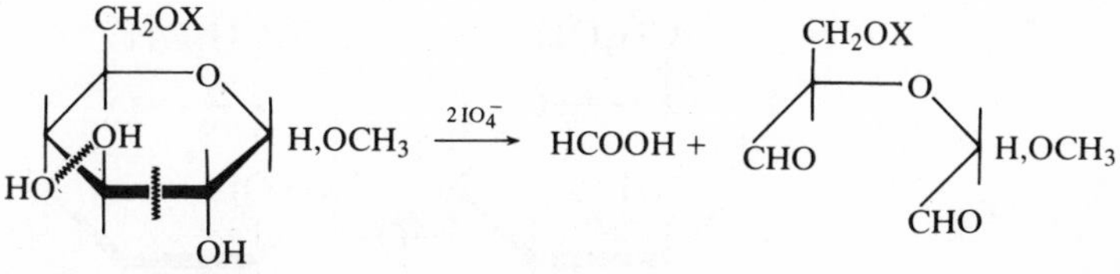

Methyl-6-substituted glucoside

(d)

FIGURE 6.7. Periodate oxidation of glucose derivatives: (a) free glucose; (b) monosubstituted glucose; (c) methyl glucoside; (d) monosubstituted methyl glucoside. ⦙ Signifies a bond broken by periodate oxidation. The dialdehydes (*J*) are very readily oxidized further by periodate, so that in the overall reaction more than 3 moles of periodate are consumed per mole of glucose

acid results. Formic acid is also produced when the anomeric carbon atom is split from an aldose. In contrast, when the carbon carrying a primary alcohol group is oxidized from a sugar, formaldehyde is obtained. The products expected from the oxidation of pyranose forms of free and monosubstituted derivatives of glucose and methyl glucoside, are shown in Figure 6.7. We see that from 2-, 3-, and 4-monosubstituted glucoses and free glucose, formaldehyde is formed. No formaldehyde is produced from substituted methylglucosides, and only the unsubstituted and 6-substituted glucosides yield formic acid.

Identification and estimation of the products of periodate oxidation of polysaccharides gives much useful information on polysaccharide structure. The reaction of carbohydrates with periodate can also be used as the basis of methods for detecting sugars after separation by chromatographic techniques.

6.2.2. The glycosidic bond

As we have seen in the previous section, the anomeric hydroxyl of a monosaccharide can react with an alcohol to form a glycoside. In theory, the 'alcohol' group could be provided by the hydroxyl group of another monosaccharide molecule, e.g.:

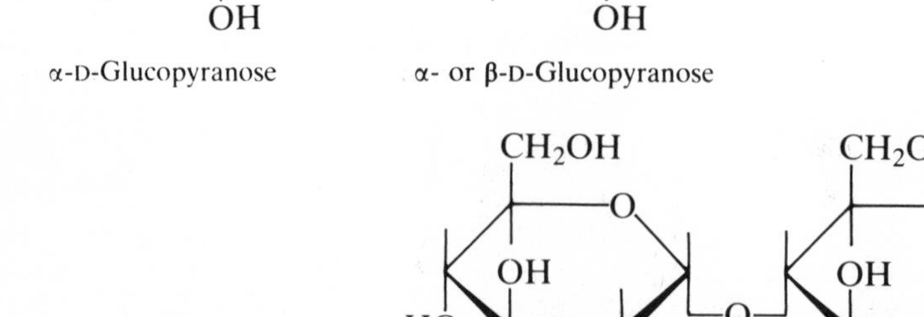

CH_2OH O   OH   HO   OH   OH + CH_2OH O   OH   HO   H,OH   OH ⟶

α-D-Glucopyranose   α- or β-D-Glucopyranose

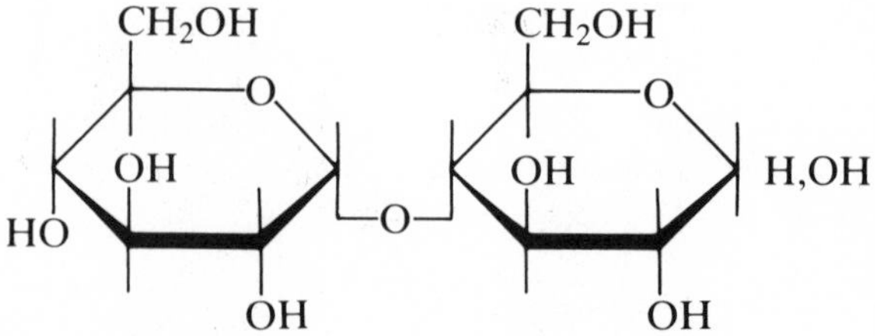

CH_2OH O   OH   HO   OH   —O— CH_2OH O   OH   H,OH   OH

Maltose

In fact, although the product exists, the process by which it is formed, either by biosynthesis or chemical synthesis, is much more complex, and is discussed in greater detail later (see Section 6.6, p. 312). The result is two monosaccharides linked by a *glycosidic bond*, and the product is known as a *disaccharide*. The hypothetical reaction has taken place between an α-hydroxyl on the anomeric carbon of one glucose residue and the hydroxyl of carbon 4 on the second glucose residue, the new bond is therefore called an α-(1 → 4)-*glycosidic bond.* The anomeric carbon of the right-hand glucose residue remains intact, and therefore the new disaccharide has still an anomeric carbon atom and is a reducing sugar.

Many disaccharides have trivial names, e.g. *maltose* for the disaccharide shown above, but also have more complex systematic names. The left-hand sugar (with the anomeric carbon involved in the glycosidic link) is described first—here it is an α-D-glucopyranosyl unit. The linkage is through oxygen (*O*- precedes the left-hand sugar name) and is a (1 → 4)-link to the right-hand sugar, a D-glucopyranose. Thus the systematic name is *O*-α-D-glucopyranosyl-(1 → 4)-*D*-glucopyranose or (*O*-α-D-Glc*p*-(1 → 4)-D-Glc*p*.

If the left-hand sugar had been β-D-glucose, a different product would result.

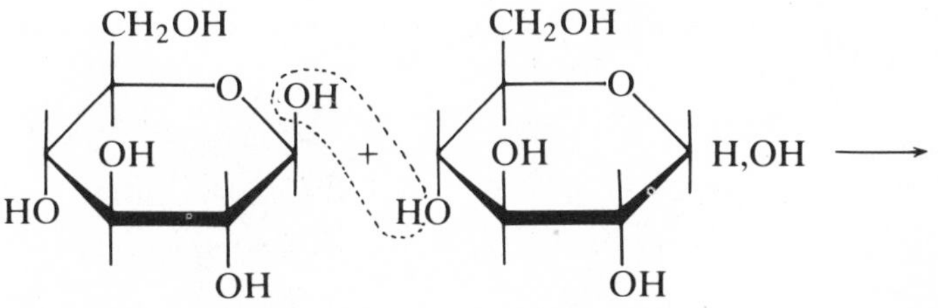

β-D-Glucopyranose + D-Glucopyranose →

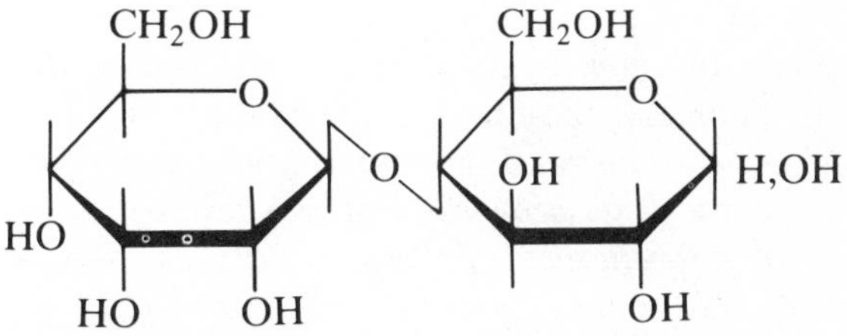

O-β-D-Glucopyranosyl-(1 → 4)-D-glucopyranose
(cellobiose)

Here the two monosaccharide units are joined by a β-(1 → 4)-glycosidic link. The difference between α- and β-links is extremely important in polysaccharides, and exercises a strong influence on the three-dimensional structure and properties of polysaccharides.

Both the disaccharides shown have a potential reducing group, on the anomeric carbon atom. In theory, then, either of these molecules could form a glycoside with a third monosaccharide unit, e.g.:

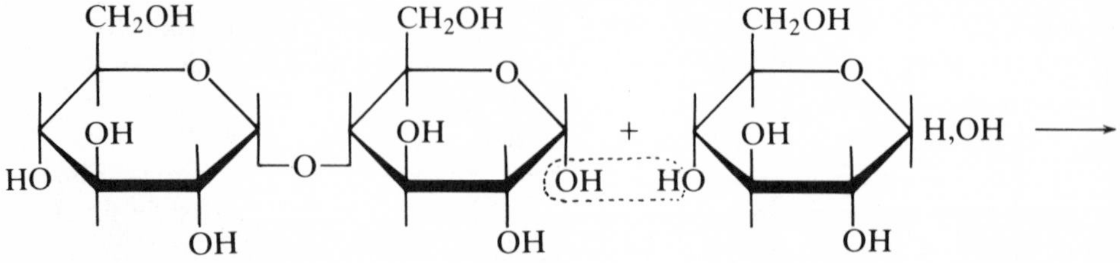

α-Maltose D-Glucopyranose

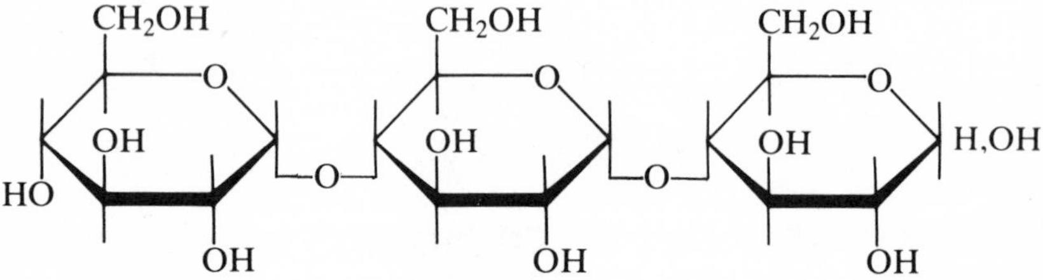

Maltotriose

O-α-D-Glucopyranosyl-(1 → 4)-*O*-(α-D-glucopyranosyl-(1 → 4)-D-glucopyranose

O-α-D-Glc*p*-(1 → 4)-*O*-α-D-Glc*p*-(1 → 4)-D-Glc*p*

Again, in practice, the product exists but is formed by a much more complex mechanism (see Section 6.6, p. 312). The product is a reducing *trisaccharide*. The right-hand residue carries the reducing group and constitutes the *reducing end* of the molecule; the left-hand glucose residue forms the *non-reducing end* of the saccharide (there is no free —OH on the anomeric carbon of this residue).

Longer chains of monosaccharides can exist, e.g. *tetrasaccharides* of four residues, *pentasaccharides* of five residues, etc. In general, these are known as *oligosaccharides. Polysaccharides* are simply large oligosaccharides, i.e. they may be thought of as condensation polymers of monosaccharides. There is no generally accepted definition of the maximum size of an oligosaccharide, or the minimum size of a polysaccharide. However, chains of 15–20 or more monosaccharides linked together would normally be considered as polysaccharides. Oligosaccharides larger than trisaccharides rarely exist in the free state in Nature, but occur, relatively commonly, linked to protein in glycoproteins.

The trisaccharide, *maltotriose*, shown above is a linear oligosaccharide, i.e. each monosaccharide is linked to only one other monosaccharide. However, it is quite possible for *two* monosaccharide residues to be linked to a single residue through non-anomeric hydroxyl groups, e.g.:

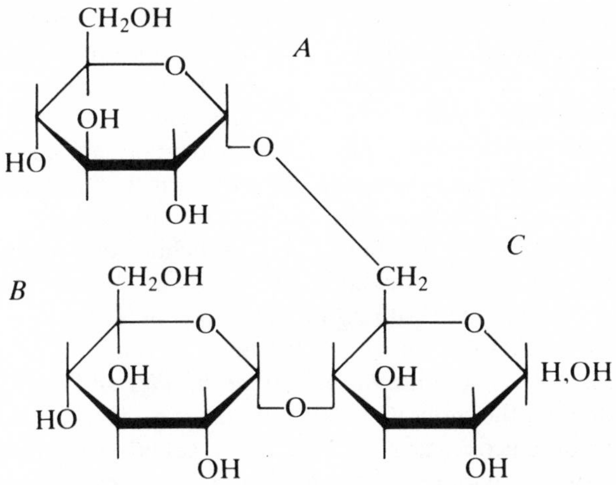

O-α-D-Glucopyranosyl-(1 → 6)-O-[α-D-glucopyranosyl-(1 → 4)]-D-glucopyranose

O-α-D-Glc$p$-(1 → 6)-O-[α-D-Glc$p$-(1 → 4)]-D-Glc$p$

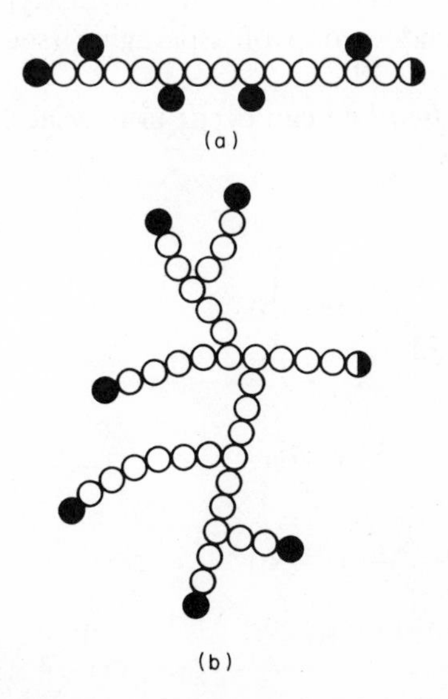

FIGURE 6.8. Possible branched structures for polysaccharides: (a) monosaccharide 'branches', (b) branch-on-branch structure; ◑ reducing end-group, ● non-reducing end-group, ○ other sugar units

The trisaccharide is then both *branched* and also reducing, as residue *C* carries a reducing group. Residues *A* and *B* are both non-reducing, i.e. this oligosaccharide has one reducing and two non-reducing ends.

Both branched and linear polysaccharides are found. Each molecule usually has one reducing group. A linear polysaccharide has one non-reducing end-group per molecule, whilst a branched polysaccharide has several. The more branched the molecule, the greater number of non-reducing end-groups it contains. This branching may consist of a few monosaccharide residues attached to a main linear chain, or a branch-on-branch structure resulting in a tree-like molecule (see Figure 6.8).

Polysaccharides, also known as glycans, may be made up of one type of monosaccharide only. These are *homopolysaccharides* and are given general names depending on the monomer they contain; e.g. starch and cellulose are both polymers of glucose, and are therefore called *glucans*; polymers of mannose are called *mannans*, etc. Polysaccharides consisting of several different monosaccharides are *heteropolysaccharides*.

6.2.3. Other bonds

Many polysaccharides occur in Nature covalently bonded to protein through the reducing end of the saccharide chain. The commonest *protein–polysaccharide links* involve the hydroxyl groups of serine or threonine, or the γ-amide group of asparagine (see Table 4.1 and Figure 6.9, p. 76).

Extensive *hydrogen bonding* can occur in polysaccharides, as they contain

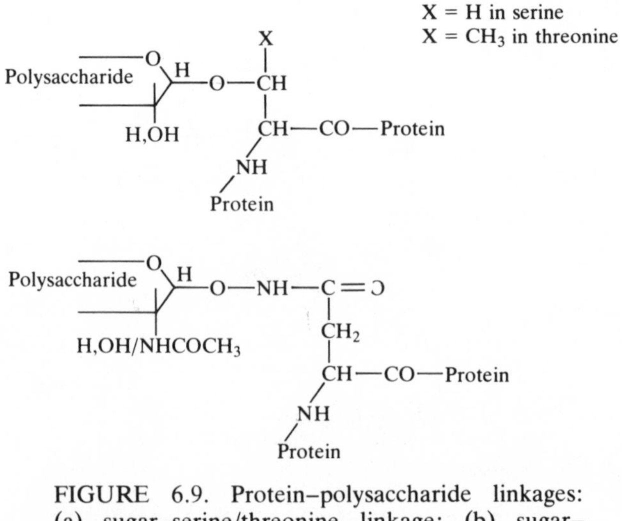

X = H in serine
X = CH_3 in threonine

Polysaccharide —O—H —O—CH(X)—CH(NH—Protein)—CO—Protein
H,OH

Polysaccharide —O—H —O—NH—C(=O)—CH_2—CH(NH—Protein)—CO—Protein
H,OH/$NHCOCH_3$

FIGURE 6.9. Protein–polysaccharide linkages: (a) sugar–serine/threonine linkage; (b) sugar–asparagine linkage—the linkage from the sugar (often glucosamine) is often β

large numbers of hydroxyl groups. Both intra- and inter-chain hydrogen bonding is believed to be important. Although this bonding may be important for maintaining three-dimensional structure in the solid, it does not dictate the structure. The conformation of a polysaccharide chain is probably determined by non-bonded interactions between groups on the sugar rings, particularly axial substituents.

In charged polysaccharides, *coulombic attractions* may be important, e.g. between amino groups and carboxyl or sulphate anions. In many cases, however, *Coulombic repulsions* may predominate if the glycan contains several carboxyl and sulphate groups, or a mixture of both.

6.3 SIZE AND COMPOSITION

6.3.1 Molecular size

The methods described in Chapter 3 can be used to estimate the size of polysaccharides. Molecular weights can vary from several thousands to many millions. Again the lower limit is arbitrary, as the distinction between a large oligosaccharide and a small polysaccharide is imprecise.

For many polysaccharides, drastic extraction procedures must be used and molecular degradation takes place. Thus many measured molecular weights should be regarded as lower limits of the size of the polysaccharide in Nature.

In general, a polysaccharide has a very wide distribution of molecular weight, i.e. it is very polydisperse. This situation is in direct contrast to that found for most proteins and nucleic acids.

For a linear homopolysaccharide, the average chain-length, ($\overline{\text{C.L.}}$), can be found simply from

$$\overline{\text{C.L.}} = \frac{\text{average molecular weight}}{\text{weight of monosaccharide residue}}.$$

When the chain-length can be determined by 'counting' the number of reducing or non-reducing end-groups in solution, then the number-average molecular weight can be calculated directly.

In contrast, the average length of unit-chain in a branched polysaccharide (i.e. the average length between branch-points) must be calculated from the number of branch-points or non-reducing ends in the molecule for usually the number of non-reducing ends equals the number of branch-points. The average concentration of branch-points, or non-reducing ends, in a sample may be determined by methylation and hydrolysis, or in some cases by periodate oxidation (see Section 6.4.1, p. 261). Then average length of unit-chain $\overline{\text{C.L.}}$ is given by:

$$\overline{\text{C.L.}} = \frac{\text{concentration of polymer in solution}}{\text{concentration of branch-points in the same solution}}.$$

6.3.2. Composition

In order to study the monomer composition, the polysaccharide must first be hydrolysed to its component monosaccharides. Sulphuric acid is commonly used for hydrolysis, although hydrochloric acid is frequently employed with glycoproteins. It is often extremely difficult to hydrolyse a polysaccharide completely to its constituent monomers without concomitant destruction of some of the monosaccharides. Thus uronic-acid-containing polymers are somewhat resistant to acid hydrolysis, and better results are obtained if the uronic acid residues are first reduced to neutral sugar residues (see Section 6.2.1, p. 248). Amino sugars released from polysaccharides during hydrolysis are readily degraded by acid, but may be protected by the inclusion of some cation-exchange resin in the hydrolysis mixture—the amino sugars become adsorbed onto the resin, and can later be washed off the resin and estimated. Because of these difficulties, it is frequently necessary to hydrolyse samples of a polysaccharide for varying times with different acid concentrations. Then, if destruction of a monosaccharide occurs, an estimate can be made of the proportion of that monomer in the polymer by extrapolation to 'zero' time of hydrolysis.

The monomers released by hydrolysis are usually separated by chromatographic methods. Nowadays, acetate or silyl derivatives would be prepared (see Section 6.2.1, p. 250), and would then be separated, identified, and estimated using gas chromatography.

Some groupings on monosaccharides, e.g. *N*-acetyl or sulphate groups, are removed from the monomers during acid hydrolysis, and so have to be determined independently.

If some information is already available on the nature of the monomers present in a polysaccharide, then enzymes (see Section 4.5.2, p. 114) of the correct specificity may be used to hydrolyse the polymer.

Methanolysis, i.e. reaction with methanol and hydrogen chloride, may also be used to give monosaccharides. In this case methylated monomers are obtained, which may be separated and identified by gas chromatography.

Unlike proteins where over 20 different monomers may be found, complex polysaccharides usually contain less than six different monomers. However, the situation is complicated by the fact that, within a polysaccharide sample, no two molecules may have an identical composition. This situation contrasts with that for pure proteins where all the molecules of a sample are usually identical.

6.4. STRUCTURE

The determination of the structure of a polysaccharide may be an exceedingly complex process, particularly in the case of some heteropolysaccharides. Firstly, it must be determined whether the polymer is linear or branched. Then the sequence of monosaccharides, the linkage positions (i.e. whether $1 \rightarrow 2$, $1 \rightarrow 3$, etc.), the linkage configurations (i.e. whether α- or β-), and the ring sizes (i.e. whether furanose or pyranose) have to be investigated. Lastly, the three-dimensional conformations of the rings and the polymer chains should be studied.

Fortunately, there are some simplifying 'rules' for polysaccharide structure. In homoglycans, more than two different types of bonds are rarely found. Even in heteroglycans, one type of monomer is often linked in the molecule by the same type of bond each time.

If two types of monosaccharide occur in a polysaccharide, they are often arranged in a regular way. In complex branched heteroglycans, hexoses and uronic acids often occur in main chains whilst pentoses may occur in side-branches; where sialic acid is found in, say, glycoproteins, it usually occurs at the non-reducing ends of chains.

6.4.1. Monosaccharide sequence and linkage

Initially, the linear or branched nature of a polysaccharide has to be studied. A simple qualitative test involves film formation by drying an aqueous polysaccharide solution spread on a glass plate: if a branched polymer is present, the film is, in general, brittle, whilst a strong film is obtained from a linear material.

Methylation studies can give information on branching in polysaccharides. Complete methylation of a polysaccharide, followed by hydrolysis, and separation and estimation of the products, constitutes the classical technique for investigating polysaccharide structure. It is important that the polymer be completely methylated (for reagents see Section 6.2.1, p. 251), and that hydrolysis does not cause monomer degradation or demethylation. The products to be expected from methylation and hydrolysis of two polysaccharides—one a linear homoglycan (cellulose), the other a branched heteroglycan (a wood hemicellulose)—are shown in Figure 6.10. Those —OH groups originally involved in glycosidic linkages or in ring formation do not become methylated, whilst tetra-*O*-methyl sugars are obtained from non-reducing chain-ends, and di-*O*-methyl sugars from branch-points. Thus no di-*O*-methyl monosaccharide is obtained from cellulose, indicating that the polymer molecules are linear. Conversely the presence of di-*O*-methylmannose in the products from the hemicellulose indicates that the polysaccharide is branched.

For cellulose an estimate of average chain-length (see Section 6.3.1) can be made from the relative proportions of 2,3,4,6-tetra-*O*-methylglucose (from a non-reducing chain-end) and 2,3,6-tri-*O*-methylglucose (from a chain interior and reducing end). The average length of unit-chain can be estimated for the branched polymer from the proportion of tetra-*O*-methyl or di-*O*-methyl monosaccharide relative to the amount of tri-*O*-methyl monosaccharide.

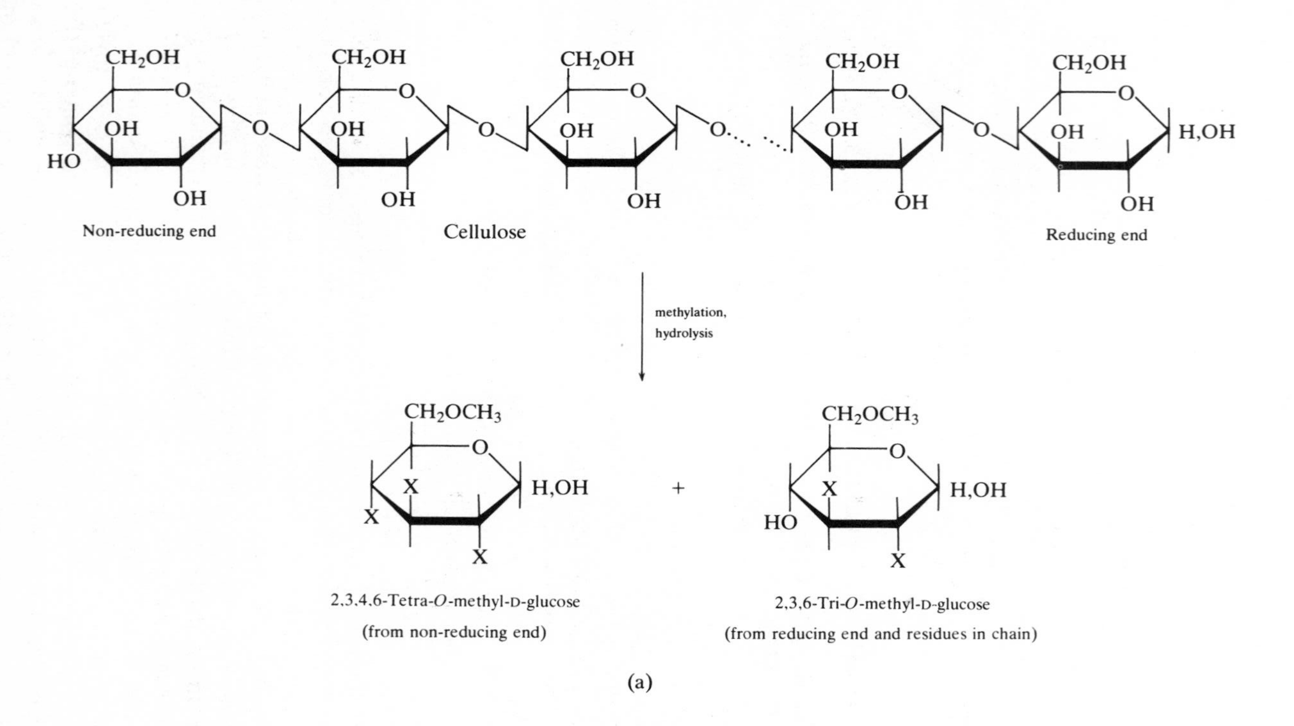

CH$_2$OH
O
OH
HO
H,OH
Non-reducing end
Cellulose
Reducing end
methylation, hydrolysis
CH$_2$OCH$_3$
X
+
2,3,4,6-Tetra-O-methyl-D-glucose
(from non-reducing end)
2,3,6-Tri-O-methyl-D-glucose
(from reducing end and residues in chain)

(a)

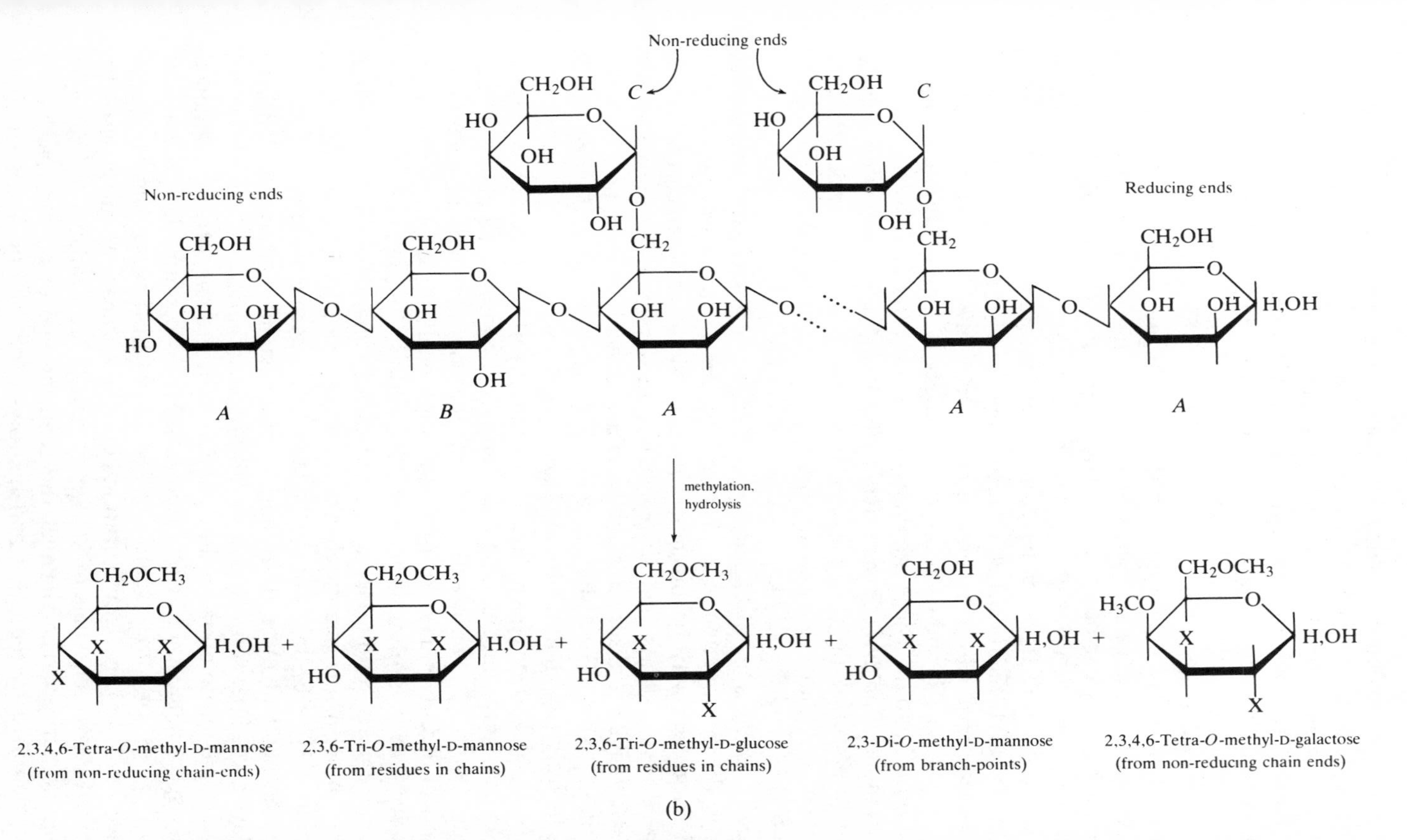

FIGURE 6.10. Products from the classical technique of methylation and hydrolysis of a polysaccharide: (a) products from cellulose (a β-(1 → 4)-glucan); (b) products from wood hemicellulose (a D-galacto-D-gluco-D-mannan); A = mannose residue; B = glucose residue, C = galactose residue, X = $—OCH_3$

A study of methylated products also yields information on ring size. For example, 2,3,4,6-tetra-*O*-methyl-D-glucose could only come from a glucose unit in the pyranose ring form at the non-reducing end of a chain (cf. Figure 6.11). However, 2,3,6-tri-*O*-methylglucose could result from glucofuranose linked (1 → 5) within a chain, or glucopyranose linked (1 → 4) within a chain. Techniques other than exhaustive methylation must be used to resolve this problem.

Methylation of the polysaccharide shown in Figure 6.10(b) indicates that mannopyranose and galactopyranose occur at non-reducing chain-ends—only mannose and galactose in the pyranose form could yield 2,3,4,6-tetra-*O*-methyl-mannose or -galactose. However, 2,3,6-tri-*O*-methyl-mannose or -glucose could result from mannofuranose or glucofuranose linked (1 → 5) within a chain (cf. Figure 6.11), or mannopyranose or glucopyranose linked by (1 → 4)-bonds. Also, the branch-points may involve a mannopyranose linked at positions 4- and 6- to other monomers, or a mannofuranose linked at positions 5- and 6-. These problems can be answered only by resorting to other methods of analysis. Exhaustive methylation, and complete hydrolysis, although a powerful tool, gives no information on monosaccharide sequence or linkage configuration (compare products from the β-linked polymer of Figure 6.10(a) and the α-linked polymer of Figure 6.11).

Partial hydrolysis to disaccharides and larger oligosaccharides can yield some of the information not provided by exhaustive methylation. This partial hydrolysis may be brought about by dilute acid—glycosidic bonds linking furanose sugars and deoxy sugars in a polysaccharide chain are particularly labile, whilst bonds involving uronic acids or 2-amino sugars are more stable. Also (1 → 4)-links are often more labile than (1 → 6)-bonds during hydrolysis with dilute mineral acid.

Enzymes of known specificity, which can hydrolyse internal glycosidic bonds in polysaccharides, can be used for partial hydrolysis, and at the same time the results give information on linkage configuration—in general an enzyme will hydrolyse an α- or β-bond of a particular type, but not both.

Acetolysis, i.e. hydrolysis using acetic anhydride with sulphuric acid, gives, under the correct conditions, acetylated oligosaccharides. These are often complementary to the oligosaccharides obtained by mineral acid hydrolysis; for example, (1 → 6)-links are particularly labile during acetolysis.

Disaccharide and even trisaccharide products may often be identified nowadays by comparison with known di- or tri-saccharides. Where this is not possible, some information on disaccharide structure may often be obtained by mass spectrometry, i.e. it is often possible to confirm the presence of pyranose rings or, for example, (1 → 6)-linked hexose sugars. Sugar sequence and linkage configuration in oligosaccharides may also be determined if *pure* enzymes of known specificity are available to hydrolyse the oligosaccharides. The sugar at the reducing end of an oligosaccharide may be identified by reduction and hydrolysis followed by examination of the sugar alcohol released. Non-reducing end-groups may be identified by

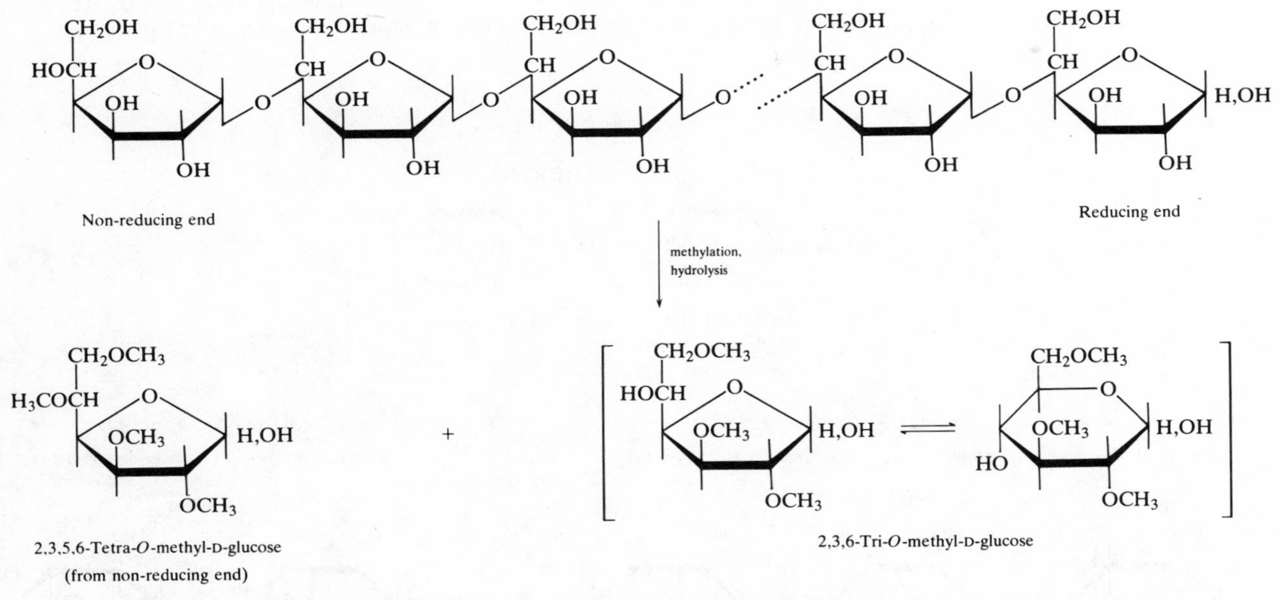

FIGURE 6.11. Products from the methylation and hydrolysis of a hypothetical polysaccharide consisting of monomers linked in the furanose ring form i.e. α-(1 → 5)-D-glucan

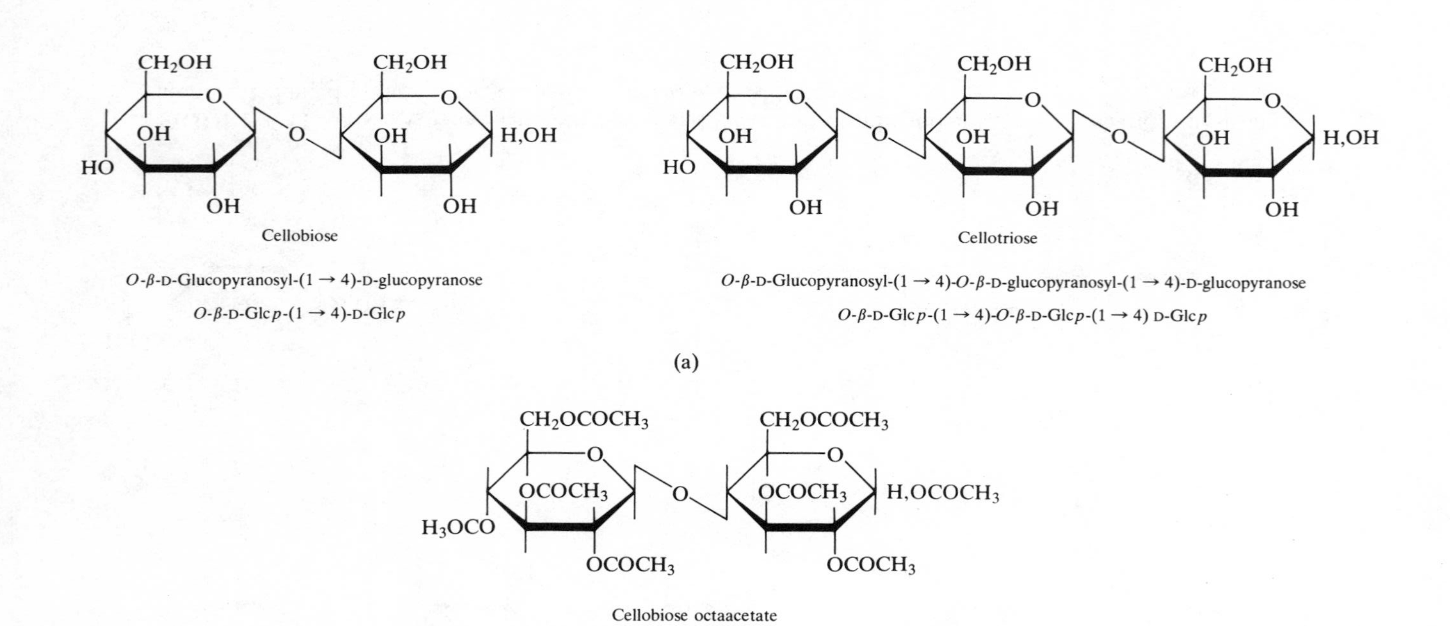

FIGURE 6.12. Cellulose degradation products: (a) oligosaccharides obtained by partial acid hydrolysis; (b) disaccharide obtained by acetolysis

methylation, as tetramethyl derivatives are obtained from hexoses at non-reducing chain-ends.

By a combination of these techniques, the sequence of sugar residues, their linkage position and configuration, and their ring sizes may be determined. For cellulose, the only di- and tri-saccharides obtained by partial acid hydrolysis are shown in Figure 6.12(a). Both these oligosaccharides are now well known, and their presence in a hydrolysate indicates that cellulose molecules consist of chains of β-(1 → 4)-linked glucopyranose units. This is further confirmed by the fact that cellulose can be hydrolysed by enzymes specific for the hydrolysis of β-(1 → 4)-linkages between glucose units. Also the only disaccharide obtained by acetolysis is cellobiose octaacetate (see Figure 6.12(b)), again confirming the structure assigned to cellulose.

For a polysaccharide such as the wood hemicellulose shown in Figure 6.10(b), the situation is more complex, but many details of the structure may be obtained. Methylation studies (see above) would show that mannopyranose and galactopyranose occur at non-reducing chain-ends, whilst glucose and mannose are found within the chains. Partial acid hydrolysis would result in the di- and tri-saccharides shown in Figure 6.13, and these would indicate that all the sugars are in the pyranose form; that galactose is always linked α-(1 → 6) to a mannose residue; and that the main chain consists of β-(1 → 4)-linked glucose and mannose units. Examination of larger oligosaccharides would indicate whether the distribution of glucose and mannose in the main chain, or of galactose-carrying mannose units, is random, or has a regular repeating pattern.

Use of enzymes specific for hydrolysing β-(1 → 4)-bonds between mannose residues or glucose and mannose residues, and others specific for α-bonds from galactose would confirm the bond configurations in the polymer.

In addition, reduction and hydrolysis of the polysaccharide shown in Figure 6.10(b) would give mannitol, the alcohol from mannose, indicating that there is a mannose unit at the reducing end of the polymer molecule.

It should be emphasized that in a sample of a polysaccharide such as the wood hemicellulose, the molecules are not likely to be all identical. Thus molecules may differ in the distribution and amount of glucose in the main chain, and in the distribution and number of galactose residues. When the structure of such a polymer is deduced, it should therefore be regarded as an 'average' structure of the molecules in a sample.

Other techniques for examining polysaccharide structure are available. For example, *immunochemical techniques* are often used in the case of bacterial polysaccharides. Antibodies (see Section 4.5.4, p. 138) can be prepared which react with a polysaccharide of known structure. If these same antibodies interact with a polysaccharide of unknown structure, it can immediately be deduced that the unknown structure must be similar to the known one. Other proteins, lectins, exist which complex strongly with

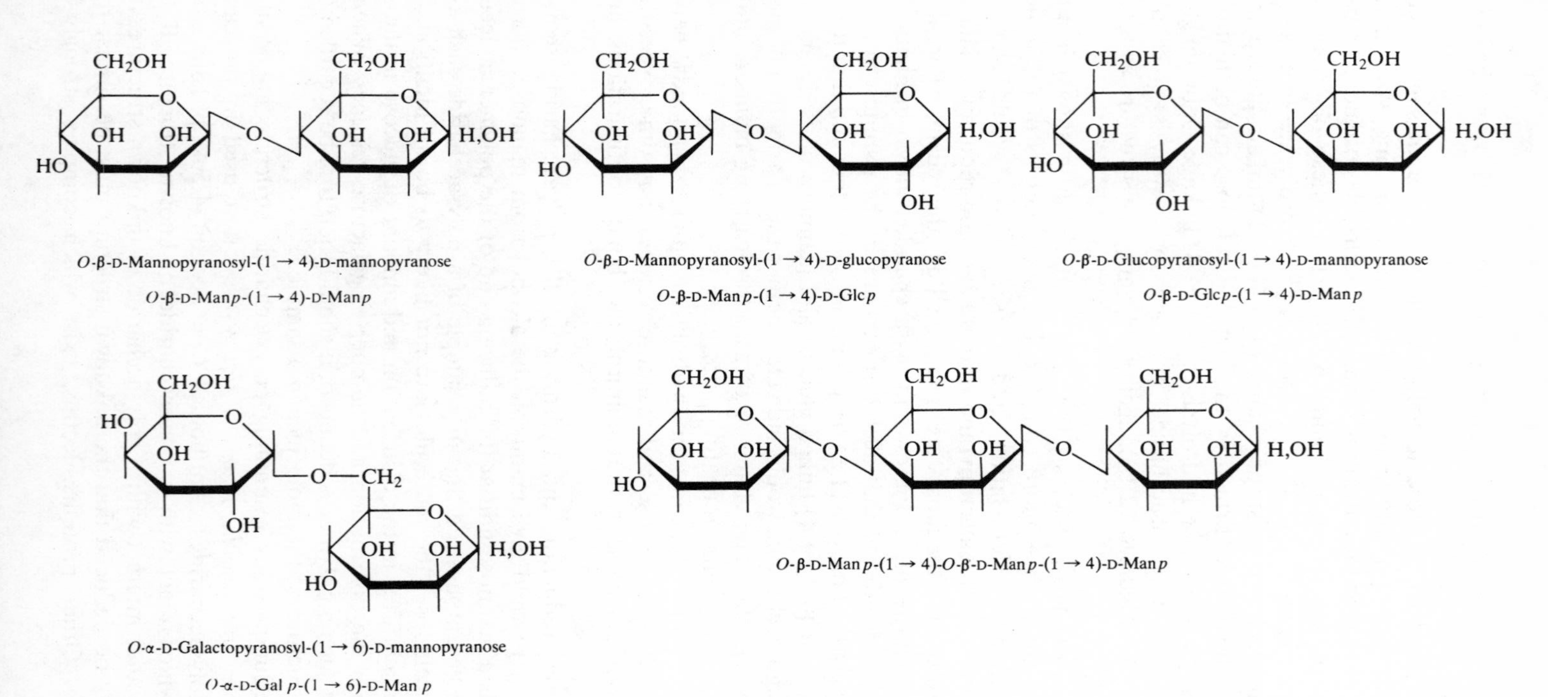

O-β-D-Mannopyranosyl-(1 → 4)-D-mannopyranose
O-β-D-Manp-(1 → 4)-D-Manp

O-β-D-Mannopyranosyl-(1 → 4)-D-glucopyranose
O-β-D-Manp-(1 → 4)-D-Glcp

O-β-D-Glucopyranosyl-(1 → 4)-D-mannopyranose
O-β-D-Glcp-(1 → 4)-D-Manp

O-α-D-Galactopyranosyl-(1 → 6)-D-mannopyranose
O-α-D-Gal p-(1 → 6)-D-Man p

O-β-D-Manp-(1 → 4)-O-β-D-Manp-(1 → 4)-D-Manp

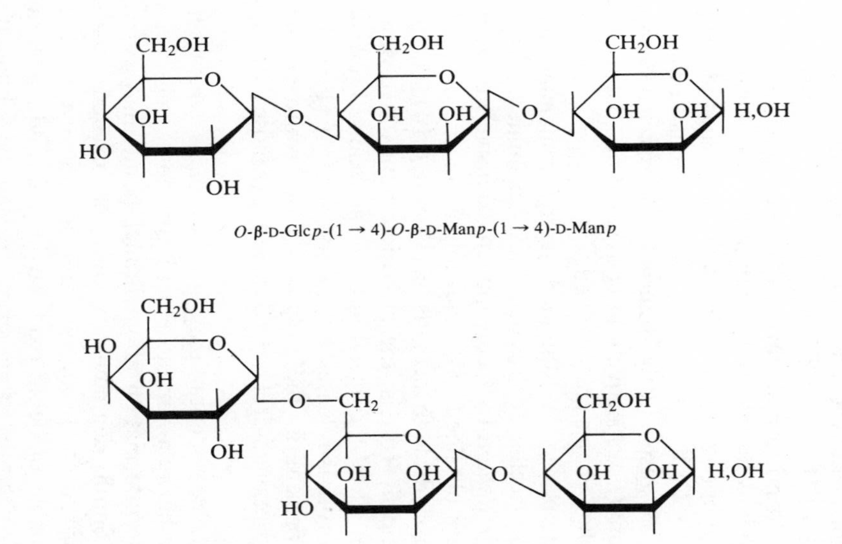

O-β-D-Glc*p*-(1 → 4)-*O*-β-D-Man*p*-(1 → 4)-D-Man*p*

O-α-D-Gal*p*-(1 → 6)-*O*-β-D-Man*p*-(1 → 4)-D-Man*p*

Other trisaccharides could also be found e.g.:

O-β-D-Man*p*-(1 → 4)-*O*-β-D-Glc*p*-(1 → 4)-D-Man*p*

and

O-β-D-Man *p*-(1 → 4)-*O*-β-D-Man *p*-(1 → 4)-D-Glc *p*

FIGURE 6.13. Di- and tri-saccharides obtained by partial acid hydrolysis of a D-galacto-D-gluco-D-mannan

specific saccharide structures, particularly at the non-reducing ends of polysaccharide chains. Thus if a polysaccharide forms a complex with a lectin of known specificity, deductions can immediately be made concerning the structure at the non-reducing chain-ends.

Another powerful tool in polysaccharide chemistry is *periodate oxidation*. Determination of the amount of formic acid and formaldehyde produced enables an estimate of the average length of unit-chain to be made. In some cases information about linkage positions can be obtained. For example, a chain of pyranose sugars, all linked either $1 \rightarrow 2$, $1 \rightarrow 3$, or $1 \rightarrow 4$, would give one molecule of formaldehyde and three molecules of formic acid, whilst a polysaccharide of $1 \rightarrow 6$-linked pyranose sugars gives a large amount of formic acid, but no formaldehyde. (To determine the amount of formic acid and formaldehyde released from a polysaccharide consider Figure 6.7, p. 252. The non-reducing end of any chain reacts like sugar *E*. The reducing ends of $1 \rightarrow 2$-, $1 \rightarrow 3$-, $1 \rightarrow 4$-, or $1 \rightarrow 6$-linked chains react like sugars *A*, *B*, *C*, or *D*, respectively. Internal monosaccharide residues react like *F*, *G*, *H*, or *I*, respectively.) The situation becomes more complicated when branched chains are involved.

Much more information can be obtained if the remaining oxidized products from the polysaccharide residues are examined. One such procedure, the *Smith degradation*, has been applied extensively for structural analyses. The dialdehydes obtained from the polysaccharide oxidation are reduced with sodium borohydride to alcohols (see Figure 6.14). The product is susceptible to mild acid hydrolysis; the glycosidic bonds involving intact sugar residues are stable, whilst the acetal links in the oxidized residues are labile. Stronger acid hydrolysis releases a 4-carbon sugar if $(1 \rightarrow 4)$-linked pyranose residues are present in the polysaccharide, whilst a 3-carbon compound, glycerol, is obtained if $(1 \rightarrow 6)$-linked pyranose residues are present.

In some cases, Smith degradation may give products which allow a distinction to be made between furanose and pyranose rings. In other polysaccharides, valuable information about branch-points may be gained. Sometimes successive Smith degradations may be carried out on a polysaccharide, i.e. unoxidized products from a first cycle of Smith degradation may be capable of being oxidized in a second cycle, and so further information on polysaccharide structure can be obtained. The products of Smith degradation are now frequently converted to silyl derivatives (see Section 6.2.1, p. 251) before being examined by gas chromatography.

Thus a variety of methods for structural analysis of polysaccharides exist, and they are often combined when a complex heteropolysaccharide is being studied.

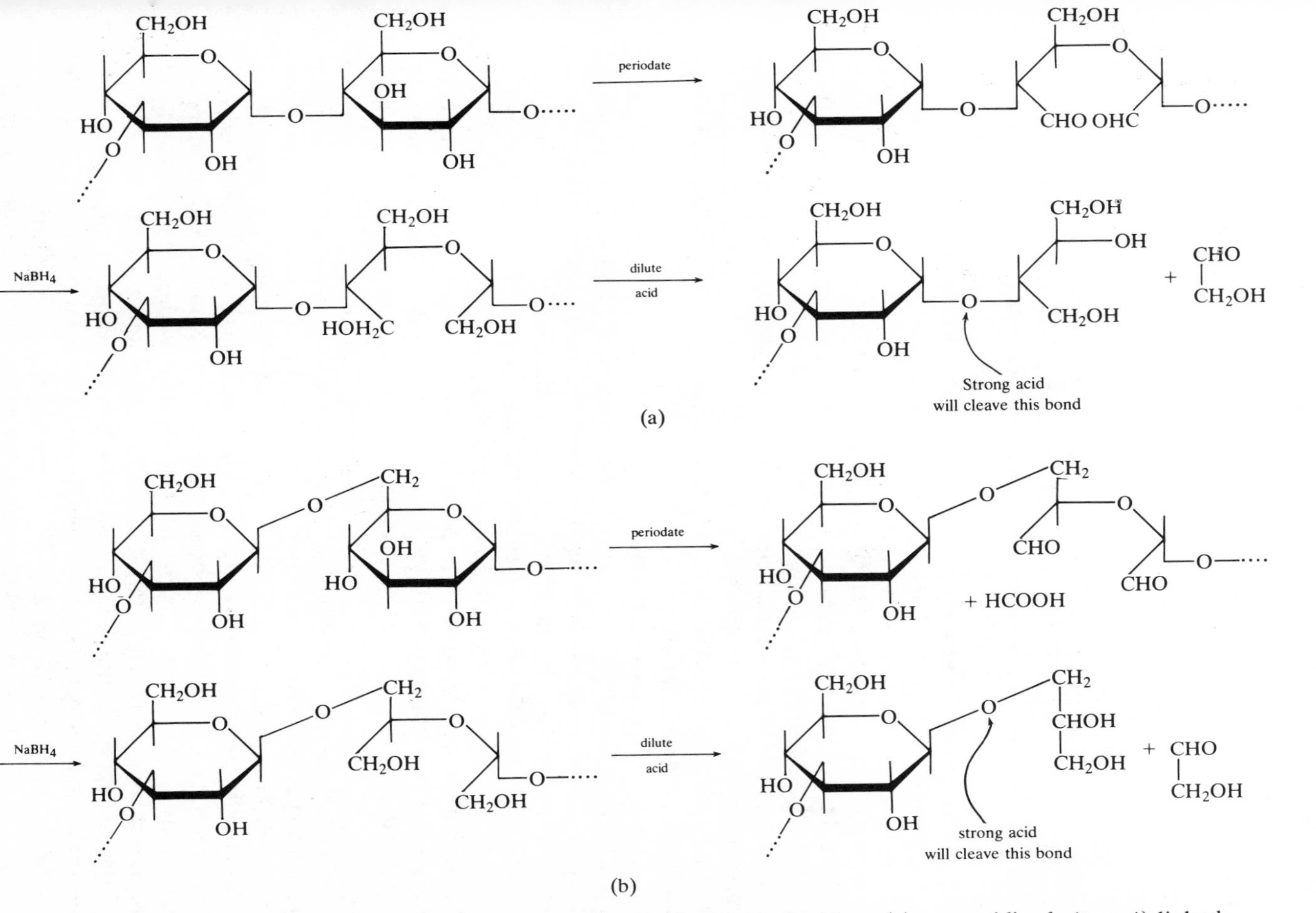

FIGURE 6.14. Smith degradation of periodate-oxidized polysaccharides: (a) an oxidized (1 → 4)-linked residue; (b) an oxidized (1 → 6)-linked residue

6.4.2. Three-dimensional structure

Here there are two principal problems to be investigated: the conformations of the rings of the monosaccharide residues, i.e. whether boat, chair, or other forms, and the relative orientations of adjacent rings, which dictates the folding of the polysaccharide chain.

Many mono- and oligo-saccharides have been crystallized and their three-dimensional structure investigated by crystallography. It has been shown that most crystalline D-hexoses occur in the chair conformation, C1, see Figure 6.4, p. 245, although the chair is somewhat distorted. From these studies, it was inferred that D-hexose residues in polysaccharides would generally exist in the C1 chair form, i.e. it was assumed that the conformations of monosaccharide residues in polysaccharides would be the same as those found in crystalline oligosaccharides. Comparative nuclear magnetic resonance studies carried out on methylated or acetylated oligosaccharides and polysaccharides in solution have indicated that this is indeed the case—chair conformations occurring commonly in oligosaccharide crystals also occur in polysaccharide chains in solution.

Investigation of polymer chain-folding is much more difficult. In solution, many polysaccharide molecules are quite flexible, and so constant changes in folding take place (see Sections 3.5 and 3.8, pp. 60 and 69). At best, only an 'average' molecular shape can be obtained.

Information on possible chain-folding may be gained more easily, at least in theory, from studies of solids where molecular motion is minimized. Electron microscopy has been used to study the size of some polysaccharide molecules, and has also indicated that these macromolecules are relatively inflexible. However, for fine structural details, X-ray crystallography is the method of choice. Unfortunately, polysaccharides do not crystallize well, because most samples are polydisperse, i.e. contain molecules of a variety of sizes. It is thus difficult for a crystal, a regular array of like molecules or atoms, to form when such a disparity exists amongst the polysaccharide chains.

In fact, single crystals of polysaccharides large enough for a detailed X-ray crystallographic study are rarely obtained. Often, partially crystalline material has been studied, i.e. solids in which there are regions of crystallinity and amorphous regions. Many polysaccharides can be drawn into fibres, and these too have been investigated by X-ray crystallography. In many cases, the number of parameters needed to describe a segment of polysaccharide chain in three dimensions is greater than the number of diffraction spots obtained, i.e. all structural details cannot be deduced from the experimental results. It is then necessary to postulate a reasonable model—usually based on known structures of oligosaccharides—and to refine this by comparison of the experimental diffraction pattern with the pattern predicted theoretically for the model.

For several polysaccharides, the size of the crystalline 'unit cell' (see

Section 3.6, p. 65) has been determined, but few other structural details are available. During studies of polysaccharide fibres, it is usually assumed that the long axis of the polymer molecule is oriented in the direction of elongation of the fibre. Distances characteristic of the repeat units of the polysaccharide chain can then often be obtained from diffraction patterns. These distances suggest that for some polysaccharides, e.g. chondroitin sulphate, the repeating unit is a disaccharide, for others, e.g. heparin, it may be a tetrasaccharide, whilst for yet others, e.g. many amylose complexes, the unit may be a section of helix.

It is now believed that, in the solid state, the chains of many polysaccharides are helical, these helices being in some cases very complex. There is evidence that the helices may be stabilized by intrachain hydrogen bonds between a monosaccharide residue and the residue directly above it on the next turn of the helix. There may also be interchain hydrogen bonding, although information on this is more difficult to obtain. For some polysaccharides, e.g. complexes of amyloses, the helices are believed to be folded back on themselves—here crystalline lamellae are formed, and the helices are thought to be oriented perpendicular to the lamellae surfaces. The helices are much longer than the thickness of the lamellae and must therefore be highly folded.

6.5. POLYSACCHARIDE STRUCTURE AND FUNCTION

Polysaccharides may be classified according to structure or function. In some cases, however, the function of a polysaccharide in Nature is not yet fully understood, and so any classification based on this property becomes difficult. It is, therefore, convenient to discuss these polymers in *groups determined by the monosaccharide composition of the main macromolecular chain.*

In general terms, polysaccharides may be considered in two large groups: *homopolysaccharides*, where the macromolecular chains are formed from one monosaccharide only, and the *heteropolysaccharides*, in which each polymer molecule is made up of two or more different monomers. Such a large variety of polysaccharides exist, particularly in plants and microorganisms, that only some of the more important examples are given here.

6.5.1. Homopolysaccharides

6.5.1.1. Glucans

These are polymers of D-glucose, and two members of this group, namely *starch* and *cellulose*, are amongst the most important of all polysaccharides from the point of view both of their abundance in Nature, and the large industries now involved in their utilization.

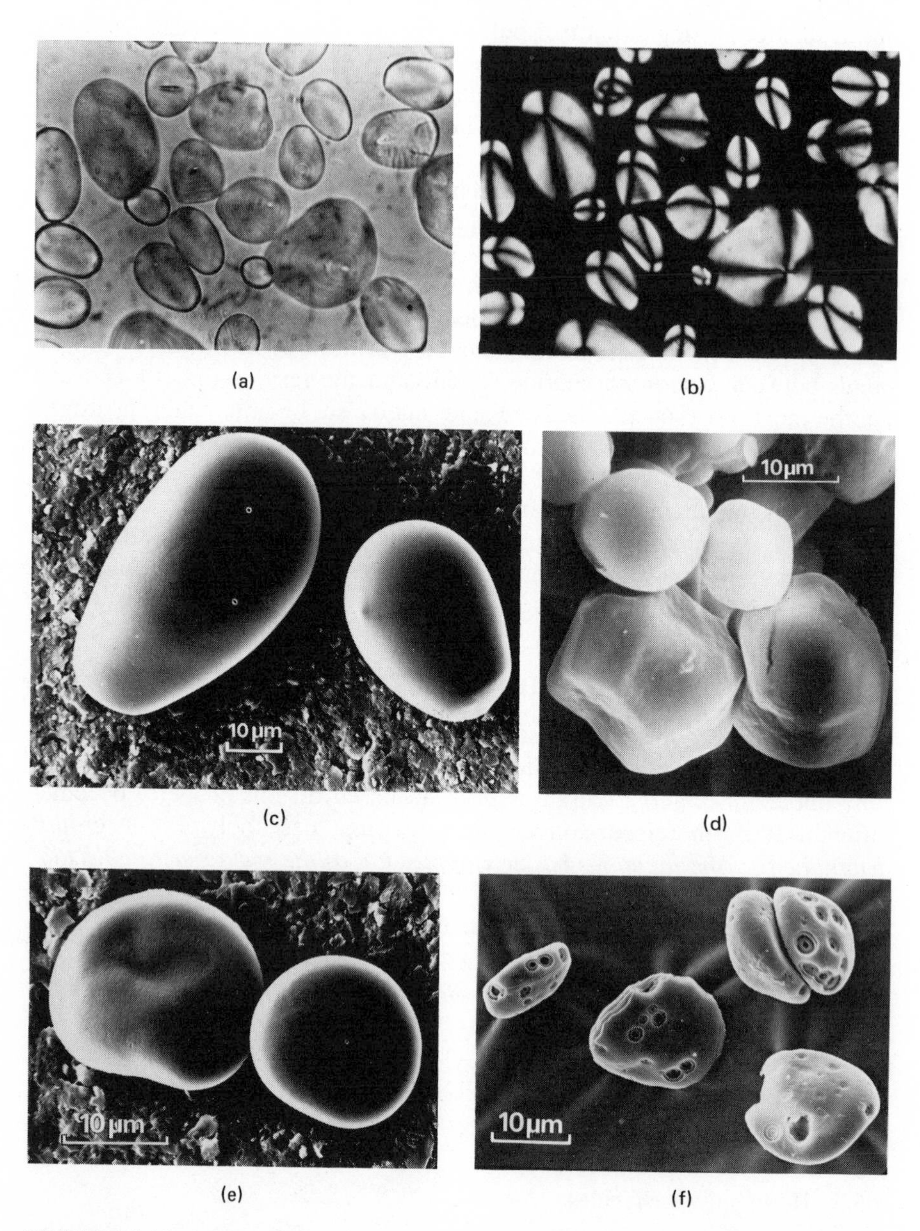

FIGURE 6.15. Micrographs of potato starch granules under (a) ordinary light, and (b) polarized light. Electron micrographs of (c) potato starch, (d) maize starch, (e) barley starch, and (f) barley starch after α-amylolysis. [(c), (e), and (f) reproduced by courtesy of Dr A. MacGregor, Canadian Grain Commission Laboratory, Winnipeg, Canada]

(*a*) *Starch and glycogen*. Starch, and the closely related glycogen, are the commonest food-reserve polysaccharides—the former in plants, the latter in animals, although some fungi also contain a glycogen-like polymer.

Starch is found in some algae and in all parts of higher plants—leaves, stems, and shoots, but mainly in the storage organs such as rhizomes, tubers, bulbs, and seeds. The amount in any one source varies from a few per cent to the 60% found in cereal seeds. Starch is, of course, an extremely important source of energy in the human diet, and indeed more than three-quarters of all food crops are cereals and starchy root crops.

Starch is characteristically laid down in the form of insoluble granules, the size, shape, and properties of which vary with the botanical source (see Figure 6.15). The granules are birefringent in polarized light, and give a diffraction pattern with X-rays. These phenomena indicate that there is a high degree of orientation and 'ordering' of the molecular structures within the granules. But in spite of very extensive investigations, much is yet unknown of how the starch granules are laid down, how the molecules are packed within the granule, or even how the starch granules are degraded when glucose is required by the plant.

Starch is considered to consist of at least two components, *amylose* and *amylopectin*. Amylose molecules are essentially linear chains of α-(1 $\rightarrow$ 4)-linked D-glucopyranose residues, whilst amylopectin has a highly branched structure consisting of chains of glucose units linked by α-(1 $\rightarrow$ 4)-bonds, these chains being cross-linked through α-(1 $\rightarrow$ 6)-bonds (see Figure 6.16). The proportion of amylose in starches can vary considerably according to the botanical source. For some starches, the so-called waxy starches, there is less than 2% amylose, whilst in others the proportion may reach 70%. Even for one source, e.g. potato starch, the amylose : amylopectin ratio differs for large and small granules, and *varies with the maturity of the plant source*.

The starch granules are insoluble in cold water, but swell in warm water—at first reversibly—until at a certain characteristic temperature, the gelatinization temperature, the swelling becomes irreversible. At this point, the granules lose their birefringence, they burst, and some starch material is leached into solution. As the temperature rises further, a starch dispersion is obtained at 100° C. (Oxygen has to be excluded from such a dispersion to avoid oxidative degradation.) Solutions of starch can be prepared also in dimethyl suphoxide, and in this solvent, disruption of the granular structure is complete.

Amylose forms insoluble complexes with substances such as thymol or *n*-butanol, and so this component can be separated from amylopectin by precipitation as a complex from a starch dispersion. This fractionation procedure enables the two main components of starches to be studied independently. Both amylose and amylopectin are readily degraded, i.e. depolymerized, by dilute acid or alkali.

Within a sample of total amylose from any one starch there exists

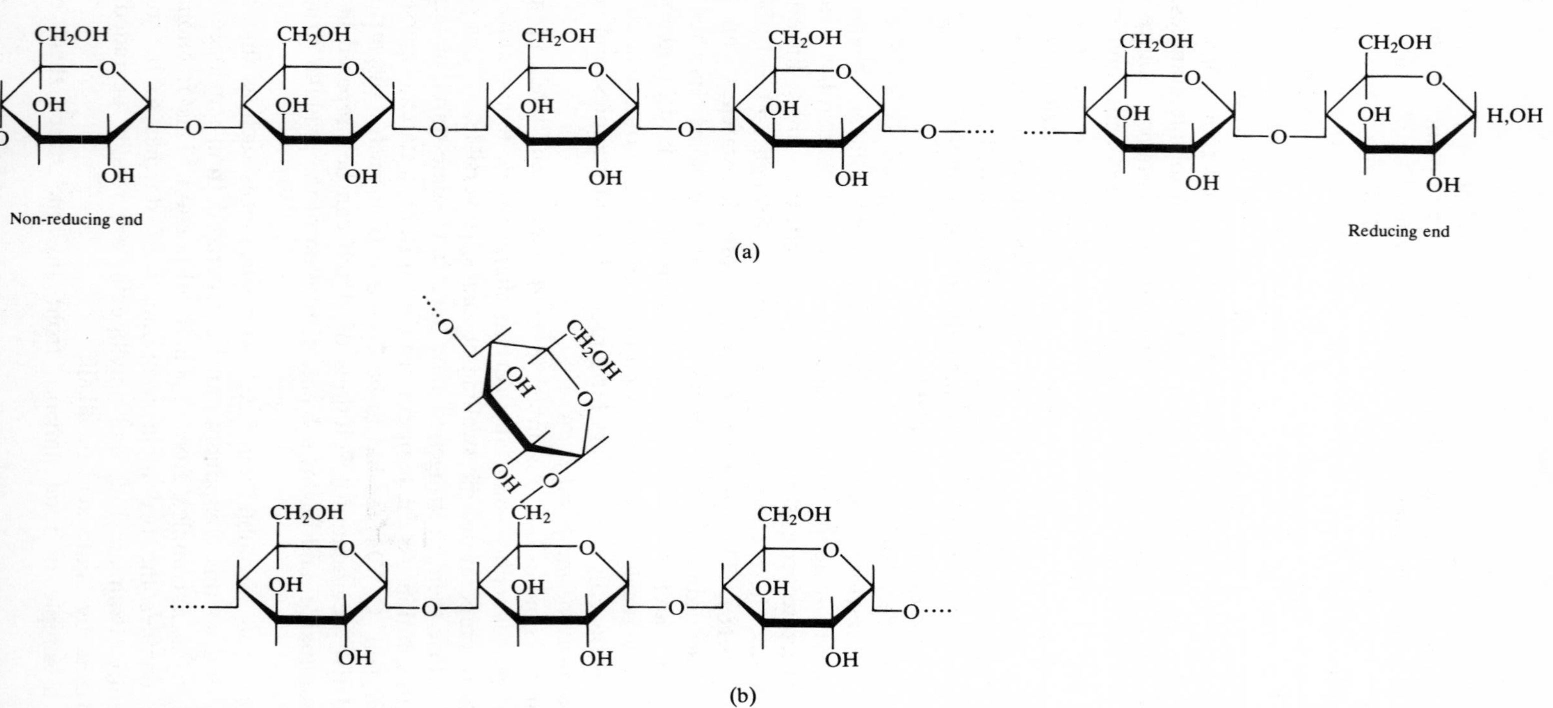

FIGURE 6.16. Polysaccharide chains of amylose and amylopectin: (a) linear amylose; (b) amylopectin—details of branch-points

heterogeneity with respect to both size and structure of the molecules: like all polysaccharides, amylose samples are polydisperse. Weight-average molecular weights up to 10^6 are obtained, but in some starches short-chain amyloses are found which form insoluble complexes with greater difficulty, and have an average molecular weight of *ca.* 10^4. Differences in structure also arise because some large amylose molecules contain a small proportion of α-(1 $\rightarrow$ 6)-linkages, i.e. branching, similar to that in amylopectin (see Fig. 6.16), occurs to a *very limited* extent.

Amylose is water-soluble, and in solution adopts a random-coil conformation. However, aqueous solutions at high concentrations, particularly involving shorter-chain amyloses, are unstable; the polysaccharide molecules associate and precipitate in an insoluble form. This process is known as *retrogradation*, but exact details of the molecular interactions or conformations involved are not yet well understood. In both amylose and amylopectin, the glucose residues probably exist in the C1 conformation (see Figure 6.4, p. 245).

An important characteristic of amylose is its ability to form a blue-coloured complex with iodine in the presence of a little iodide. Indeed, this interaction is the basis of the well-known starch–iodine reaction, which can be used to estimate the percentage of amylose in a starch sample. Insoluble amylose–iodine complexes give an X-ray diffraction pattern, which can be interpreted to indicate that the amylose adopts a helical conformation, with six glucose residues per turn of the helix of pitch 8 Å. The iodine atoms are thought to lie along the hollow core of the helix parallel to the helical axis, and intrachain hydrogen bonds involving hydroxyl groups of the glucose residues stabilize the helical structure. Other insoluble amylose complexes, e.g. the amylose–*n*-butanol complex, probably involve a similar conformation of the polymer chains.

Amylopectin also is water-soluble, but the molecules are on average much larger than those of amylose. Amylopectin samples are very polydisperse, and weight-average molecular weights are often in the range 10^6–10^7. Much structural heterogeneity exists because the length and position of the branched chains in the molecule are probably (at least partially) randomly determined. Molecules are believed to have a branch-on-branch structure (cf. Figures 6.8(b) and 6.17), but not all chains may extend to the surface of the molecule. In potato amylopectin, for example, there is one non-reducing chain-end per 20–25 glucose residues in the molecules, but in amylopectins from other sources the average length of unit-chain (number of glucose residues/non-reducing chain-end) may be greater or smaller. This average chain-length of 20–25 for the potato polysaccharide suggest that on average there will be 20–25 glucose units between branch-points, but in actual molecules the distance between branch-points can vary from 4 to 60 glucose residues. Thus it can be seen that in a preparation of amylopectin it is improbable that any two molecules are identical. This situation is very different from that found for most proteins and nucleic acids.

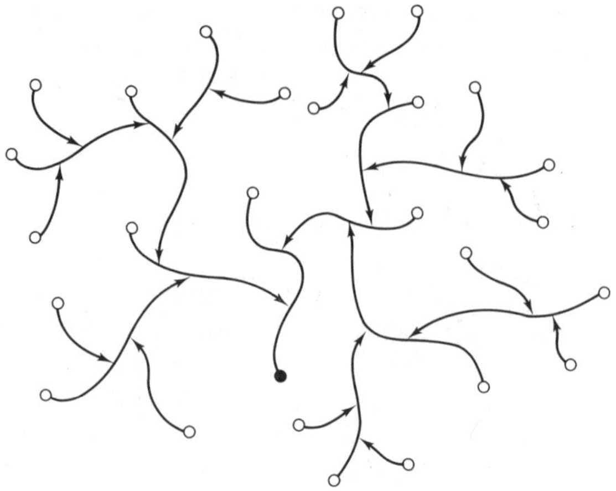

FIGURE 6.17. Structure of amylopectin: ● = reducing chain-end; ○ = non-reducing chain-end

In solution, the amylopectin molecules appear to be essentially extended, but do not retrograde easily, as do those of amylose. However, gels can form in very concentrated solutions, and it is this property which makes starch useful as a thickening agent in the food industry. Amylopectin also reacts with an iodine solution, but much more weakly than does amylose, to give a reddish-purple coloured complex.

Glycogen is generally cited as the food-reserve polysaccharide of animals, but it occurs also in many fungi and some yeasts. Even certain plants, such as sweet corn, synthesize a polysaccharide with many of the structural characteristics of glycogen.

In animals, glycogen is found principally in the liver and muscle, although it occurs in many other organs. It is laid down in a particulate form, and may be extracted essentially undamaged in dimethyl sulphoxide or water. (Earlier extraction methods using alkali resulted in extensive degradation of the polymer.)

Glycogen molecules have a multiply-branched structure rather similar to that of amylopectin (Figure 6.17), but larger numbers of chains in a molecule have branches attached to them. Again, as with amylopectin, not all chains extend to the surface. The average distance between branch-points is less, and average chain-lengths of 10–14 glucose units are common. Structural heterogeneity occurs in a glycogen sample, because of differences in length and position of attachment of branch-chains within the molecules.

Extracted glycogen is usually very polydisperse with a weight-average molecular weight of 10^8 or more. Electron microscopy of liver glycogen reveals simple particles, so-called *β-particles*, and larger, complex *α-particles*—some of which may attain a molecular weight of 10^9. Current evidence suggests that glycogen particles contain a small amount of protein, to which the polysaccharide chains are covalently attached. β-Particles may consist of single glycogen molecules of molecular weight 1–2 × 10^7; a small number of β-particles may be attached to a protein unit which is itself capable of forming disulphide bonds (see Section 4.2.3, p. 83) with other similar protein units. Thus large α-particles are simply associations of β-particles on protein, the disulphide bonds of the protein providing the 'glue' to hold the large complexes together. Certainly α-particles can be dissociated by reagents which rupture disulphide bonds but leave polysaccharides unaffected. Details of the protein structure or the nature of the linkage of glycogen to the polypeptide chains are not yet known.

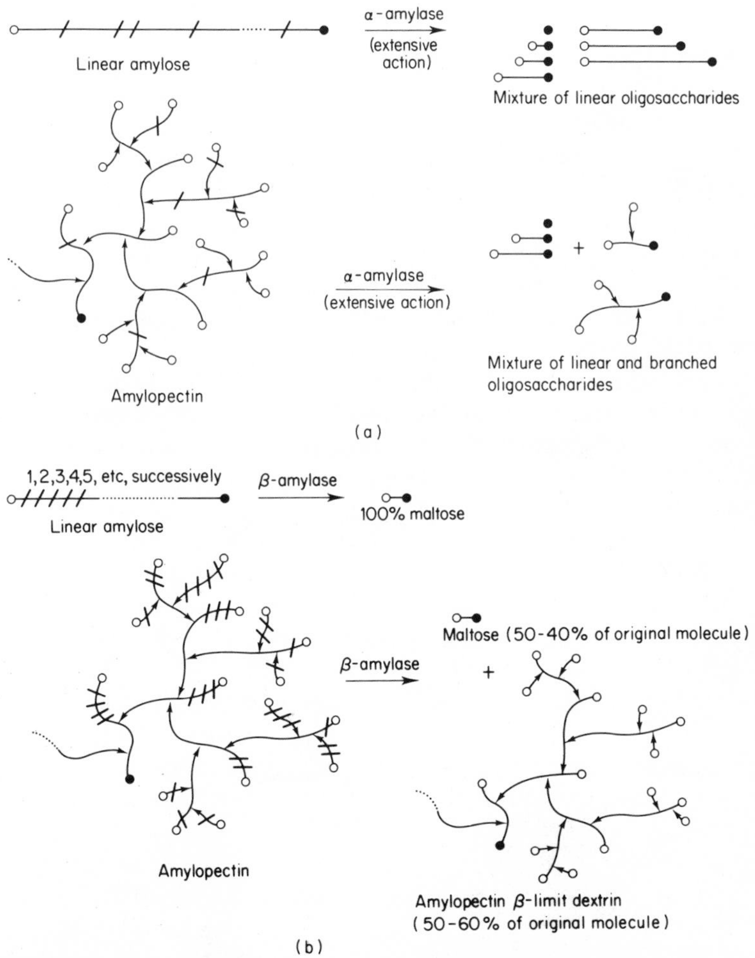

FIGURE 6.18. Action of α- and β-amylases on the starch components: (a) α-amylolysis; (b) β-amylolysis; ● = reducing end of α-(1 → 4)-linked glucan chain; ○ = non-reducing end of α-(1 → 4)-linked glucan chain; → = α-(1 → 6)-linkage; / = point of attack by enzyme

Glycogen reacts weakly with iodine to give a yellow-orange colour. It is believed that the lengths of unbranched chain in glycogen are too short to give the stronger colours of the amylose–iodine, or even amylopectin–iodine, complex. (A minimum of 50 linear, contiguous glucose residues may be necessary to give the blue colour characteristic of the amylose–iodine reaction.)

Starch and glycogen are synthesized (see Section 6.6, p. 314) in living organisms at times of high glucose availability, and are readily degraded to the monomer under conditions when glucose is required to supply energy for the organism.

In plants, two main types of enzymes are available which catalyse the hydrolysis of α-(1 → 4)-bonds in glucans, and hence are important in starch breakdown. These are the *α-amylases*, which cause an essentially random hydrolysis of the glycosidic bonds, and so depolymerize both amylose and amylopectin extensively, and the *β-amylases*, which in contrast degrade starch from non-reducing end-groups with the release of β-maltose. [Oligosaccharides in the α-configuration (see Figure 6.3, p. 244) are produced by the action of α-amylases, while β-maltose is obtained by β-amylolysis—hence the names of the amylases.] Whereas α-amylases can bypass α-(1 → 6)-linkages and hydrolyse starch readily to oligosaccharides, β-amylase cannot. Thus the action of β-amylase on starch gives complete hydrolysis of linear amylose molecules, but incomplete degradation of molecules containing branch-points (see Figure 6.18). Other enzymes exist in plants to facilitate the hydrolysis of α-(1 → 6)-bonds in glucan chains, or the production of glucose from the oligosaccharide products of α-amylase action on starch. Amylases present in various regions of the human digestive tract are important for digesting the starches in our foodstuffs.

In animal muscle and liver, glycogen is broken down when glucose demand is high by an enzyme, *phosphorylase*, which catalyses the reaction:

Non-reducing end of a glycogen chain

Repeat

Glucose-1-phosphate

Glycogen chain shortened by one glucose unit

The enzyme exists in two forms, an active and a less active form. The relative proportion of the two is kept under strict control, so that only when glucose is required by the cell does the active form predominate. Debranching enzymes must also be present to remove α-(1 → 6)-linked sugar units, which hinder the action of phosphorylase.

The importance of phosphorylase and other enzymes involved in glycogen breakdown in man is emphasized by the occurrence of several hereditary disorders, known as glycogen storage diseases (often fatal for the individual concerned), where unusually large amounts of glycogen, sometimes of abnormal structure, accumulate in the liver or muscles. It has been shown that these disorders result from a complete lack, or severe decrease in amount, of one or other glycogen-degrading enzyme.

(*b*) *Cellulose and plant cell-walls. Cellulose* is the most abundant, naturally occurring organic substance, and the polysaccharide functions as the main structural material of all higher plants. The polymer constitutes approximately one-third of the mass of higher plants, but the purest natural form, seed hairs of the cotton plant, contains 95% or more of cellulose. The polymer is also synthesized by some algae and bacteria.

Cellulose, like starch, is a polymer of D-glucopyranose units. In this case, the monosaccharide residues are linked by β-(1 → 4)-bonds, and the molecules are unbranched. Because of the β-configuration of the intermonomer links, the glucose units effectively alternate up and down in the chain (see Figure 6.19). The monosaccharide residues are believed to exist in the C1 conformation. Linear chains of α-(1 → 4)-linked glucose units in amylose (see previous section) are quite flexible, but in contrast, the β-(1 → 4)-bonded chains of cellulose are essentially rigid and straight. As a result, cellulose molecules can readily align themselves side-by-side, and this arrangement is stabilized by intra- and inter-chain hydrogen bonding between —OH groups on the glucose residues. In fact, intermolecular bonding is so strong that cellulose is insoluble in water, and even in strong sodium hydroxide. The polymer can, however, be dissolved in cuprammonium hydroxide solution through complex formation.

Cellulose occurs in higher plants in close association with—or covalently bonded to—other polysaccharides, some protein and, in woody tissues, lignin (see Chapter 7). Delignification and removal of contaminating polysaccharides usually involves very drastic conditions such as heating with alkali, or with sulphur dioxide in bisulphite solution. These procedures leave cellulose as an insoluble residue, but some molecular degradation takes place. Thus many estimations of the molecular weight of the resultant cellulose must be regarded as giving a lower limit for the molecular weight of the polymer. Cellulose samples are usually polydisperse, and weight-average molecular weights of approximately 10^6 are generally obtained with careful extraction.

In vivo, cellulose molecules aggregate to form *microfibrils* of diameter 35 Å; the polymeric chains probably extend along the length of the

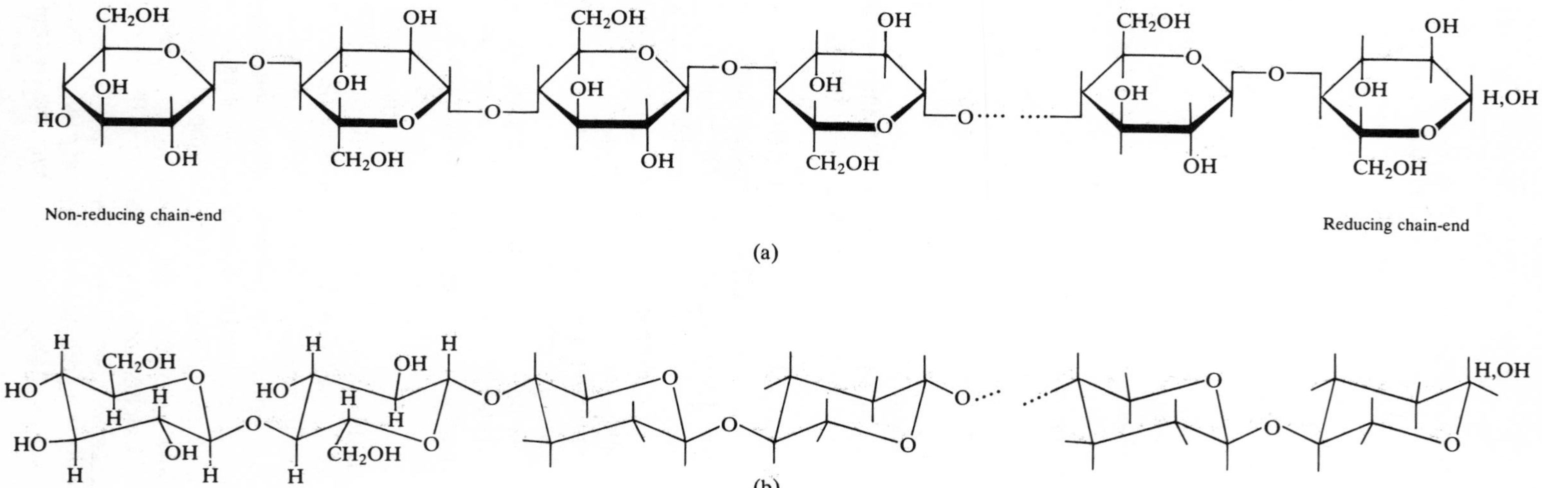

Non-reducing chain-end

Reducing chain-end

(a)

(b)

β-D-Glc *p*-(1 → 4)-β-D-Glc *p*-(1 → 4)-β-D-Glc *p* . . . . -β-D-Glc *p*

(c)

FIGURE 6.19. Structure of cellulose: (a) using Haworth convention; (b) showing the most stable ring conformation; (c) shorthand form—abbreviations as in Figure 6.6, p. 246 (compare Figure 6.12, p. 226—it is conventional to write polysaccharide sequences without the —O— shown)

microfibrils. These then associate to form thicker *fibrils*, which in turn aggregate giving the *cellulose fibres*. The latter act as the 'skeletons' of higher plants. Cellulose fibres give X-ray diffraction patterns, suggesting a high degree of order within them. The fibres are not completely crystalline—in native cellulose an apparent value of around 70% crystallinity is obtained. Various theories of the arrangement of molecules within the microfibrils have been suggested. In one of these, it is proposed that there are regions of perfect three-dimensional order, separated from each other by so-called amorphous regions where the molecules are arranged at random. As the crystalline regions are shorter than the length of a cellulose chain, there would be alternating ordered and disordered regions along an individual macromolecule. In another theory, dislocation of chain-ends within a microfibril would be sufficient to account for the amorphous regions (see Figure 6.20). From the fibre X-ray diffraction pattern, it has

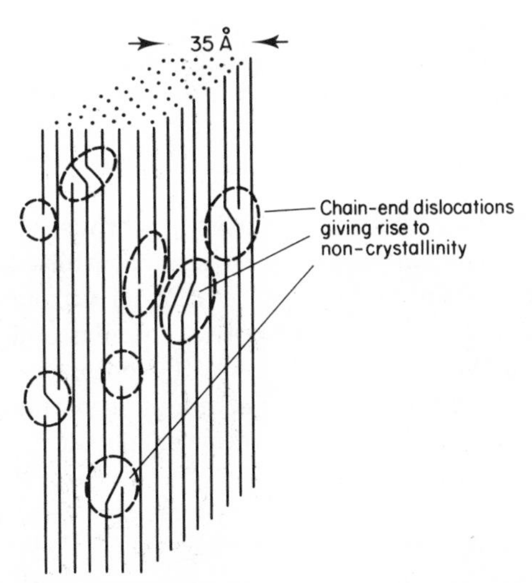

FIGURE 6.20. Possible structure for a cellulose microfibril

been calculated that there is a repeating unit 10.3 Å in length along the fibre axis. Although a helical structure has been proposed for cellulose chains in the microfibril, other possible arrangements are parallel or antiparallel arrays of extended molecules stabilized by hydrogen bonding (an antiparallel arrangement of molecules is shown in Figure 6.21).

Cellulose provides the effective framework for cells in plant tissues. Plant cells can experience very high osmotic pressures, and so the cell walls must be strong—both to withstand pressure and support the weight of the plant.

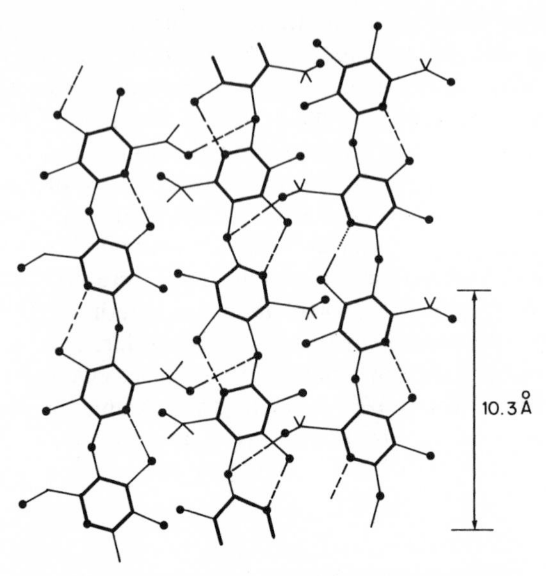

FIGURE 6.21. Possible arrangement of chains in the crystalline region of a cellulose microfibril

In Nature, support structures often consist of fibrous substances embedded in an amorphous matrix, the fibres resist tension, whilst the matrix helps to resist compression. In plants, the matrix material surrounding the cellulose fibres contains a complex mixture of polysaccharides and protein, the exact nature of the biopolymers depending on the botanical source. In woody plants, lignin (see Chapter 7) also is of major importance in cell-walls.

Plant cells grow and differentiate—their shape being maintained by the cell-walls. The properties of the walls are in turn determined by the biopolymers involved, and by the 'super molecular' arrangement within the walls, i.e. the arrangement and binding of the different component molecules. Although much information is now available concerning the structures of these components, little is known of the organization and binding of the different biopolymers with respect to each other within the cell-walls.

When plant cells divide, the first component of the new cell-wall to be laid down appears to be a watery pectin layer, followed by deposition of cellulose fibres. Indeed the properties of these fibres may be influenced by the associated pectin materials. The first complete wall to form round the new cell is known as the *primary cell-wall*, and contains on average some 20% cellulose. In this structure, the cellulose fibres are loosely interwoven. The other components of this primary wall are mainly polysaccharides—pectins (see Section 6.5.2.4, p. 303) and hemicelluloses (see Section 6.5.2.2, p. 296)—and a little protein which probably cross-links some of the amorphous polysaccharides to form a network. It is now considered likely that many of the wall components are covalently bonded together, thus explaining the difficulty of extracting cell-wall polysaccharides in mild solvents, such as water. Some of the amorphous polysaccharides probably associate closely with the cellulose fibres, and monosaccharides other than glucose are found as components of the fibres.

Primary cell-walls can be readily extended in one dimension, allowing for cell enlargement, and hence plant growth. Once the extension of a cell is completed, a new thicker, inner cell-wall, the *secondary cell-wall*, is formed. It is this secondary cell-wall which gives the ability to withstand high pressures.

In many plants, three distinct layers can be distinguished by electron microscopy of the secondary cell-wall (see Figure 6.22). In these layers, the

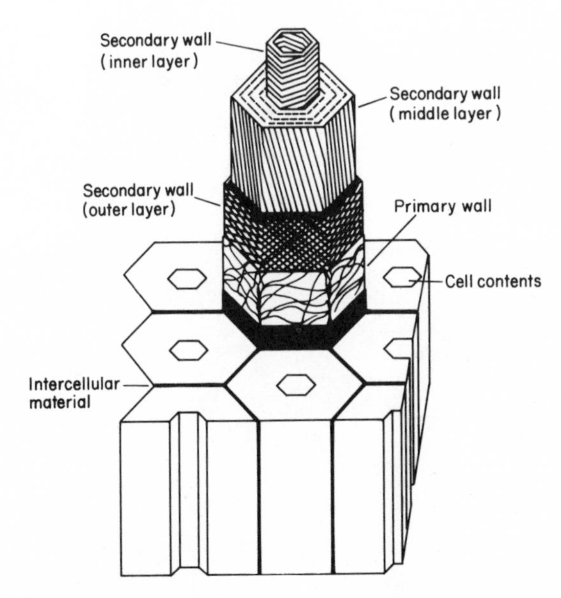

FIGURE 6.22. Schematic representation of a plant cell-wall

cellulose fibres are woven together in different ways. For example, in soft woods, parallel bundles of fibres crossing each other at a fixed angle are found in the thin outer layer of the secondary cell-wall; in the broad middle layer, parallel fibres run at a slight angle to the cell axis; whilst the fibres of the inner layer are arranged in a flat helix. Again, as in the primary cell-wall, the amorphous material consists of a complex mixture of polysaccharides with some protein. The amount of lignin (see Chapter 7), a very important

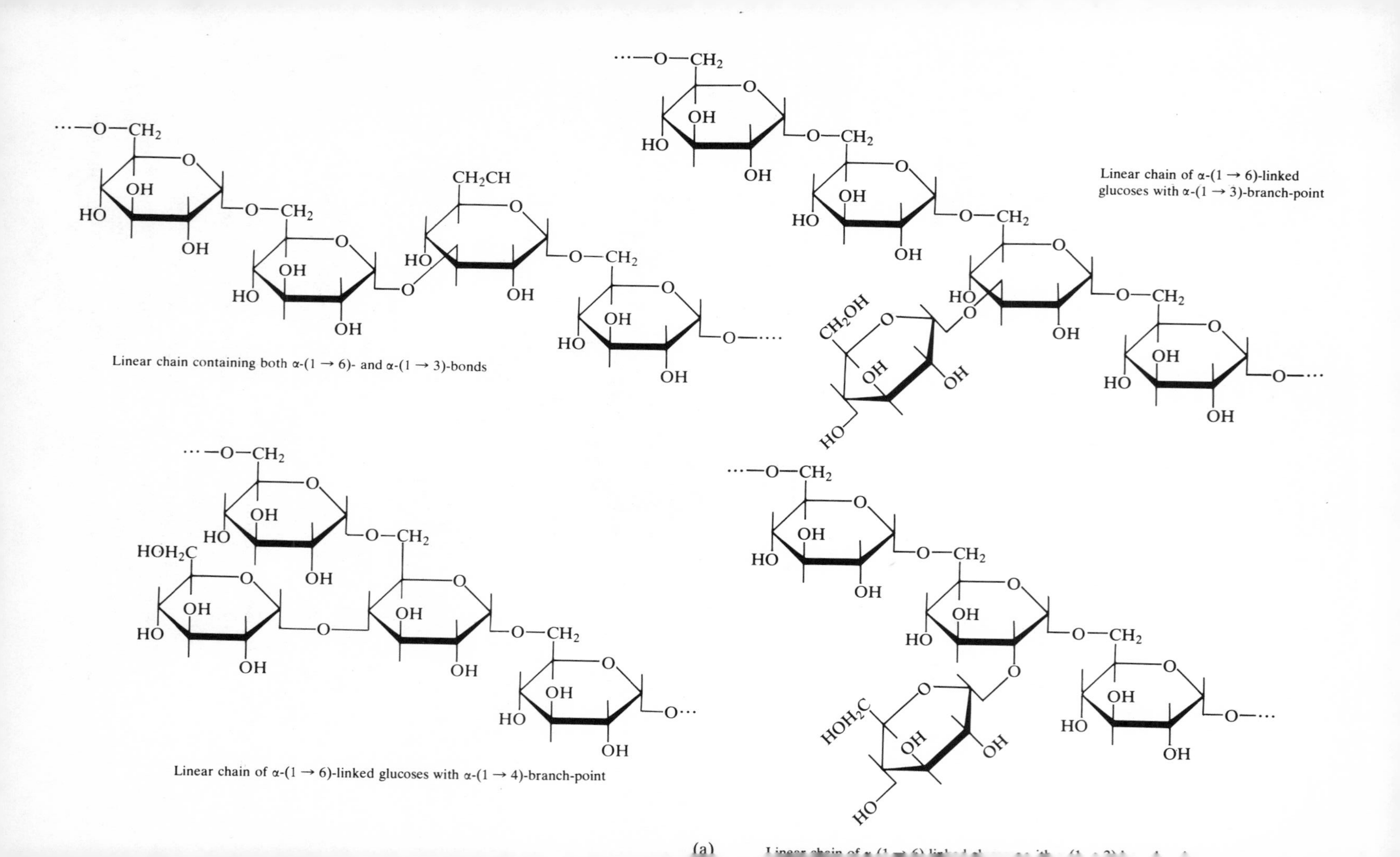
Linear chain containing both α-(1 → 6)- and α-(1 → 3)-bonds
Linear chain of α-(1 → 6)-linked glucoses with α-(1 → 3)-branch-point
Linear chain of α-(1 → 6)-linked glucoses with α-(1 → 4)-branch-point
(a)

··· -α-D-Glc*p*-(1 → 6)-α-D-Glc*p*-(1 → 3)- α -D-Glc*p*-(1 → 6)-α-D-Glc*p*- ···

··· -α-D-Glc*p*-(1 → 6)-α-D-Glc*p*-(1 → 6)-α-D-Glc*p*-(1 → 6)-α-D-Glc*p*- ···

3 ↑ 1

α-D-Glc*p*

··· -α-D-Glc*p*-(1 → 6)-α-D-Glc*p*-(1 → 6)-α-D-Glc*p*- ···

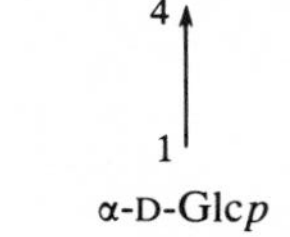

α-D-Glc*p*

··· -α-D-Glc*p*-(1 → 6)-α-D-Glc*p*-(1 → 6)-α-D-Glc*p*- ···

α-D-Glc*p*

FIGURE 6.23. Partial structures of dextrans: (a) using Haworth formulae; (b) shorthand form

component of the secondary wall, varies with the botanical source, and is highest in woody plants where mechanical rigidity is required.

(*c*) *Dextrans.* These are extracellular polysaccharides synthesized by certain bacteria when grown on sucrose; they probably act as a protective coating for the bacteria concerned.

The polymers consist essentially of branched chains of α-(1 → 6)-linked D-glucopyranose residues, the branches being formed by the presence of α-(1 → 2)-, α-(1 → 3)-, or α-(1 → 4)-bonds in the molecules (see Figure 6.23). Many different dextrans exist, the structure depending on the strain of bacteria which synthesizes the polysaccharide. The monosaccharide residues are probably in the C1 conformation.

In some dextrans, α-(1 → 3)-bonded glucose units may form part of a linear chain (see Figure 6.23), whilst in others these bonds give branched structures. Where α-(1 → 2)- or α-(1 → 4)-links are found, they always consitute branch-points. The positions of branch-points are probably randomly determined, sometimes adjacent to each other, mostly further apart. Frequently branches consist of single glucose units (see Figure 6.8(a), p. 257), although branches of four or five glucose units are also common. Larger branches are also known, and such molecules probably have the branch-on-branch structure of Figure 6.8(b). Dextrans are usually quite polydisperse, with weight-average molecular weights of up to 5×10^5.

These polysaccharides may be water-soluble or -insoluble, but both types are capable of forming *gels*, in which it is thought that linear segments of molecular chains associate, possibly in a helical conformation, to give junction zones. (In the presence of aqueous solvents, many polysaccharides can form the network structures which entrap solvent molecules: such structures are known as gels. In these structures, there are amorphous regions where the polymer chains are probably in a random-coil conformation. These regions are held together by junction zones where there is a high degree of ordering of the molecular chain segments involved. Differences between polysaccharide gels are determined by the nature and number of junction zones.)

Dextran-synthesizing bacteria can flourish in the human mouth, particularly on sucrose-containing food particles which become trapped between the teeth. The polysaccharides formed by the bacteria can form part of dental plaque, and thus are implicated in tooth decay.

Sugar cane and sugar beet, after harvesting, can also become infected by dextran-synthesizing bacteria. The resultant dextran interferes with sugar refining, choking filters and pipelines. Thus, on the whole, dextran, unlike starch and cellulose, cannot be considered as being useful to man.

(*d*) β-*Glucans.* We have already discussed one β-glucan, cellulose, but there are many other polymers of β-glucose, some acting as food-reserve polysaccharides, with others having a structural rôle.

The food-reserve β-glucans are found in some microorganisms and some seaweeds. *Laminaran* from brown seaweed, for example, is mainly a β-(1 → 3)-linked D-glucopyranose polymer, with some β-(1 → 6)-branch-points.

The structural β-glucans are usually linear, and are found in the cell-walls of plants, and some microorganisms. Water-soluble β-glucans can be obtained from oat and barley grains. Indeed these polysaccharides may be the main structural components of the cell-walls, for little, if any, cellulose is present. The molecules are linear chains containing both β-(1 → 3)- and β-(1 → 4)-bonds, in the ratio 1:3. The presence of (1 → 3)-links in the molecule makes the polymer chain more flexible than the rigid chain of cellulose where all interglucose bonds are of the β-(1 → 4)-type. The molecules of cereal β-glucans contain partial sequences of the type:

···-(1 → 3)-β-D-Glc*p*-(1 → 4)-β-D-Glc*p*-(1 → 4)-β-D-Glc*p*-(1 → 4)-
-D-Glc*p*-(1 → 3)-β-D-Glc*p*-···

i.e. cellotriose or cellobiose units linked by β-(1 → 3)-bonds. Some regions of the polymer chains, however, may contain contiguous β-(1 → 3)-linked glucose residues. Number-average molecular weights are probably approximately 2–4 × 10^4, and the polysaccharide may exist in the cell-walls covalently bonded to protein.

Lichenan, a β-glucan from Iceland moss, bears a certain similarity to the cereal β-glucan in that it too contains β-(1 → 3)- and β-(1 → 4)-linked glucose units, in this case in equal numbers. Again, the molecular chains are linear rather than branched, with a minimum molecular weight of 70,000.

6.5.1.2. Fructans

Fructans are polymers of D-fructofuranose units, and are important as short-term, energy-reserve polysaccharides of grasses and plants of the daisy family. They are water-soluble and easily hydrolysed, and hence can readily supply fructose for the plant's energy requirements. Fructans are also elaborated by some microorganisms.

Inulin is a typical fructan; the main inulin of dahlia has short linear molecules of β-(2 → 1)-linked fructose residues (see Figures 6.24(a)). The molecules are believed to be synthesized on a sucrose 'primer' (see Section 6.6, p. 316), and so have a glucose residue at one end of the polysaccharide chain. The number-average molecular weight is generally around 6,000.

Other plant fructans, the *levans* from grasses for example, consist of linear chains of β-(2 → 6)-linked fructose units (see Figure 6.24(b)). Again, a glucose residue 'blocks' the reducing end of the molecules, i.e. fructans are non-reducing polysaccharides. Number-average molecular weights range from 5 to 50 × 10^3.

Bacterial fructans, also called levans, have β-(2 → 6)-bonded fructose chains which can contain β-(2 → 1)-branching links, i.e. a partial structure

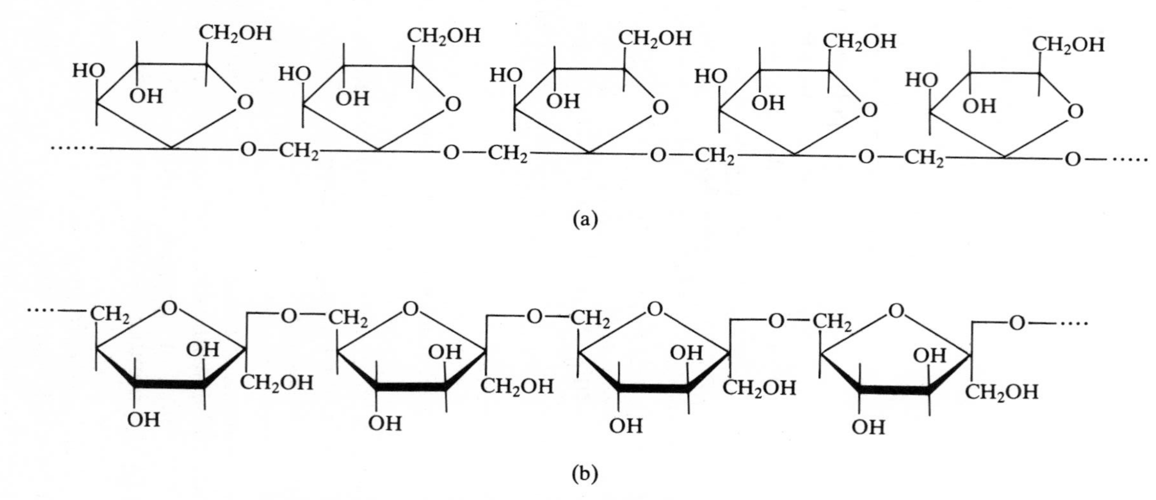

FIGURE 6.24. Partial structure of fructans: (a) inulin; (b) levan

could be

$$
\begin{array}{l}
\cdots\text{-}(2 \rightarrow 6)\text{-}\beta\text{-D-Fru}f\text{-}(2 \rightarrow 6)\text{-}\beta\text{-D-Fru}f\text{-}(2 \rightarrow 6)\text{-}\beta\text{-D-Fru}f\text{-}(2 \rightarrow 6)\text{-}\beta\text{-D-Fru}f\text{-}(2 \rightarrow 6)\text{-}\beta\text{-D-Fru}f \\
\qquad\qquad\quad {}^{1}\!\uparrow \qquad\qquad\qquad\qquad\qquad\qquad\qquad\qquad\qquad {}^{1}\!\uparrow \\
\qquad\qquad\quad {}_{2}| \qquad\qquad\qquad\qquad\qquad\qquad\qquad\qquad\qquad {}_{2}| \\
\cdots\text{-}\beta\text{-Fru}f\text{-}(2 \rightarrow 6)\text{-}\beta\text{-Fru}f \quad \cdots\text{-}(2 \rightarrow 6)\text{-}\beta\text{-Fru}f\text{-}(2 \rightarrow 6)\text{-}\beta\text{-Fru}f\text{-}(2 \rightarrow 6)\text{-}\beta\text{-Fru}f
\end{array}
$$

These bacterial polysaccharides are sometimes very large and highly branched, with weight-average molecular weights of over 10^6.

6.5.1.3. Mannans

Mannans are found in plants, particularly in certain seeds, and in some microorganisms such as yeasts and algae. These polysaccharides usually fulfill a structural rôle, but in seeds may act as a food reservoir.

A food-reserve mannan is present in tagua palm seeds, the so-called ivory nuts. The polysaccharide is insoluble in water, but exists in two forms in the seeds, one soluble in alkali, and the other, a microfibrillar form, readily soluble only in cuprammonium hydroxide (cf. cellulose, Section 6.5.1.1). This second form may have a structural rôle in the seed, for a very similar polysaccharide occurs as a structural fibre in certain algae. The polysaccharide consists essentially of linear chains of β-(1 → 4)-linked D-mannopyranose residues (see Figure 6.25) which are more flexible than cellulose chains—because the difference in configuration at C-2 of the sugar rings (cf. Figure 6.25 and 6.19(a), p. 282) has a profound effect on intra- and/or inter-molecular bonding. X-ray diffraction measurements on mannan fibres indicate a repeat distance of 10.3 Å, and there is believed to be both intra- and inter-chain hydrogen bonding.

Polysaccharides containing mannan only are relatively rare in plants. Although mannans may exist in plant cell-walls, heteropolysaccharides containing mannose are much more common, and will be discussed in Section 6.5.2.2, p. 297).

Some yeasts contain a cell-wall mannan in which the main chains consist of α-(1 → 6)-linked D-mannopyranose residues with side-chains of α-(1 → 2)- and α-(1 → 3)-bonded mannose units. In these polysaccharides, some mannose units carry phosphate groups on C-6 of the sugar ring, and the polysaccharides may occur in the cell-wall covalently bonded to protein.

6.5.1.4. Xylans, arabinans, galactans, and galacturonans

These homopolysaccharides of xylose, arabinose, galactose, and galacturonic acid residues, respectively, are grouped together because they may occur as components of plant cell-walls, but are relatively rare. Much more common are the heteropolysaccharides containing two or more

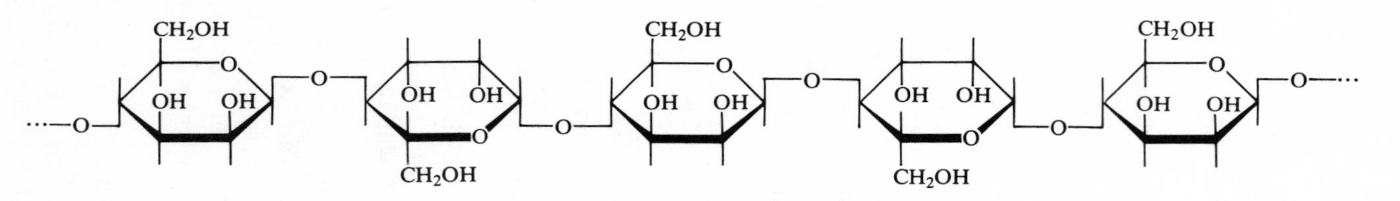

FIGURE 6.25. Partial structure of a mannan

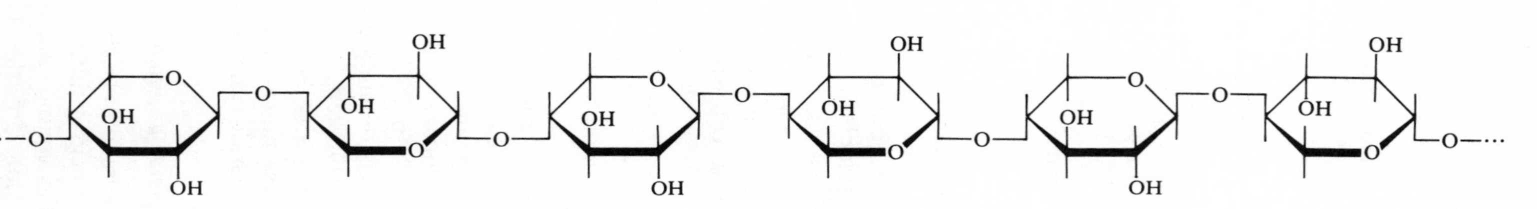

FIGURE 6.26. Partial structure of a xylan

monosaccharides e.g. arabinoxylans, arabinogalactans, etc. (see Section 6.5.2, p. 296).

Polysaccharides with xylose main chains are an important component of 'hemicelluloses', the alkali-soluble polymers closely associated with cellulose, particularly in the secondary cell-walls of higher plants. Xylose homopolymers are uncommon in plants, but such a *xylan* can be extracted from esparto grass, and has been shown to consist essentially of linear chains of β-(1 → 4)-linked D-xylopyranose residues (see Figure 6.26). A small number of branches linked at the 3-position of a xylose unit may occur in the polymer. The macromolecule is more flexible than cellulose and in this case, the lack of the —CH_2OH group on C-5 must mean that some of the hydrogen bonding, which gives cellulose its rigidity, cannot now take place.

Fibres of β-(1 → 4)-linked xylans have been examined by X-ray diffraction, and repeat distances of 14.8 Å along the fibre axis have been found. A helical conformation has been proposed for the polymer chains within the fibre.

Some green algae contain a xylan in the fibre form as an important structural component. In contrast to the polysaccharide from higher plants, this algal polymer consists of β-(1 → 3)-linked D-xylose residues. X-ray diffraction indicates a repeat distance of 6.1 Å along the fibre axis, and again a helical arrangement of chains is implicated. Other algae, however, contain β-(1 → 4)-linked xylans, whilst yet others have xylans containing both (1 → 3)- and (1 → 4)-bonds.

Arabinans occur in plants as a component of the pectic substances of cell-walls. Most arabinose-containing polysaccharides, however, are heteropolysaccharides, or are covalently bonded to polysaccharides containing other monosaccharide units. In this second case, it becomes difficult to define the individual polysaccharide. If large homopolymers exist covalently bonded to other different polymers, the complex perhaps extending over the whole cell-wall, do we define this as a single polysaccharide, or as a number of distinct polysaccharides making up a supermolecular arrangement? Indeed, as the structures of more and more 'simple' biopolymers become known, research work is moving towards the elucidation of supermolecular structure—the arrangement of many polymers to give a microscopically recognizable, subcellular unit. Studies of polysaccharides are, in this respect, lagging behind those of proteins and nucleic acids, because of difficulties in determining when a pure polysaccharide has been obtained, and because of the complex heterogeneity in molecular architecture.

One arabinan which has been studied has been shown to have main chains of α-(1 → 5)-linked L-arabinofuranose residues. Single α-L-arabinofuranose residues are attached, as side-chains, by (1 → 3)-bonds to about 50% of the main-chain sugar units. In some arabinans, arabinose residues of the backbone may also carry 'branching' arabinose units at C-2, or at both C-2 and C-3. (Polysaccharides consisting of pentose sugars such as xylose or arabinose are often referred to as *pentosans*.)

Galactans are found with arabinans as part of the pectic substances of plant cell-walls and intercellular material. Again, galactose homopolymers are not as common as galactose-containing heteropolysaccharides. Polymers of β-(1 → 4)-linked D-galactopyranose units have been found in plants, as have β-(1 → 3)-bonded galactose chains. These polysaccharides often have side-chains of β-(1 → 6)-linked galactose units. Molecular weights are usually in the region of 10^4.

Galactans are found extensively in seaweeds, but these polysaccharides contain two types of galactose residue, and will be discussed in Section 6.5.2.3, p. 300.

A simple *galacturonan* consists of chains of α-D-galacturonic acid linked by (1 → 4)-bonds. Such a substance is *pectic acid*, an important pectin. Pectins are located mainly in primary cell-walls and intercellular material of plants, and are particularly abundant in fruits. As fruits ripen, the pectin polysaccharides are degraded, which leads to the softening of the fruit concerned.

In many pectins, the galacturonic acid is present, at least in part, as the methyl ester. Much commoner than simple galacturonans are pectins containing other monosaccharides in addition to galacturonic acid in the main chains (see Section 6.5.2.4, p. 303).

The molecular weights of many of these polysaccharides in their native state are not known. Most must be extracted from plants in alkali, which causes degradation as the polymers dissolve. Number-average molecular weights of isolated polymers are usually $1–4 \times 10^4$.

6.5.1.5. Chitin

Chitin is an important structural polysaccharide which often replaces cellulose in the cell-walls of lower plants. It is found not only in most fungi and in some algae, but also in the animal kingdom in the cuticles of arthropods and molluscs. Chitin rarely occurs alone, but is usually associated with protein to which it may be covalently bonded. Arthropod exoskeletons, in addition, contain inorganic salts such as calcium carbonate. Chitin, like cellulose, is inert and insoluble in water, and is thus difficult to isolate without degradation. It is soluble in concentrated mineral acids, but this makes meaningful molecular weight determinations difficult. *In vivo*, chitin molecules may be as large as those of cellulose, and certainly, chitin provides a rigid structure like cellulose.

Chitin is a polymer of β-(1 → 4)-linked *N*-acetyl-D-glucosamine residues, and thus is similar to cellulose, but an acetamido group replaces the hydroxyl group on C-2 of the glucose ring (cf. Figures 6.27 and 6.19(a), p. 282).

Fibres of chitin can be formed, but the exact three-dimensional structure of the polysaccharide in Nature may depend on its interaction with the

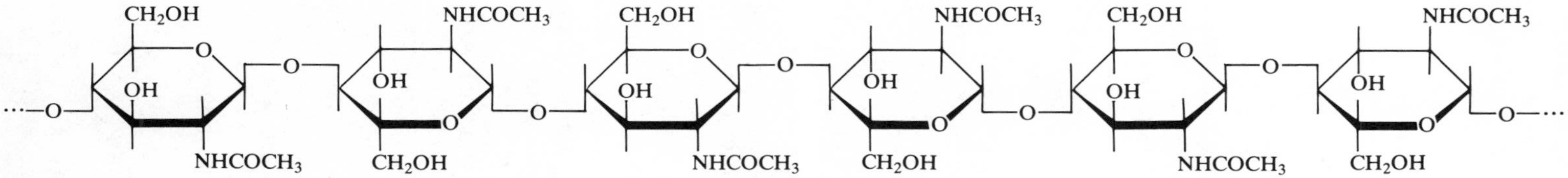

FIGURE 6.27. Partial structure of chitin

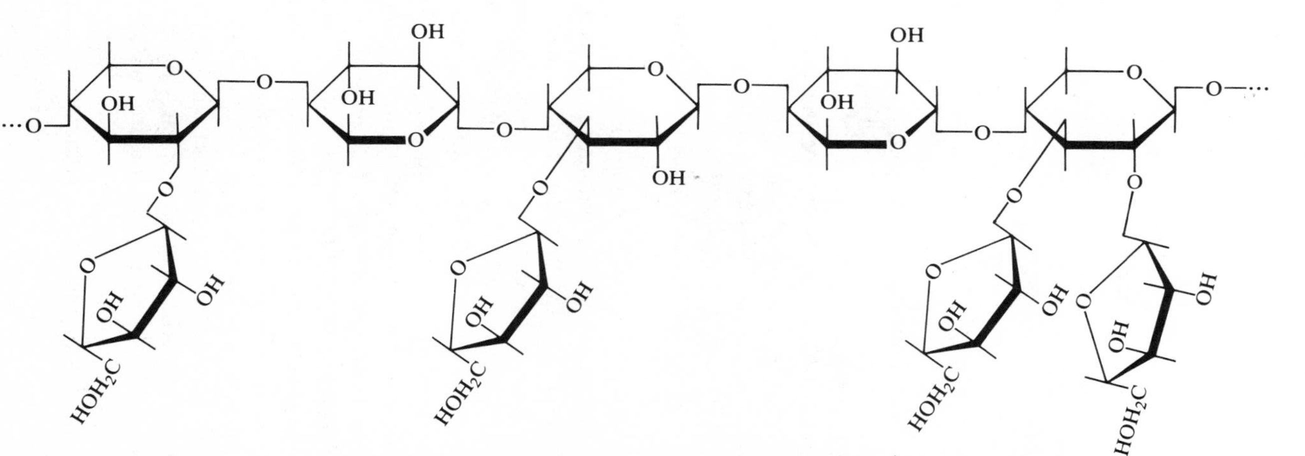

FIGURE 6.28. Partial structure of an arabinoxylan

matrix material, i.e. the protein and/or salts surrounding it. X-ray diffraction of chitin fibres indicates a 10.3 Å repeat distance along the fibre axis, which is very similar to that for cellulose. It is likely that intra- and intermolecular hydrogen bonds stabilize the three-dimensional structure.

6.5.2. Heteropolysaccharides

Here polysaccharides containing two or more different monosaccharide residues are grouped according to the components of the *main* polymeric chain. Many of these polymers are structural components of plant cell-walls. As yet, little is known of the way in which they participate in cellular architecture, or how their own molecular structure is adapted for this purpose.

6.5.2.1. Polysaccharides with a xylose backbone

Xylans occur in all land plants, and are particularly abundant in annual plants and hard woods, where they form an important component of the hemicelluloses of the secondary cell-wall.

The main chains of both *glucuronoxylans* and *arabinoxylans* consist of β-(1 → 4)-linked xylose residues as discussed earlier (see Section 6.5.1.4, p. 293, and Figure 6.26). Glucuronoxylans have, in addition, side-chains of single 4-*O*-methyl-α-D-glucuronic acid residues linked by (1 → 2)-bonds to the polymer backbone. Approximately one in ten of the main-chain xylose units carries a branch-point, and some of the xylose units themselves may be acetylated. The distribution of glucuronic acid residues is probably random, but stretches of contiguous acidic residues are not believed to occur. Number-average molecular weights of 4×10^4 have been found, but it is very likely the polymer was partially degraded on isolation. One glucuronoxylan has been prepared in a sufficiently crystalline form to obtain an X-ray diffraction pattern, and a repeat length of 15 Å was found.

Arabinoxylans, along with β-glucans (see Section 6.5.1.1, p. 288), make up the major components of the cell-walls of some cereal grains. Most arabinoxylans have single L-arabinofuranose residues linked by α-(1 → 2)- or α-(1 → 3)-bonds to the main xylose chain, but more complex side-chains are known. The nature and distribution of the side-chains depends on the botanical source of the polymer. In one barley arabinoxylan, several xylose residues are doubly branched, carrying arabinose units at both C-2 and C-3 of the xylose ring (see Figure 6.28). The distribution of arabinose side-chains in the polymer may be random.

In a highly-branched arabinoxylan, such as that from wheat flour, the following type of distribution might be expected:

```
—Xylp—Xylp—Xylp—Xylp—Xylp—Xylp—Xylp—Xylp—Xylp—
   |           |     |           |           |
  Araf        Araf  Araf        Araf        Araf
```

Occasionally, arabinose side-chains occur on two or three contiguous xylose units, but never on four or more. These branched sections alternate every 20–25 main-chain units with short sections (2–5 units) of unbranched xylose chain. The unbranched stretches may be important for the binding of arabinoxylans to other cell-wall polysaccharides—arabinoxylans with longer unbranched sections can bind cellulose, whilst those with short (two-unit) unbranched stretches cannot.

In softwoods, the major xylan contains both L-arabinose and 4-*O*-methyl-D-glucuronic acid side-chains.

6.5.2.2. *Polysaccharides with glucose and mannose backbones*

Xyloglucans are found in some seeds and in hardwood cell-walls. They form part of the primary cell-wall, and are closely associated with cellulose fibres. The backbone of the molecules is that of cellulose, i.e. β-(1 → 4)-linked glucose residues, but most main-chain residues carry side-chains of single xylose units, or trisaccharides containing fucose, galactose and xylose. Thus a partial structure is given by:

```
   α-D-Xylp               α-D-Xylp               α-D-Xylp
        1|                     1|                     1|
         ↓                      ↓                      ↓
        6                      6                      6
....β-D-Glcp-(1 → 4)-β-D-Glcp-(1 → 4)-β- D-Glcp-(1 → 4)-β-D-Glcp-(1 → 4)-β-D-Glcp-....
                                                              6↑
                                                               |
                                                              1|
                    α-L-Fucp-(1 → 2)-β-D-Galp-(1 → 2)-α-D-Xylp
```

The glucose units of the polymer backbone may be hydrogen-bonded to cellulose, and the xyloglucan molecules would then align themselves along cellulose fibres with the side-chains pointing away from the fibres.

Galactomannans, on the other hand, may not have a structural rôle in plants. They are found in seed mucilages (i.e. gummy substances), and may act as food reserves, or help to retain water in the seeds. These macromolecules are essentially linear chains of β-(1 → 4)-linked D-mannopyranose residues with side-chains of single D-galactose units joined to the polymer backbone by α-(1 → 6)-bonds. The galactose: mannose ratio varies with the botanical source of the gum, being commonly in the range 1:1 to 1:5. There may be a random distribution of galactose side-chains, and the solubility and viscosity depend on the galactose:mannose ratio. The higher the ratio, the more water-soluble is the polymer. (Because very viscous solutions can be obtained, these polysaccharides are often used as thickening agents in the food industry.)

Galactomannan preparations are usually very polydisperse, with number-average molecular weights up to 3×10^5. It has been possible to study galactomannan fibres by X-ray diffraction, and a fibre repeat distance of 10.3 Å has been found, i.e. very similar to that of cellulose. An extended ribbon-like structure has been proposed for the polysaccharide in the fibre.

Glucomannans are found in some seeds and bulbs, where they may act as food reserves. In hardwoods, however, glucomannans (classed as hemicelluloses) play a structural rôle. These molecules are linear, and contain both D-glucopyranose and D-mannopyranose residues in the main chain, linked by β-(1 → 4)-bonds. The glucose:mannose ratio varies with the source of the polysaccharide, but is, in general, around 1:2 in hardwoods. The distribution of mannose and glucose along the chains may be partly random; up to six contiguous residues of mannose have been found, but not more than two contiguous glucose units. Number-average molecular weights in the range 2 to 8×10^4 have been reported.

Galactoglucommannans, related in structure to galacto- and gluco-mannans, occur in softwood cell-walls, where they are classed as hemicelluloses. The polymer backbone is like that of glucomannans—a chain of β-(1 → 4)-linked glucose and mannose residues—but like the galactomannans, side-chains of single galactose units are present, attached to main-chain mannose units by α-(1 → 6)-bonds. Some of the mannose residues may be partially acetylated. The glucose:mannose ratio is usually 1:3, but the galactose:mannose ratio can vary from 1:15 to 1:3. Again, the higher the galactose content, the more water-soluble is the polymer. Thus the presence of galactose side-chains must prevent strong intermolecular bonding. A typical partial structure would be:

```
  α-D-Galp-                                                                                   α-D-Galp-
      1|                                                                                          1|
       ↓6                                                                                          ↓6
....(1 → 4)-β-D-Manp-(1 → 4)-β-D-Manp-(1 → 4)-β-D-Manp-(1 → 4)-β-D-Glcp-(1 → 4)-β-D-Manp-
```

6.5.2.3. Polysaccharides with galactose backbones

These may be considered in two sub-groups, i.e. polysaccharides containing unmodified galactose residues, and those containing, in addition, such monomers as anhydrogalactose and/or galactose sulphates.

(a) Arabinogalactans: galactans containing unmodified galactose residues. Two major types of arabinogalactans are found in plants. The first, associated with pectic materials, occurs in seeds such as the lupin and soya bean. These polysaccharides have main chains of β-(1 → 4)-linked D-galactopyranose residues with side-chains of single, or disaccharide,

arabinofuranose units. A partial structure may be written as:

$$
\cdots\text{-}(1\rightarrow 4)\text{-}\beta\text{-D-Gal}p\text{-}(1\rightarrow 4)\text{-}\beta\text{-D-Gal}p\text{-}(1\rightarrow 4)\text{-}\beta\text{-D-Gal}p\text{-}(1\rightarrow 4)\text{-}\beta\text{-D-Gal}p\text{-}(1\rightarrow 4)\text{-}\cdots
$$

(Second Galp: 3 ← 1 L-Araf; fourth Galp: 3 ← 1 L-Araf, which is substituted at 5 ← 1 L-Araf)

Number-average molecular weights of up to 10,000 have been obtained.

The second, more common type occurring in plants are the highly-branched arabinogalactans with a β-(1 → 3)-linked D-galactopyranose backbone. Such polymers have been extracted from wheat flour, coffee beans, sycamore cells, and larch-wood. These polysaccharides probably form an important, if minor, part of the structure of cell-walls, and there is evidence suggesting the molecules are covalently bonded to protein *in vivo*. Side-chains, in this case, may contain galactose as well as arabinose units. Thus an arabinogalactan from larch can be represented by the partial structure:

$$
\text{-}(1\rightarrow 3)\text{-}\beta\text{-D-Gal}p\text{-}(1\rightarrow 3)\text{-}\beta\text{-D-Gal}p\text{-}(1\rightarrow 3)\text{-}\beta\text{-D-Gal}p\text{-}(1\rightarrow 3)\ \beta\text{-D-Gal}p\text{-}(1\rightarrow 3)\text{-}\beta\text{-D-Gal}p\text{-}(1\rightarrow 3)\text{-}\beta\text{-D-Gal}p\text{-}
$$

(Side chains, each attached 1 → 6 to the backbone: (1 → 3)-β-D-Galp carrying 6 ← 1 β-D-Galp; β-D-Galp carrying 6 ← 1 β-D-Galp; R; β-D-Galp; β-D-Galp carrying 6 ← 1 β-D-Galp)

R = L-Ara*f*
or L-Ara *f*-(1 → 3)-β-L-Ara*f*

In larch, the galactose:arabinose ratio of the polysaccharide is approximately 6:1, and number-average molecular weights up to 10^5 have been found.

Plant gums are exuded as viscous fluids, either spontaneously, or at the site of an injury to the plant. This fluid then becomes dehydrated to give hard, clear nodules consisting mainly of polysaccharide. The function of the exudate may simply be to protect the plant from further damage. Some of these plant gums are polysaccharides based on galactose chains.

The polysaccharides of the gums of tropical trees all possess complex, highly-branched structures, and, as a group, are amongst the most complicated biopolymers known. Most likely a gum polysaccharide consists of a number of closely related molecular species, differing in the nature and distribution of side-chains.

The principal polysaccharide of *gum arabic* has main chains of β-(1 → 3)-linked D-galactopyranose residues with side-chains containing rhamnose, arabinose, galactose, and glucuronic acid. The molecules may be built up from repeating units, and weight average molecular weights as high as 1 to 2×10^6 have been obtained. A possible partial structure is given in Figure 6.29.

(*b*) *Carrageenans and agar: galactans containing modified galactose residues.* The name *carrageenan* covers a range of sulphated polysaccharides found in various species of red seaweeds, where they probably play an important structural rôle. The polysaccharides are essentially linear galactans, where the monomers are joined by alternating (1 → 4)- and (1 → 3)-bonds. They can be regarded, in general, as made up of repeating

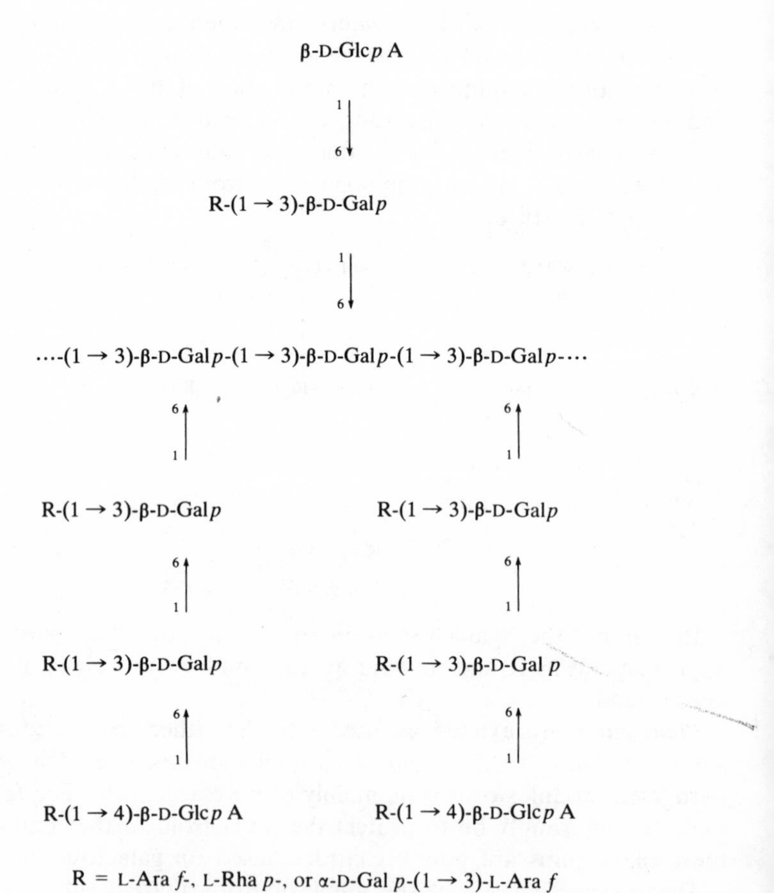

R = L-Ara *f*-, L-Rha *p*-, or α-D-Gal *p*-(1 → 3)-L-Ara *f* or less frequently β-L-Ara *p*-(1 → 3)-L-Ara *f*

FIGURE 6.29. Partial structure of gum arabic

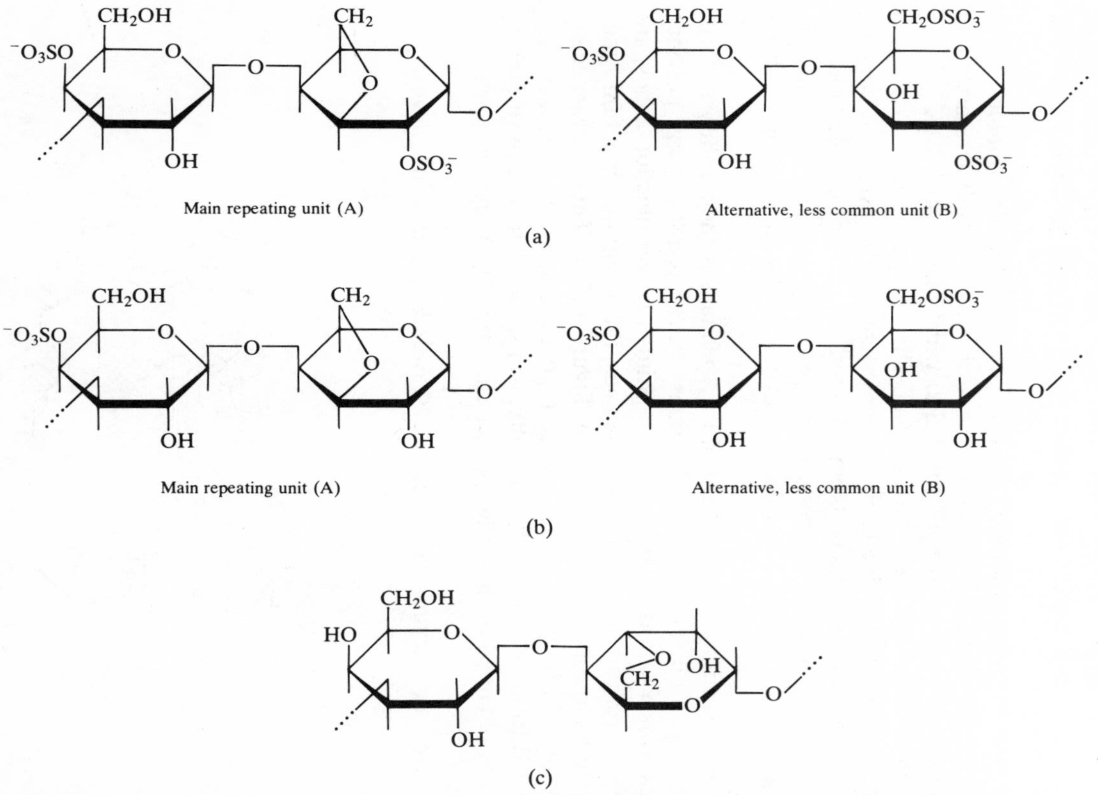
CH2OH
CH2
−O3SO
O
O
O
O
OH
OSO3−
Main repeating unit (A)
CH2OH
CH2OSO3−
−O3SO
O
O
O
OH
O
OH
OSO3−
Alternative, less common unit (B)
(a)
CH2OH
CH2
−O3SO
O
O
O
O
OH
OH
Main repeating unit (A)
CH2OH
CH2OSO3−
−O3SO
O
O
O
OH
O
OH
OH
Alternative, less common unit (B)
(b)
CH2OH
HO
O
O
O
OH
CH2
O
O
OH
(c)

FIGURE 6.30. Partial structures of carrageenans and agarose: (a) ι-carrageenan; (b) κ-carrageenan; (c) agarose

disaccharide units. In two industrially important carrageenans, the disaccharide repeating unit consists of a β-galactopyranose residue linked by a β-(1 → 4)-bond to a 3,6-anhydro-D-galactopyranose unit (see Figure 6.30). The two polysaccharides, ι- and κ-carrageenan differ in their sulphate content. In both polymers, some of the anhydrogalactose units may be replaced by sulphated galactose residues; thus galactose-2, 6-disulphate is found in ι-carrageenan, and galactose-6-sulphate in κ-carrageenan. Molecular weights of 10^5 to 10^6 are common.

Both ι- and κ-carrageenan are believed capable of forming double helices consisting of two parallel, staggered polysaccharide chains. There are three disaccharide repeats per turn of the helix, which has a pitch of 27 Å for ι-carrageenan, and 25 Å for κ-carrageenan. The sulphate groups are located on the outside of the helix, and the helix is stabilized by hydrogen bonding between the constituent chains. Uninterrupted stretches of double helix are thought possible only for sections of chain containing anhydrogalactose. When a galactose sulphate replaces an anhydro residue, a 'kink' in the helix must result.

An extremely important property of carrageenans is their ability to form gels. As discussed earlier (see Sections 6.5.1, p. 288) long polymer molecules, which can associate at certain points to form junction zones, are necessary for gel formation. For carrageenans, it is likely that junction zones involve stretches of double helix (see Figure 6.31). Thus carrageenans having little anhydrogalactose (i.e. more repeat *B* relative to repeat *A*—see Figure 6.30) form gels only with difficulty, whilst more stable gels of greater strength are obtained from similar polysaccharides with a high proportion of anhydrogalactose.

In Nature, an enzyme exists in red seaweeds which can convert the

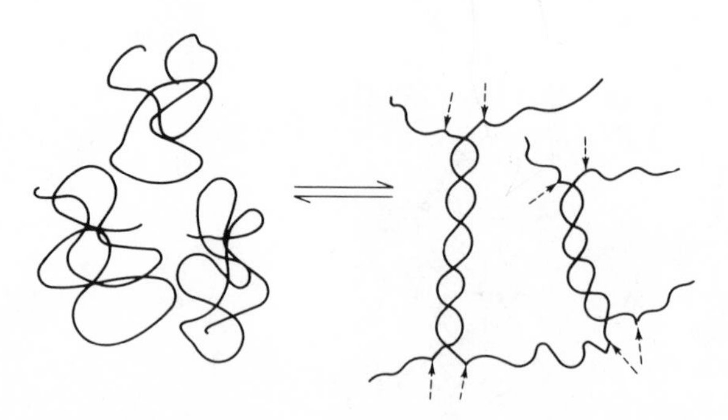

FIGURE 6.31. Junction zones of carrageenan gels. The dashed arrows indicate galactose sulphate or disulphate residues replacing anhydrogalactose units: these residues break up the regular helical structure

galactose-6-sulphate of κ-carrageenan to 3,6-anhydrogalactose. This would be expected to increase helix formation, and allow a stiffer gel to be formed. In fact, it has been found that in seaweeds growing where there is strong wave action, the carrageenan has a high proportion of anhydrogalactose. Thus it seems that the plant can control its structure to some extent, and react to stress, i.e. strong wave action, by stiffening the carrageenan gel it contains.

Because of their gelling properties, carrageenans are widely used as food thickeners and emulsion stabilizers in the food industry.

The name *agar* refers to a family of polysaccharides, again from red seaweed, which contain alternating residues of β-D-galactopyranose and 3,6-anhydro-α-L-galactopyranose (see Figure 6.30). (Thus an agar is similar to a carrangeenan with anhydro-L-galactose substituting for the anhydro-D-galactose of the carrageenans.) The members of this family differ in extent of sulphation and methylation, and some may contain glucuronic acid residues. Agar forms strong gels even when very dilute, and these gels are widely used as a support for culturing bacteria.

Agarose is the agar polysaccharide with the greatest gelling tendency, and contains no glucuronic acid residues. The macromolecules are believed to be capable of forming a double helix of pitch 19 Å, the two chains being parallel and staggered. It is therefore likely that the junction zones of agarose gels are stretches of double helix. Agarose gels are widely used in laboratories for gel-permeation chromatography and gel electrophoresis (see Section 2.3.1.3, p. 17 and 2.3.2.2, p. 23).

6.5.2.4. *Polysaccharides with backbones containing uronic acid residues*

Here we shall discuss some plant polysaccharides, and leave the animal polymers which contain both uronic acids and amino sugars until the next section.

In the primary cell-walls and intercellular material of land plants are located the pectic substances, polysaccharides which can be extracted with hot water, EDTA, or dilute acid. These polysaccharides are mainly galactans (see Section 6.51, p. 294), arabinans (see Section 6.5.1, p. 293), arabinogalactans (see Section 6.5.3, p. 298), and polymers of galacturonic acid—the *pectic acids*. There is some evidence that they are covalently-bonded together in the cell-wall. Indeed, pectic acids have often been reported to have 'side-chains' which are long stretches of galactose or arabinose residues, i.e. galactans and/or arabinans. Although we have already mentioned pectic acid as a homopolymer (see Section 6.5.1, p. 294), the polysaccharide usually contains some rhamnose units in the main chain. Indeed the backbone of a pectic acid probably consists of stretches of contiguous α-(1 → 4)-linked D-galacturonic acid residues interrupted by a trisaccharide containing two rhamnose and one galacturonic acid units. Side-chains consisting of fucose and xylose, or galactose and xylose, are common. Thus a partial structure may be written as:

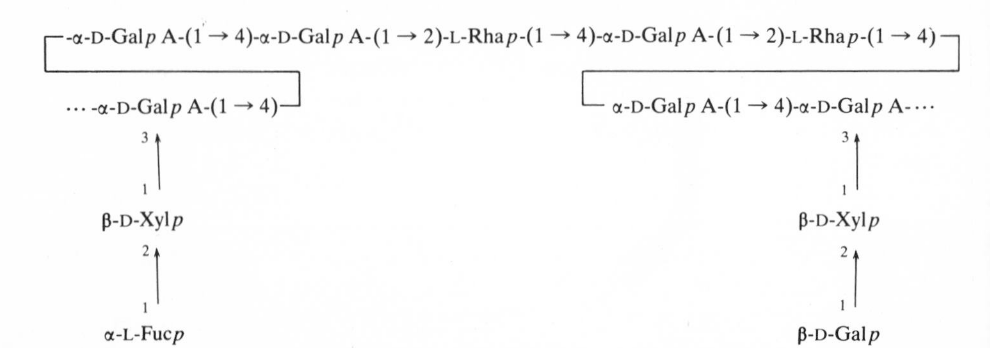

Weight-average molecular weights may be as high as 10^8. In cell-walls, galactans may be linked to pectic acids through the rhamnose residues.

Such polymers are found in cell-walls when rapid growth is taking place. They confer elasticity on the cell-wall, in order to accommodate the changes in cell volume then occurring. They are polydisperse with respect to size, nature and distribution of side-chains, and also in the extent of esterification. *Pectinic acids*—the polysaccharides with a high proportion of methyl ester groupings—are located more in intercellular material, whilst pectic acids with little esterification are concentrated in the primary cell-wall.

Pectins are capable of forming gels, particularly in the presence of sucrose, and so are very important in the jam and jelly industry.

Various models have been proposed for plant cell-walls, in which attempts have been made to explain the interrelations of the different polysaccharides, but to date none are completely successful.

Some gum exudates contain polysaccharides which are very similar to the pectic acids. For example, *tragacanthic acid*, the principal polysaccharide of *gum tragacanth*, has a structure which closely resembles that shown above for pectic acid, but the polymer backbone may not contain rhamnose. (The high viscosity of tragacanthic acid makes it useful as a food thickener.)

In some brown seaweeds, a polysaccharide is found in cell-walls, closely associated with cellulose, which can be considered the marine plant equivalent of pectic acid. This is *alginic acid*, a (1 → 4)-linked polymer of β-D-mannuronic acid and α-L-guluronic acid (see Figure 6.32). The main chain is thought to consist of blocks of contiguous mannuronic or gluronic acid residues. It has been found that the stretches of guluronic acid units give a more rigid structure. In the seaweeds, more rigid tissues contain a higher proportion of guluronic acid than flexible tissues. (Alginic acids form stable gels, and are widely used in the food industry.)

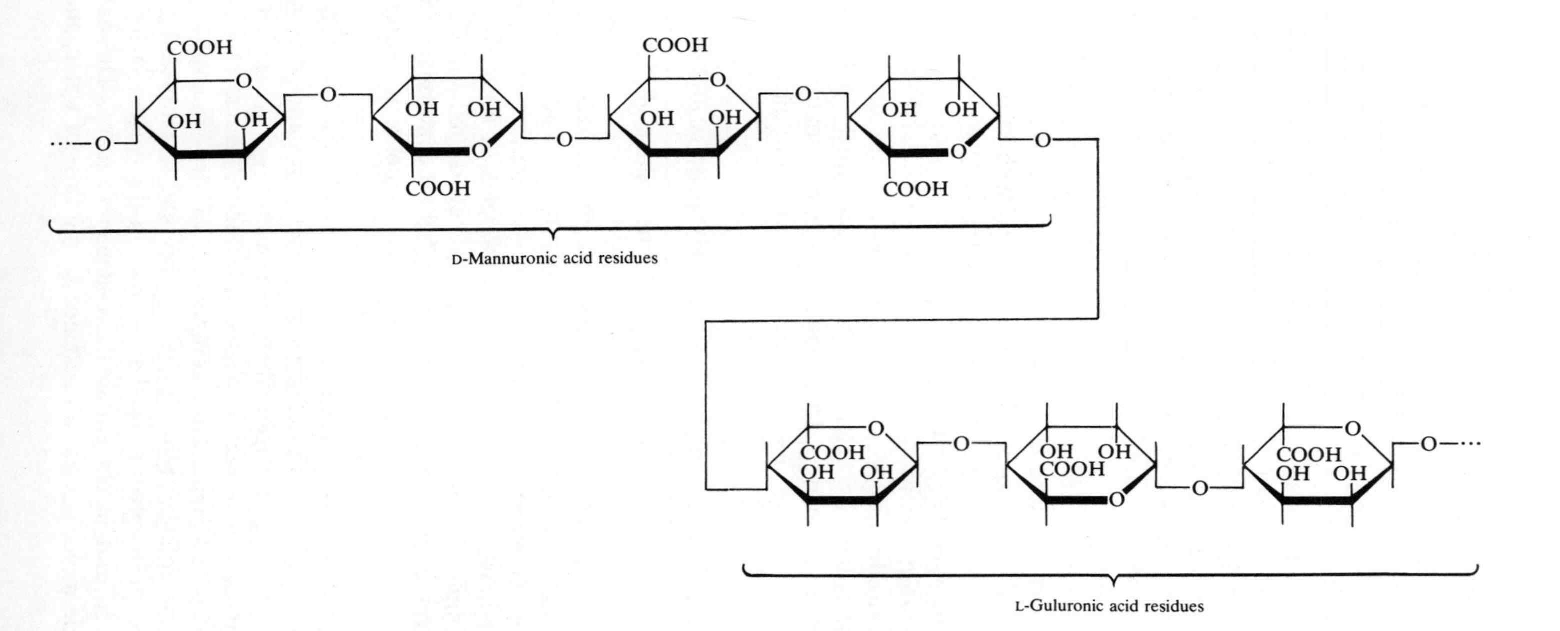

FIGURE 6.32. Partial structure of alginic acid

6.5.2.5. Glycosaminoglycans

These are polysaccharides containing amino sugar residues, and are mostly of animal origin, although some are synthesized by bacteria. Most glycosaminoglycans consist of linear molecules of repeating disaccharides (see Figure 6.33).

Hyaluronic acid is found in skin, connective tissues, umbilical cord, and the synovial fluid of joint. The weight-average molecular weight can be very large, over 10^7, and solutions of hyaluronic acid are very viscous. X-ray diffraction studies indicate that the macromolecules can adopt a helical conformation, but it is not known whether they do so in Nature. The polymer molecules are capable, however, of forming gels, and this property is believed to be important for the natural function of hyaluronic acid. In the synovial fluid of joints, for example, it probably acts as a lubricant, and in cartilage it may also function, along with chondroitin sulphates, as a shock absorber. In disease states, such as osteoarthritis, the hyaluronic acid of joints is partly degraded, resulting in a marked loss of elasticity of the synovial fluid.

Chondroitin sulphates are found in cartilage, bone, and skin, but not as free polysaccharides. They exist as proteoglycan complexes where the polysaccharide is linked covalently to protein through the reducing ends of the polysaccharide chains. This linkage region contains a particular monosaccharide sequence which is found in other protein–polysaccharide complexes i.e.

$$\cdots\text{-}\beta\text{-D-Glc}p\,\text{A-}(1\rightarrow 3)\text{-}\beta\text{-D-Gal}p\text{-}(1\rightarrow 3)\text{-}\beta\text{-D-Gal}p\text{-}(1\rightarrow 4)\text{-}\beta\text{-D-Xyl}p\text{— Serine residue of protein}$$

The proteoglycan of cartilage consists of protein (approx. 10%) and keratan sulphate (see below), as well as chondroitin sulphate (mainly the 4-sulphate in humans) covalently linked to this protein. Individual chondroitin sulphate chains have molecular weights of up to 5×10^4, and the whole complex has a molecular weight of several million. Chondroitin sulphate molecules are capable of adopting a helical conformation in fibres, but whether they do so in the proteoglycan complex in Nature is not known.

The function of this proteoglycan in cartilage can be regarded as analogous to that of the non-cellulosic polysaccharides and protein of plant cell-walls. In cartilage, collagen fibres (see section 4.5.5.1, p. 146) supply the tensile strength provided in plants by cellulose fibres, whilst cartilage proteoglycan is an important component of the amorphous matrix surrounding the collagen fibres, giving rigidity and incompressibility to the structure. The proteoglycan may have a rôle in the correct laying down of collagen fibres, and in cementing the collagen in place. The network of polysaccharide molecules has a resistance to the flow of water, and so may

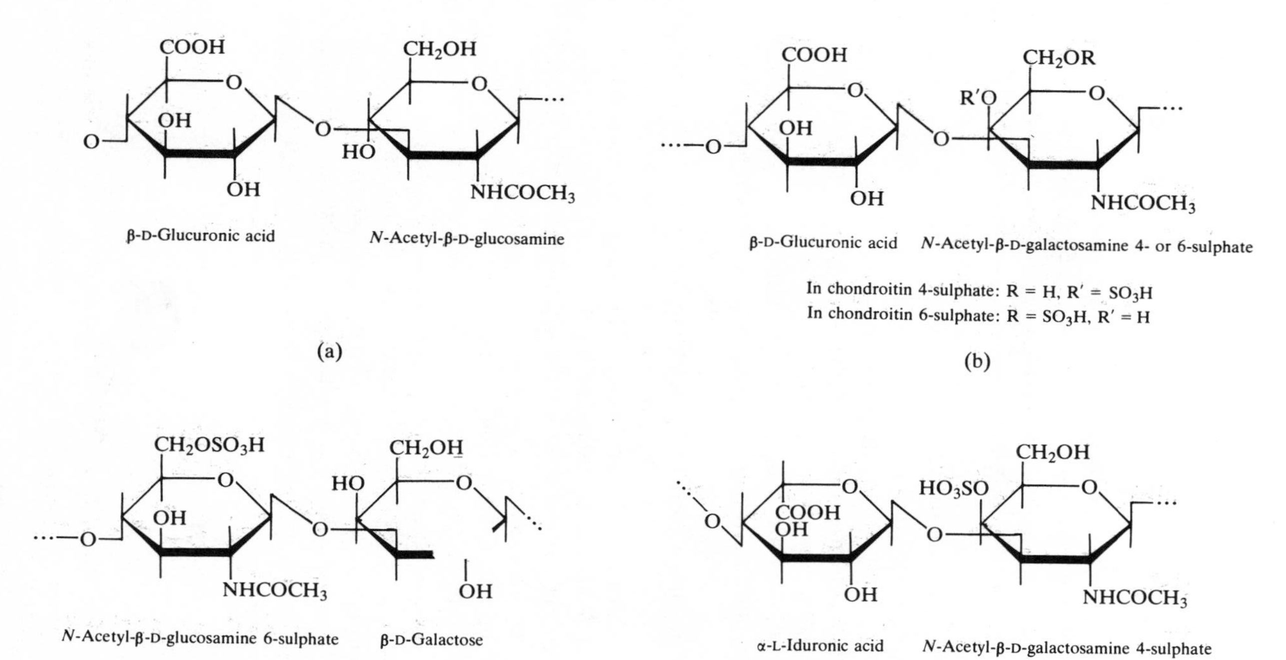

FIGURE 6.33. Repeating disaccharides of some glycosaminoglycans: (a) hyaluronic acid; (b) chondroitin 4- or 6-sulphate; (c) keratan sulphate; (d) dermatan sulphate

help to keep cartilage hydrated. Also, such a system can undergo plastic deformation on compression—the water is squeezed out slowly to a neighbouring uncompressed region, and returns slowly when compression is removed. Thus the protein–polysaccharide network can act as a shock absorber.

Keratan sulphate, the second polysaccharide component of cartilage proteoglycan, is also found in the cornea of the eye. The polymer chains are usually smaller than those of chondroitin sulphate, with molecular weights up to 2×10^4. The reducing end of a keratan sulphate chain is linked to protein through a sequence of monosaccharide residues different to that shown above for chondroitin sulphate, involving instead a link through mannose and an amino sugar to an asparagine or serine residue of the protein.

Dermatan sulphate occurs in skin, tendon, and arterial walls, where it is found as part of proteoglycan complex. Molecules of dermatan sulphate are about the same size as those of chondroitin sulphate, and are capable of adopting a helical conformation. The linkage region of polysaccharide to protein resembles that for chondroitin sulphate, and glucuronic acid can substitute for some iduronic acid within the chains. Dermatan sulphate-protein complexes may interact with collagen fibres in tissues in a manner analogous to chondroitin sulphate proteoglycans.

Heparin, found in lung, liver, and arterial walls, is rather more complex in structure, for it contains D-glucuronic acid, L-iduronic acid, and D-glucosamine residues (see Figure 6.34). The glucosamine residues may be *N*-acetylated or *N*-sulphated. Because heparin is fairly resistant to acid hydrolysis, it is not yet certain whether the glucuronic acid residues in the polymer are α- or β-anomers. The glucuronic acid and iduronic acid residues are not distributed randomly, but occur in blocks. The chains usually have molecular weights of up to 20,000, and heparin is found both as the free polysaccharide and covalently-linked to protein in animal tissues. The protein–polysaccharide linkage region is similar to that found for chondroitin sulphate proteoglycans. Heparin acts as an anticoagulant—an inhibitor of blood clotting—and although it is widely used for this purpose in medicine, it is not certain this is the sole function of heparin in Nature.

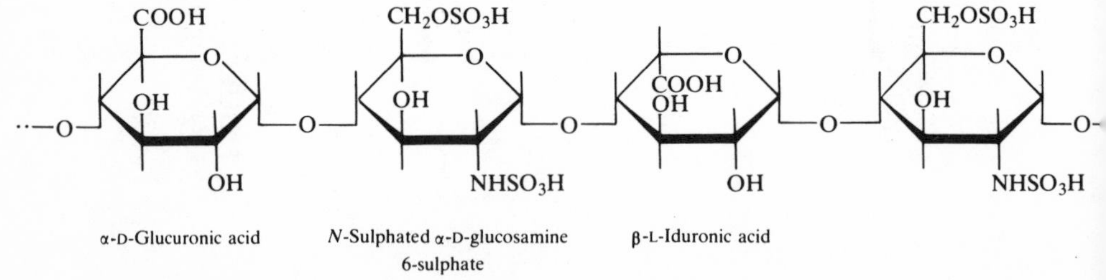

COOH CH$_2$OSO$_3$H CH$_2$OSO$_3$H

OH OH COOH OH OH

OH NHSO$_3$H OH NHSO$_3$H

α-D-Glucuronic acid — *N*-Sulphated α-D-glucosamine 6-sulphate — β-L-Iduronic acid

FIGURE 6.34. Partial structure of heparin

6.5.2.6. Some bacterial polysaccharides and related polymers

The cell-walls of many bacteria often contain polysaccharides which, like those of gum exudates, are extremely complex. Very many different capsular polysaccharides have been found, their structure depending on the bacterial source. The *pneumococcal capsular polysaccharides* (produced by different strains of *Streptococcus pneumoniae*), for example, have been widely studied, and many types occur—some with linear molecules, others with branched structures, some containing phosphate groups in the main chain, some with uronic acid residues, and others with amino sugars. At this stage, no generalization about the structure of pneumococcal polysaccharides can be made. One complex example is shown in Figure 6.35(a).

Gram-negative bacteria have *lipopolysaccharide* at their surface, i.e. polysaccharide covalently linked to lipid. The polysaccharide usually

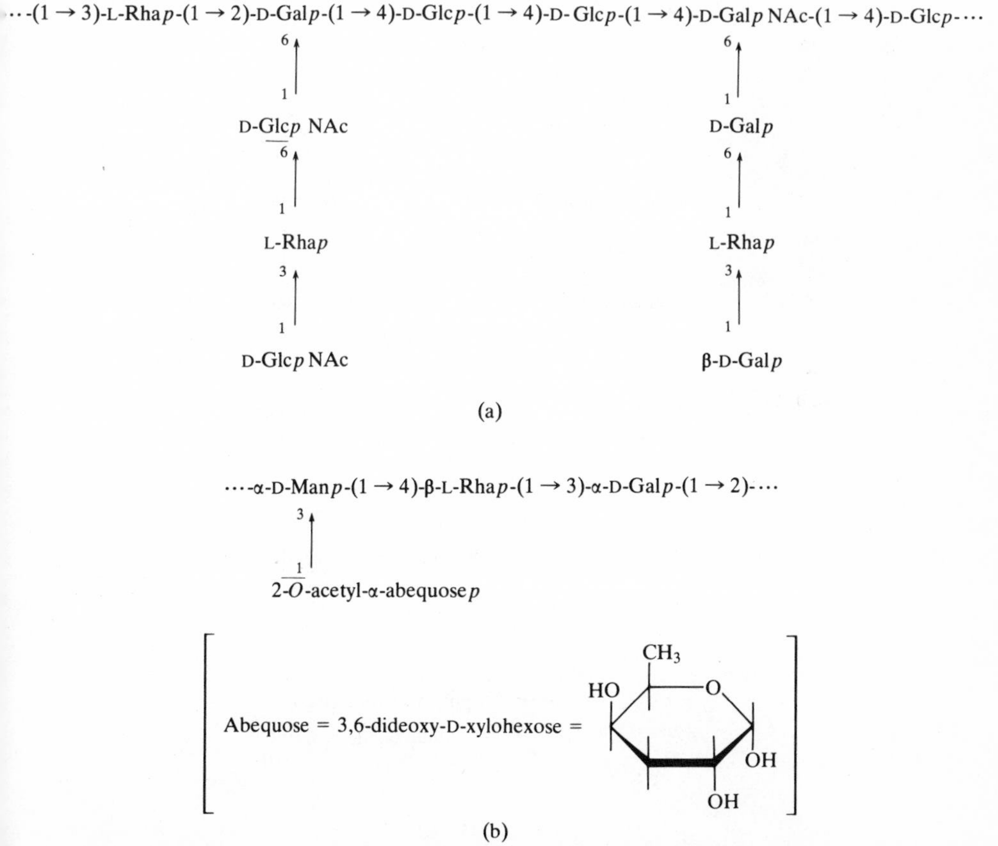

···-(1 → 3)-L-Rha*p*-(1 → 2)-D-Gal*p*-(1 → 4)-D-Glc*p*-(1 → 4)-D-Glc*p*-(1 → 4)-D-Gal*p* NAc-(1 → 4)-D-Glc*p*-···

Branch on D-Gal*p* (6 ← 1): D-Glc*p* NAc (6 ← 1) L-Rha*p* (3 ← 1) D-Glc*p* NAc

Branch on D-Gal*p* NAc (6 ← 1): D-Gal*p* (6 ← 1) L-Rha*p* (3 ← 1) β-D-Gal*p*

(a)

····α-D-Man*p*-(1 → 4)-β-L-Rha*p*-(1 → 3)-α-D-Gal*p*-(1 → 2)-····

Branch on α-D-Man*p* (3 ← 1): 2-*O*-acetyl-α-abequose*p*

[Abequose = 3,6-dideoxy-D-xylohexose =]

(b)

FIGURE 6.35. Repeat units of some bacterial polysaccharides: (a) complex repeat of pneumococcal Type 7 capsular polysaccharide; (b) outer chain repeat of lipopolysaccharide of *Salmonella typhimurium*

consists of a repeating tetra- or penta-saccharide unit (which varies with the nature of the bacterial source, see Figure 6.35(b)), linked, through an inner core of some half dozen monosaccharide residues, to lipid. The lipopolysaccharide itself may form part of a larger protein–polysaccharide–lipid complex.

Peptidoglycan, also known as *murein*, is the material which maintains the

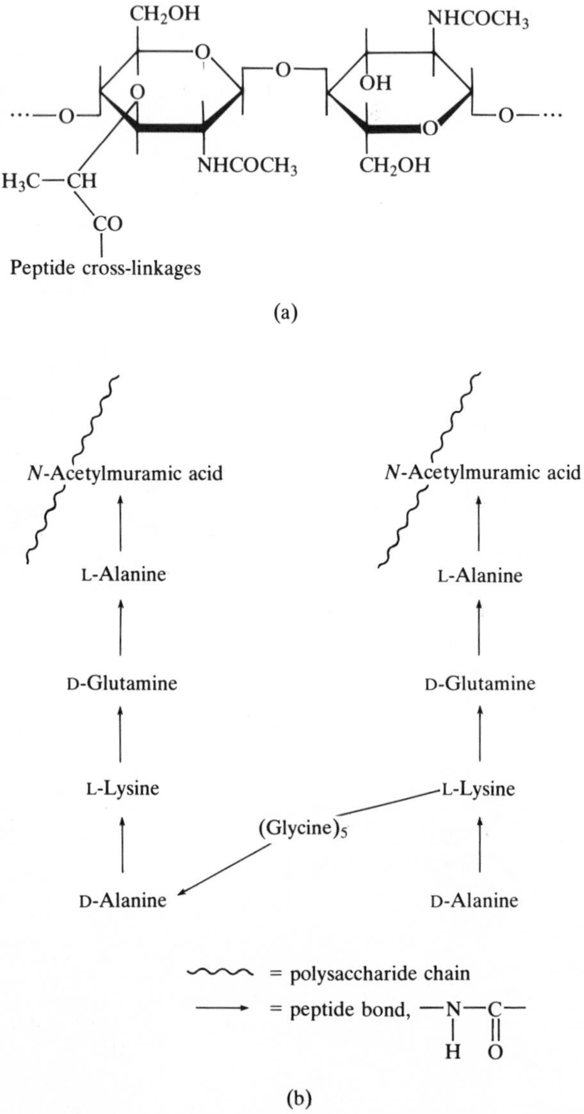

CH_2OH NHCOCH$_3$ O OH ···—O— —O—··· H_3C—CH NHCOCH$_3$ CH_2OH CO Peptide cross-linkages

(a)

N-Acetylmuramic acid → L-Alanine → D-Glutamine → L-Lysine → D-Alanine

N-Acetylmuramic acid → L-Alanine → D-Glutamine → L-Lysine → D-Alanine

(Glycine)$_5$

〰 = polysaccharide chain

→ = peptide bond, —N(H)—C(=O)—

(b)

FIGURE 6.36. A typical peptidoglycan: (a) repeat unit of the polysaccharide chain; (b) peptide cross-link

shape of most bacteria. It consists essentially of polysaccharide strands cross-linked by peptides to give a very large 'molecule', which could conceivably cover an entire bacterium. The polysaccharide chains contain β-(1 → 4)-linked alternating residues of *N*-acetylglucosamine and *N*-acetylmuramic acid (the 3-*O*-carboxyethyl derivative of *N*-acetylglucosamine) as shown in Figure 6.36(a). These chains are usually 10–50 sugar residues long, and are the natural substrate for the enzyme lysozyme (see Section 4.5.2.3, p. 121). The nature of the cross-linking peptides depends on the bacterial source, and one possible example is given in Figure 6.36(b). (Note that D-amino acid residues occur in peptidoglycans.) In gram-negative bacteria, peptidoglycan constitutes a smaller proportion of the cell-wall than in gram-positive bacteria, but even in the latter, other polymers such as protein or teichoic acids are important components of the walls.

Teichoic acids are closely related to polysaccharides, and are most often polymers of glycerol or ribitol phosphate, but some teichoic acids contain hexose phosphates and neither glycerol nor ribitol. These polymers are

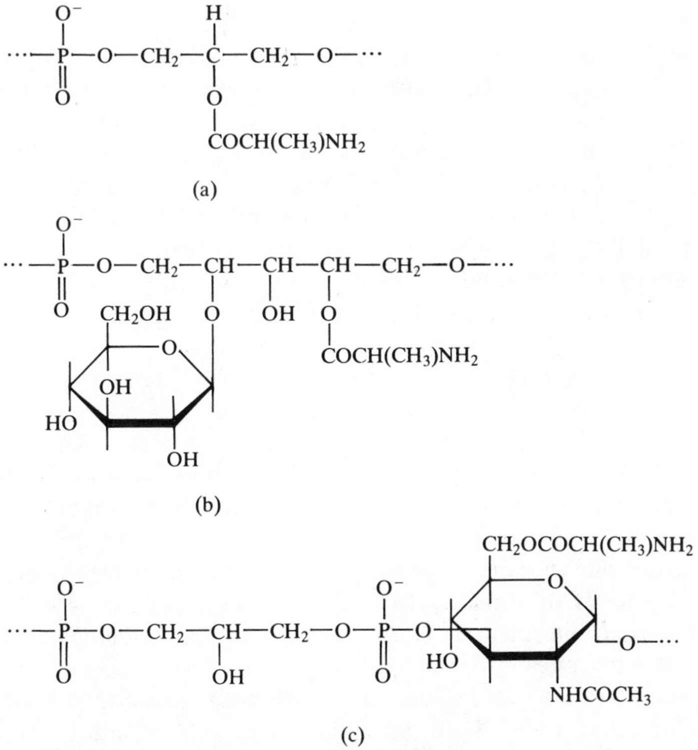

$$\cdots\!-\!P(=O)(O^-)\!-\!O\!-\!CH_2\!-\!CH(OCOCH(CH_3)NH_2)\!-\!CH_2\!-\!O\!-\!\cdots$$

(a)

$$\cdots\!-\!P(=O)(O^-)\!-\!O\!-\!CH_2\!-\!CH(O\text{-glucosyl})\!-\!CH(OH)\!-\!CH(OCOCH(CH_3)NH_2)\!-\!CH_2\!-\!O\!-\!\cdots$$

(b)

$$\cdots\!-\!P(=O)(O^-)\!-\!O\!-\!CH_2\!-\!CH(OH)\!-\!CH_2\!-\!O\!-\!P(=O)(O^-)\!-\!O\!-\!\text{hexosamine}(CH_2OCOCH(CH_3)NH_2,\ HO,\ NHCOCH_3)\!-\!O\!-\!\cdots$$

(c)

FIGURE 6.37. Repeat units of teichoic acids: (a) glycerol phosphate backbone; (b) ribitol phosphate backbone; (c) glycerol phosphate–hexosamine phosphate backbone

found in bacterial cell-walls and membranes. Ribitol teichoic acids occur in cell-walls only, whilst the glycerol biopolymers are components of both walls and membranes. In cell-walls, teichoic acids exist bonded through phosphate bridges to muramic acid residues of peptidogylcan. Three types of teichoic acid are illustrated in Figure 6.37.

6.6. BIOSYNTHESIS

In general, polysaccharides are synthesized from *nucleoside diphosphate sugars*, compounds which are themselves formed by the reaction of a monosaccharide phosphate with a nucleoside triphosphate. Each stage in synthesis is catalysed by a specific enzyme, and the formation of uridine 5′-(D-glucopyranosyl pyrophosphate), i.e. uridine diphosphate glucose, or UDP-glucose for short, is shown in Figure 6.38. Not all nucleoside sugars are formed directly from the sugar phosphate and nucleoside triphosphate; some are formed instead by conversion of, say, UDP-glucose. For example, UDP-galacturonic acid synthesis can be summarized as:

$$\text{UDP-glucose} \xrightarrow[\text{at C-6 of glucose}]{\text{oxidation}} \text{UDP-glucuronic acid} \xrightarrow[\text{at C-4}]{\text{epimerization}} \text{UDP-galacturonic acid}$$

Most monosaccharides are utilized for polysaccharide synthesis in the form of UDP-sugars. However, starch is probably synthesized from adenine 5′-(D-glucopyranosyl pyrophosphate), i.e. ADP-glucose, although UDP-glucose can also be utilized. The guanosine derivative, GDP-glucose, has been implicated in both cellulose and glucomannan formation. Mannose and mannuronic acid may in general, be transferred to polymer from the GDP-sugar. The bacterial wall polymer, teichoic acid, is synthesized from the cytidine derivative of glycerol or ribitol. (For the structures of adenosine, guanosine, cytidine, and uridine see Figure 5.2, p. 170).

The monosaccharide is then transferred from the nucleotide to the non-reducing end of a growing polysaccharide chain. The reaction is catalysed by an appropriate transferase enzyme. Generally, an oligosaccharide primer is required, to which the monosaccharide residue can be added from the nucleotide. Details of the synthesis of primer have not yet been elucidated.

During the biosynthesis of some polysaccharides, transfer is believed to take place directly from the nucleotide sugar to the polymer. In other cases, however, a lipid intermediate is involved. In this case, a monosaccharide is transferred from nucleotide to lipid to form an oligosaccharide–lipid, and the oligosaccharide is later added as a unit to the growing polysaccharide. The requirement for a lipid intermediate has been most convincingly demonstrated for bacterial polysaccharides and many glycoproteins.

The lipid intermediates studied to date are polyprenols of the type $H\text{-}[CH_2{-}C(CH_3){=}CH{-}CH_2]_n\text{-}OH$, where n = 5–24, and one or more

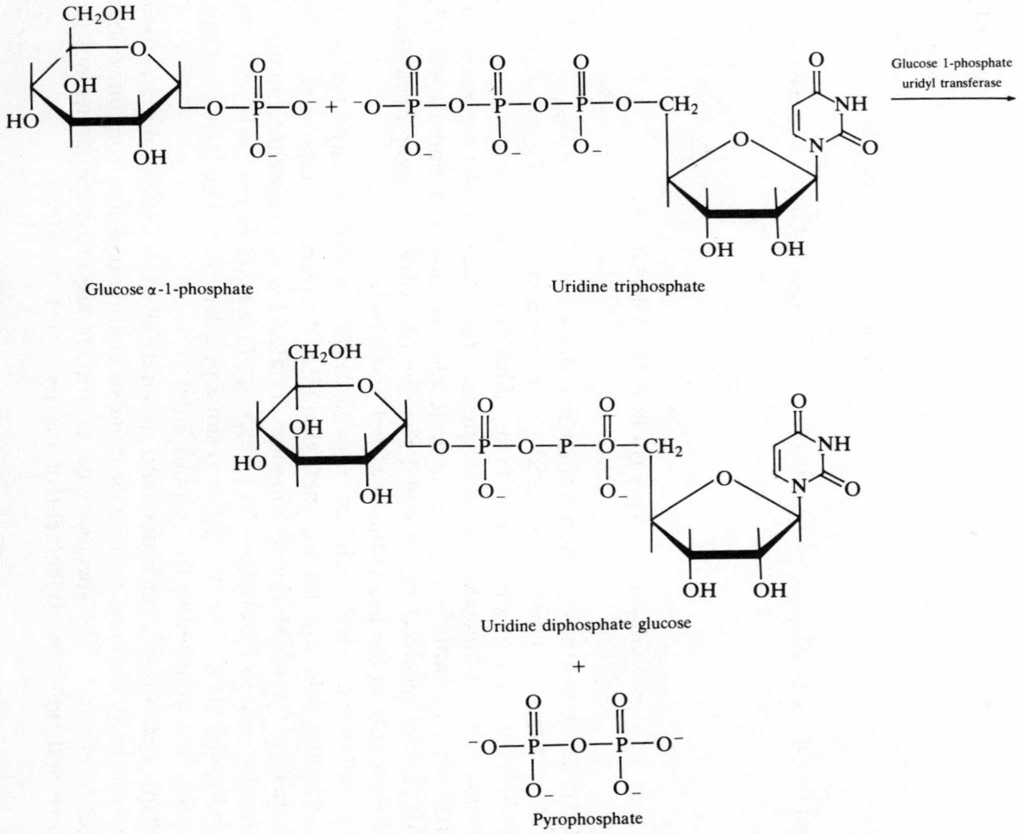

Glucose 1-phosphate uridyl transferase

Glucose α-1-phosphate + Uridine triphosphate → Uridine diphosphate glucose + Pyrophosphate

FIGURE 6.38. Nucleoside diphosphate sugar formation

isoprene residues may be saturated, usually the residue adjacent to the hydroxyl group. These polyprenols are found in a wide range of living organisms, from bacteria to mammals, and they are involved in polysaccharide biosynthesis as the mono- or di-phosphates.

In general, plant polysaccharides are thought to be synthesized by direct monosaccharide transfer from the nucleotide to the polymer. The rôle, if any, of polyprenols in plant polysaccharide biosynthesis is at present very uncertain. Thus the biosynthesis of a *linear portion of a glucan chain* in starch may be represented as in Figure 6.39. Amylose can be synthesized by this mechanism, but a further reaction catalysed by a branching enzyme is required for amylopectin formation. Such branching enzymes catalyse scission of an oligosaccharide from the non-reducing end of an existing polymer chain, and reattach the oligosaccharide by an α-(1 → 6)-bond, i.e.:

····-α-D-Glc*p*-(1 → 4)-α-D-Glc*p*-(1 → 4)-α-D-Glc*p*- (1 → 4)-α-D-Glc*p*-(1 ⇸ 4)-α-D-Glc*p*-··

scission

↓ α-1,4-glucan: α-1,4-glucan 6-glucosyl transferase

····-Glc*p*-(1 → 4)-α-D-Glc*p*-(1 → 4)-α-D-Glc*p*-(1 → 4)-α-D-Glc*p*-(1 → 4)-···

6 ↑ 1

··· α-D-Glc*p*-(1 → 4)-α-D-Glc*p*-(1 → 4)-α-D-Glc*p*-(1 → 4)-α-D-Glc*p*

Glycogen is synthesized in a similar manner from UDP-glucose.

Other linear homopolysaccharides are formed as in Figure 6.39 in reactions involving the appropriate nucleotide sugar and transferase enzyme. For branched homopolymers, the factors determining the distribution of branches within a molecule are not well understood. (In 1970, Leloir received the Nobel Prize for his work on sugar nucleotides and their rôle in the biosynthesis of polysaccharides.)

In polysaccharides such as *pectinic acid*, where methyl esters of galacturonic acid are found, methylation is believed to take place after polymer synthesis (Methyl groups are donated by *S*-adenosylmethionine, a molecule where $HOOC—CH_2(NH_2)—(CH_2)_2—\overset{+}{S}(CH_3)—$ replaces the —OH group on C-5 of the ribose residue of adenosine. The 'active' methyl group is that attached to the sulphur atom.)

Plant heteropolysaccharides and animal glycosaminoglycans are also synthesized by stepwise transfer of monosaccharide residues from nucleotide sugar to polymer. The specificity of the transferase enzymes determines the nature and point of attachment of the incoming monomer.

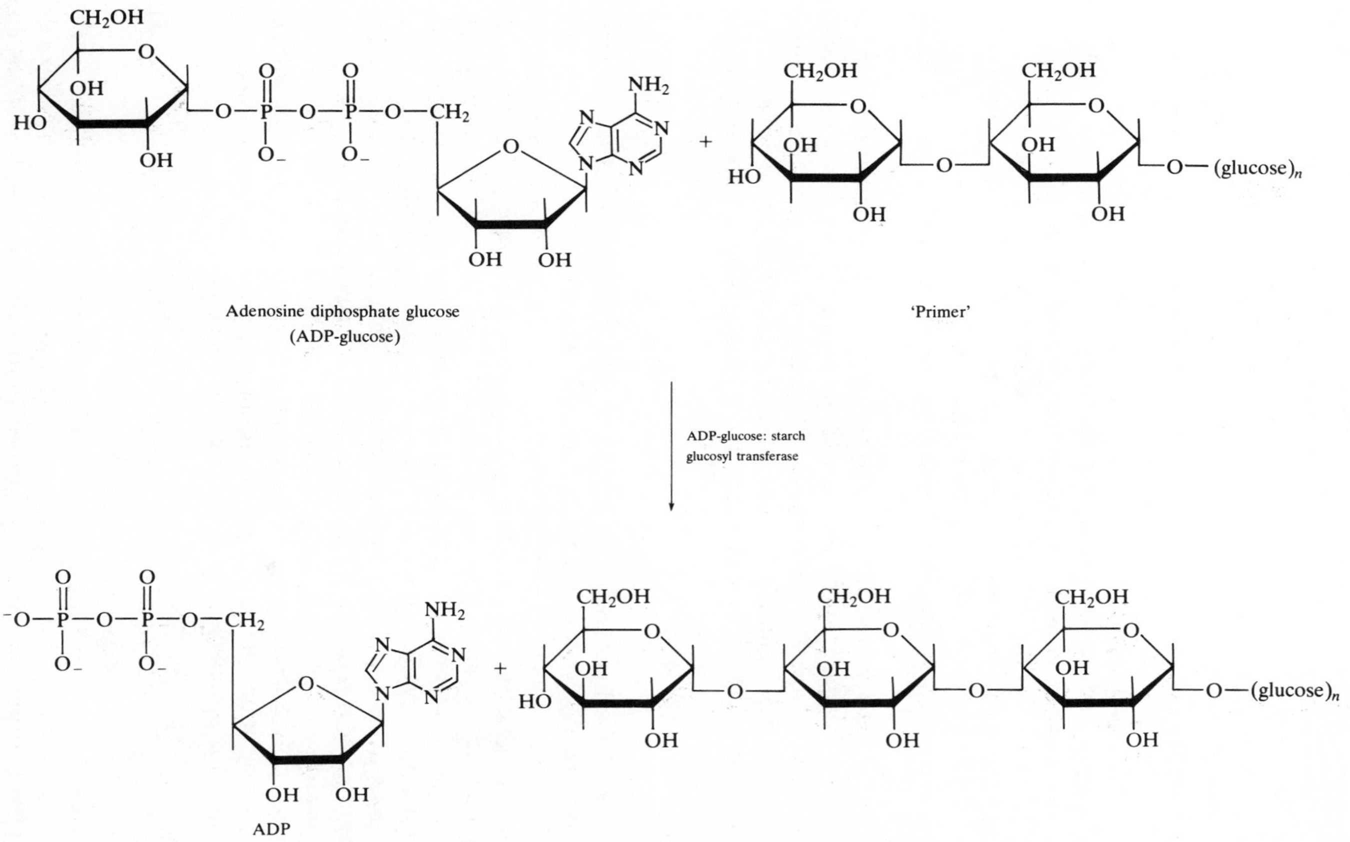

CH_2OH OH HO OH O—P—O—P—O—CH_2 O^- O^- NH_2 N OH OH

Adenosine diphosphate glucose
(ADP-glucose)

+ CH_2OH OH HO OH O CH_2OH OH OH O—(glucose)$_n$

'Primer'

ADP-glucose: starch
glucosyl transferase

$^-$O—P—O—P—O—CH_2 O^- O^- NH_2 N OH OH

ADP

+ CH_2OH OH HO OH O CH_2OH OH OH O CH_2OH OH OH O—(glucose)$_n$

FIGURE 6.39. Lengthening of a polysaccharide chain: amylose synthesis

Most glycosaminoglycans are synthesized in the form of protein-polysaccharides, i.e. the polysaccharide chains are initiated and grow on completed protein chains. Thus for a chondoitin sulphate–protein complex, xylose is first transferred enzymically from UDP-xylose to a serine residue of the protein. Then a galactose residue from UDP-galactose is attached to the xylose, and a second galactose residue is added to the non-reducing end of the galactosyl–xylosyl–serine chain (for the sequence of protein-polysaccharide linkage region see Section 6.5.2, p. 306). Next, glucuronic acid is transferred to the growing chain, and finally *N*-acetylglucosamine and glucuronic acid residues are added alternately to the non-reducing end of the polysaccharide, both from UDP-derivatives. The monosaccharides are added in the correct order by the transferases concerned—these enzymes are specific for both the sugar being transferred from UDP, and the residue already on the non-reducing end of the growing chain. Thus the transferases determine the monosaccharide sequence in the polysaccharide.

After the beginning of chain synthesis of a polysaccharide such as *chondroitin sulphate*, sugar residues in the chain are sulphated by specific sulphotransferases. These enzymes catalyse the transfer of a sulphate group from adenosine-3-phosphate-5-phosphosulphate. (Adenosine-3-phosphate-5-phosphosulphate is an adenosine molecule carrying a phosphate group instead of a hydroxyl group at ribose C-3, and a $^{-}O-\overset{O}{\underset{O}{S}}-O-\overset{O}{\underset{O^{-}}{P}}-O-$ group in place of the hydroxyl group at C-5 of the ribose ring.) All the enzymes involved occur as a large membrane-bound complex.

Heparin is thought to be synthesized by addition of D-glucuronic acid and *N*-acetyl glucosamine residues. After polysaccharide formation, the glucosamine is deacetylated and subsequently sulphated at the amino group. L-Iduronic acid units are produced from some of the D-glucuronic acid residues by enzymic epimerization at C-6. Further sulphation at C-6 of the glucosamine, or C-2 of the uronic acid residue, can take place. The L-iduronic acid residues of dermatan sulphate may be formed by a similar mechanism.

Some polysaccharides may be synthesized by transglycosylation without the intervention of nucleoside diphosphate sugars. Thus *dextrans* and *fructans* are probably formed directly from sucrose:

$$\underset{\text{Sucrose}}{\text{Glc—Fru}} + \underset{\text{Dextran}}{\text{Glc—|Glc|—Glc—}\cdots} \longrightarrow \text{Glc—Glc—Glc—Glc—}\cdots + \text{Fru}$$

or

$$\underset{\text{Sucrose}}{\text{Glc—Fru}} + \underset{\text{Fructan}}{\text{Fru—Fru—Fru—}\cdots} \longrightarrow \text{Fru—Fru—Fru—Fru—}\cdots + \text{Glc}$$

For fructans the 'new' monosaccharide residue is transferred from sucrose to the non-reducing end of the growing polysaccharide chain, whilst for dextrans the opposite occurs.

The biosynthesis of the *bacterial polysaccharides* discussed earlier (Section 6.5.2, p. 309) is more complex than the syntheses outlined above. Lipid intermediates are involved in the production of capsular and lipo-polysaccharides, peptidoglycans, and teichoic acids. Similar intermediates also participate in the synthesis of the carbohydrate portion of many glycoproteins. One of the commonest of these intermediates is undecaprenol phosphate:

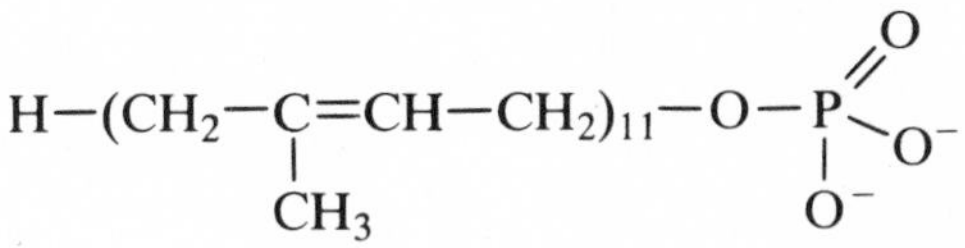

$$\mathrm{H{-}(CH_2{-}\underset{\underset{CH_3}{|}}{C}{=}CH{-}CH_2)_{11}{-}O{-}\underset{\underset{O^-}{|}}{P}(=O){-}O^-}$$

The repeat oligosaccharide of a capsular or lipo-polysaccharide is synthesized on a lipid phosphate or diphosphate by transfer of individual monosaccharide residues from nucleoside diphosphate sugars to the non-reducing end of the growing oligosaccharide. Thus steps in the synthesis of a capsular polysaccharide of the type $\mathrm{\left[Gal{-}\underset{\underset{Glc\,A}{|}}{Man}{-}Gal\right]_n}$ would be:

$$\mathrm{UDP\text{-}galactose + \bar{O}{-}\overset{\overset{O}{\|}}{\underset{\underset{O^-}{|}}{P}}{-}O{-}undecaprenol \longrightarrow}$$

$$\mathrm{Gal{-}O{-}\overset{\overset{O}{\|}}{\underset{\underset{O^-}{|}}{P}}{-}O{-}\overset{\overset{O}{\|}}{\underset{\underset{O^-}{|}}{P}}{-}O{-}undecaprenol + UMP}$$

$$\mathrm{GDP\text{-}mannose + Gal{-}O{-}\overset{\overset{O}{\|}}{\underset{\underset{O^-}{|}}{P}}{-}O{-}\overset{\overset{O}{\|}}{\underset{\underset{O^-}{|}}{P}}{-}O{-}undecaprenol \longrightarrow}$$

$$\mathrm{Man{-}Gal{-}O{-}\overset{\overset{O}{\|}}{\underset{\underset{O^-}{|}}{P}}{-}O{-}\overset{\overset{O}{\|}}{\underset{\underset{O^-}{|}}{P}}{-}O{-}undecaprenol + GDP}$$

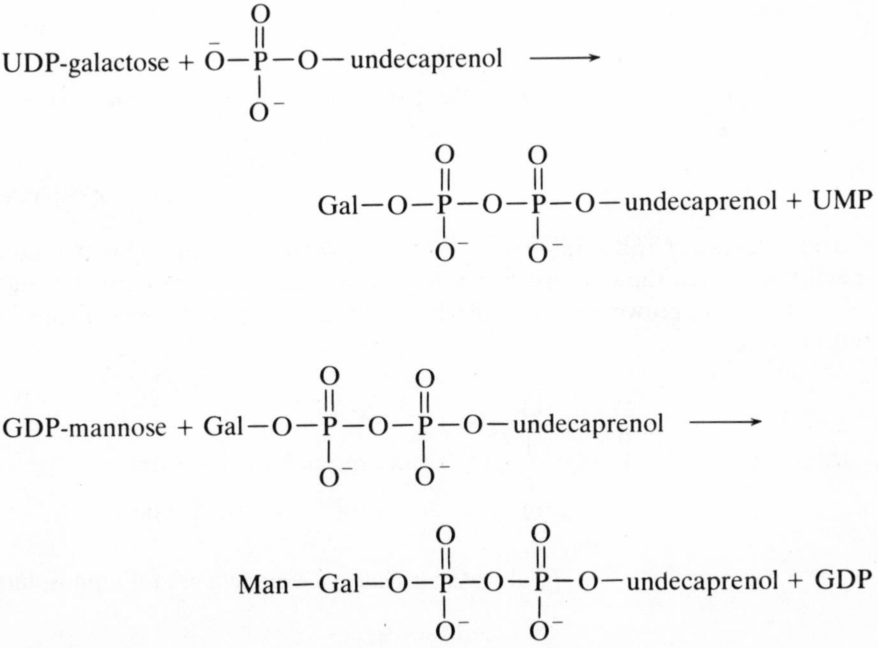

$$\text{UDP-Glc A} + \text{Man}-\text{Gal}-\text{O}-\underset{\underset{\mathrm{O^-}}{|}}{\overset{\overset{\mathrm{O}}{\|}}{\mathrm{P}}}-\text{O}-\underset{\underset{\mathrm{O^-}}{|}}{\overset{\overset{\mathrm{O}}{\|}}{\mathrm{P}}}-\text{O}-\text{undecaprenol} \longrightarrow$$

$$\underset{\underset{\text{Glc A}}{|}}{\text{Man}}-\text{Gal}-\text{O}-\underset{\underset{\mathrm{O^-}}{|}}{\overset{\overset{\mathrm{O}}{\|}}{\mathrm{P}}}-\text{O}-\underset{\underset{\mathrm{O^-}}{|}}{\overset{\overset{\mathrm{O}}{\|}}{\mathrm{P}}}-\text{O}-\text{undecaprenol} + \text{UDP}$$

$$\text{UDP}-\text{Gal} + \underset{\underset{\text{Glc A}}{|}}{\text{Man}}-\text{Gal}-\text{O}-\underset{\underset{\mathrm{O^-}}{|}}{\overset{\overset{\mathrm{O}}{\|}}{\mathrm{P}}}-\text{O}-\underset{\underset{\mathrm{O^-}}{|}}{\overset{\overset{\mathrm{O}}{\|}}{\mathrm{P}}}-\text{O}-\text{undecaprenol} \longrightarrow$$

$$\text{Gal}-\underset{\underset{\text{Glc A}}{|}}{\text{Man}}-\text{Gal}-\text{O}-\underset{\underset{\mathrm{O^-}}{|}}{\overset{\overset{\mathrm{O}}{\|}}{\mathrm{P}}}-\text{O}-\underset{\underset{\mathrm{O^-}}{|}}{\overset{\overset{\mathrm{O}}{\|}}{\mathrm{P}}}-\text{O}-\text{undecaprenol} + \text{UDP}$$

Two or more of these oligosaccharide–lipid complexes may react together to give a lengthened oligosaccharide attached to lipid e.g.:

$$2\ \text{Gal}-\underset{\underset{\text{Glc A}}{|}}{\text{Man}}-\text{Gal}-\text{O}-\underset{\underset{\mathrm{O^-}}{|}}{\overset{\overset{\mathrm{O}}{\|}}{\mathrm{P}}}-\text{O}-\underset{\underset{\mathrm{O^-}}{|}}{\overset{\overset{\mathrm{O}}{\|}}{\mathrm{P}}}-\text{O}-\text{undecaprenol} \longrightarrow$$

$$[\text{Gal}-\underset{\underset{\text{Glc A}}{|}}{\text{Man}}-\text{Gal}]_2-\text{O}-\underset{\underset{\mathrm{O^-}}{|}}{\overset{\overset{\mathrm{O}}{\|}}{\mathrm{P}}}-\text{O}-\underset{\underset{\mathrm{O^-}}{|}}{\overset{\overset{\mathrm{O}}{\|}}{\mathrm{P}}}-\text{O}-\text{undecaprenol}$$

$$+ \text{ undecaprenol diphosphate}$$

This oligosaccharide–lipid complex may then pass through the cell membrane, when the oligosaccharide is transferred to the non-reducing end of a primer or growing polysaccharide, and undecaprenol diphosphate is released, i.e.:

$$[\text{Gal}-\text{Man}-\text{Gal}]_2-\text{O}-\underset{\underset{\mathrm{O^-}}{|}}{\overset{\overset{\mathrm{O}}{\|}}{\mathrm{P}}}-\text{O}-\underset{\underset{\mathrm{O^-}}{|}}{\overset{\overset{\mathrm{O}}{\|}}{\mathrm{P}}}-\text{O}-\text{undecaprenol} + [\text{Gal}-\underset{\underset{\text{Glc A}}{|}}{\text{Man}}-\text{Gal}]_n$$

$$\longrightarrow [\text{Gal}-\underset{\underset{\text{Glc A}}{|}}{\text{Man}}-\text{Gal}]_{n+2} + \text{ undecaprenol diphosphate}$$

The repeat unit of a teichoic acid is also synthesized on a lipid phosphate carrier by transfer of monomers from the appropriate nucleoside diphosphate derivative. In this case, a molecule such as glycerol phosphate or ribitol phosphate (i.e. with a phosphate group still attached) is transferred to the lipid intermediate. The complete repeating unit is then added to the growing polymer and lipid phosphate is released.

Peptidoglycan synthesis is even more complicated, for both monosaccharide and amino acid residues must be incorporated into the polymer.

Firstly the uridine diphosphate derivative of *N*-acetylmuramic acid is formed. Then, for the peptidoglycan shown on page 310, five amino acid residues are added in stepwise fashion, to give a UDP-muramyl pentapeptide i.e. UDP-Mur N Ac—L-alanine—D-glutamic acid—L-lysine—D-alanine—D-alanine. The linkage of D-glutamic acid to L-lysine is unusual, for the γ-COOH group of glutamic acid is involved, i.e.:

$$\text{to L-alanine} \longleftarrow \text{NH}-\underset{}{\overset{\text{COOH}}{\overset{|}{\text{CH}}}}-\text{CH}_2-\text{CH}_2-\text{CO} \longrightarrow \text{to L-lysine}$$

The muramyl pentapeptide is transferred to a lipid intermediate:

$$\text{UDP-Mur NAc}-\text{pentapeptide} + \text{phospholipid} \longrightarrow \underset{\text{pentapeptide}}{\underset{|}{\text{Mur NAc}}}-\text{diphospholipid} + \text{UMP}$$

Next, an *N*-acetylglucosamine residue is added to the non-reducing end of the muramic acid unit:

$$\text{UDP-GlcNAc} + \underset{\text{pentapeptide}}{\underset{|}{\text{Mur NAc}}}-\text{diphospholipid} \longrightarrow \beta\text{-Glc NAc-}(1 \rightarrow 4)-\underset{\text{pentapeptide}}{\underset{|}{\text{Mur NAc}}}-\text{diphospholipid}$$

The glycine residues (see Figure 6.36(b)) are then added to the growing repeat unit by transfer from a glycine-*t*-RNA complex. The transfer ribonucleic acids involved here differ from those participating in 'normal' protein biosynthesis (see Section 5.5.5, p. 213). Instead of the usual GT′ψC base sequence (see Figure 5.11, p. 192), there is a GUGC sequence, which may prevent interaction of these 'special' *t*-RNAs with ribosomes, and so stop them from becoming involved in protein synthesis. Thus we have:

β-Glc NAc-(1 → 4)-Mur NAc — diphospholipid
|
L-Ala
|
D-Glu
|
L-Lys
|
D-Ala
|
D-Ala

$\xrightarrow{\text{5 glycyl-}t\text{-RNA}}$

t-RNA
+
β-Glc NAc-(1 → 4)-Mur NAc — diphospholipid
|
L-Ala
|
D-Glu
|
$(Gly)_5$ — L-Lys
|
D-Ala
|
D-Ala

(for abbreviations of names of amino acids see Section 4.2, p. 76).

This completely assembled repeating unit is transferred to a growing polysaccharide chain of a peptidoglycan molecule and diphospholipid is released. Lastly, a cross-linking reaction takes place where the last D-alanine residue is hydrolysed from the pentapeptide chain and a glycine bridge is completed:

Polysaccharide		Polysaccharide		Polysaccharide		Polysaccharide
|		|		|		|
L-Ala		L-Ala		L-Ala		L-Ala
|		|		|		|
D-Glu		D-Glu		D-Glu		D-Glu
|	+	|	→	|		|
— L-Lys		$(Gly)_5$ — L-Lys		— L-Lys		L-Lys
|		|		|		|
D-Ala		D-Ala		D-Ala — $(Gly)_5$		D-Ala
-----|---		|		+		|
D-Ala				D-Ala		

(Pencillin prevents proper cross-linking, and hence kills growing bacteria. It has a greater effect on gram-positive than gram-negative bacteria as the former are surrounded by a thicker layer of peptidoglycan.)

6.7. CHEMICAL SYNTHESIS

To date, few syntheses of polysaccharides have been achieved by purely chemical reaction.

Some polymers such as amylose can be synthesized in the laboratory by the use of an enzyme, phosphorylase, which catalyses the reversible reaction:

$$n \text{ Glucose-1-phosphate} \rightleftharpoons [\text{glucose}]_n + n\text{-phosphate.}$$

Thus, in the presence of excess glucose-1-phosphate and with the removal of liberated phosphate, a synthetic amylose can be formed.

Phosphorylase is found in many plants, but there is some controversy as to whether its natural function involves synthesis or degradation of amylose. A similar enzyme in mammals is accepted as being important for glycogen breakdown.

Non-enzymic synthesis of polysaccharides is much more difficult, however, because of the number of reactive groups on a monosaccharide residue. Thus all hydroxyl groups, except those participating in the desired reaction, must be protected by blocking groups. Reactions have been developed for oligosaccharide synthesis, and some of these have been utilized to give small polysaccharides.

The Koenigs–Knorr reaction was amongst the first developed which was capable of yielding oligosaccharides. Here a sugar halide reacts with a free hydroxyl on another monosaccharide residue to give a disaccharide, with simultaneous inversion of configuration at the anomeric carbon (see Figure 6.40). Acetyl groups are often used to protect hydroxyl groups not participating in disaccharide formation, but benzyl groups may be used for the same purpose. Repetition of the Koenigs–Knorr reaction produces larger oligosaccharides. In these reactions α-halides yield β-residues in the oligosaccharide. It is more difficult to obtain oligosaccharides containing α-sugar units. However, the presence of mercuric salts in solvents such as nitromethane tends to promote the formation of α-glycosides if the starting material is a β-halide. Insertion of a group such as a nitro group at C-2 of the β-halide also favours the formation of an α-glycoside. In addition α-glycosides may be obtained from α-halides if the halide ion is included in the reaction mixture.

In the reactions shown in Figure 6.40, the new oligosaccharides contain (1 → 6)-bonds. Other intermonomer bonds can be formed by utilizing an isopropylidene anhydro derivative of one monosaccharide; for example, in the reaction shown in Figure 6.41, a (1 → 4)-bond results.

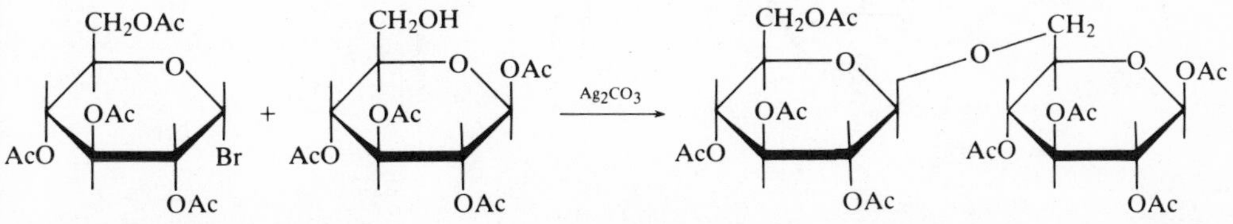

Tetra-*O*-acetyl-α-D-glucopyranose bromide

FIGURE 6.40. Example of oligosaccharide synthesis using Koenigs–Knorr reaction

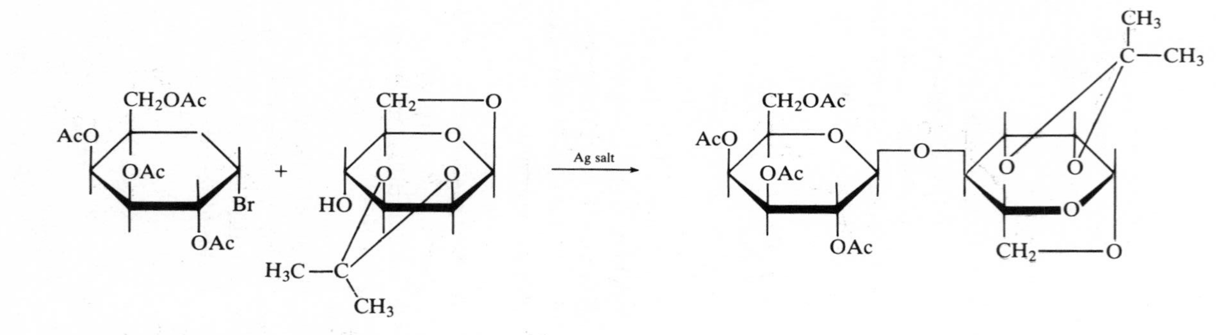

Tetra-*O*-acetyl-
α-D-galactopyranose bromide

1,6-Anhydro-2,3-*O*-isopropylidene-
D-mannopyranose

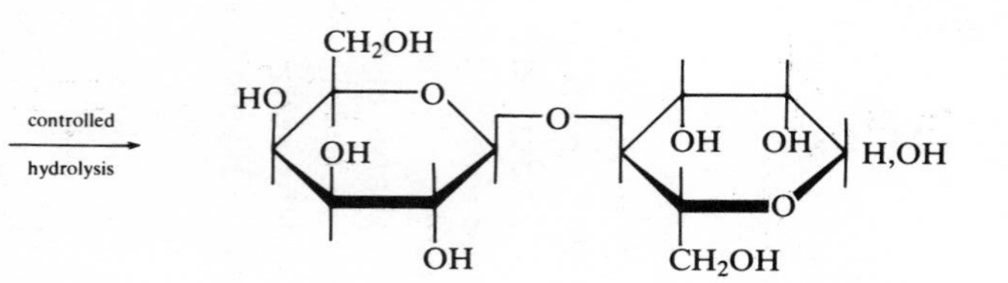

4-*O*-β-D-Galactopyranosyl-D-mannopyranose

FIGURE 6.41. Synthesis of a (1 → 4)-linked oligosaccharide

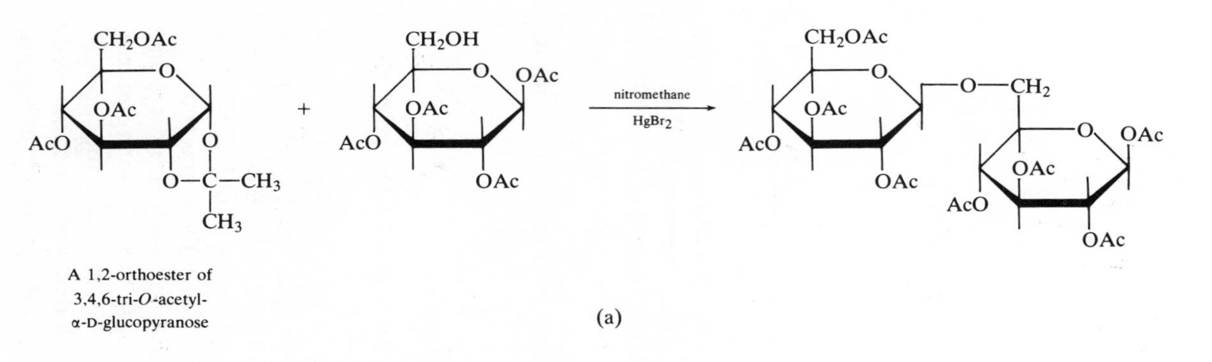

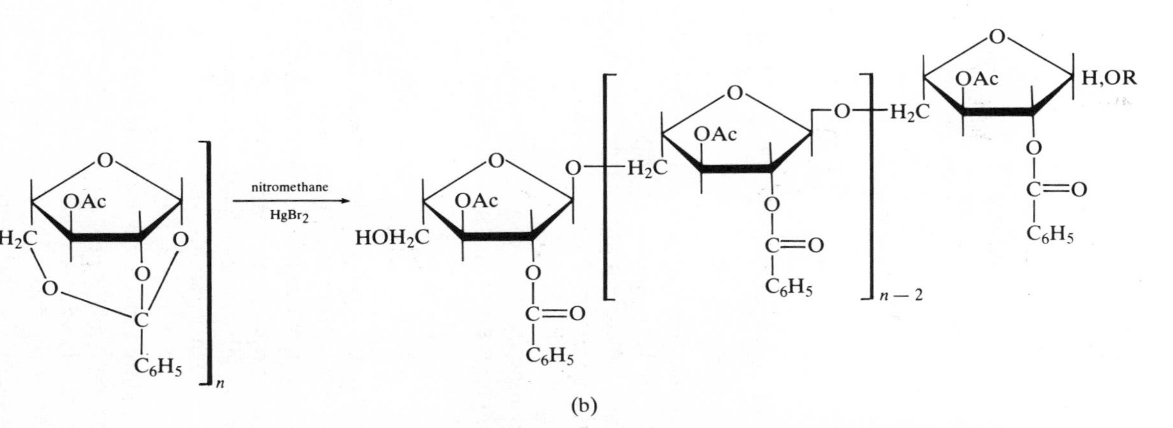

FIGURE 6.42. Use of orthoesters in oligosaccharide synthesis: (a) disaccharide formation from an orthoester; (b) polysaccharide formation from a tricyclic orthoester

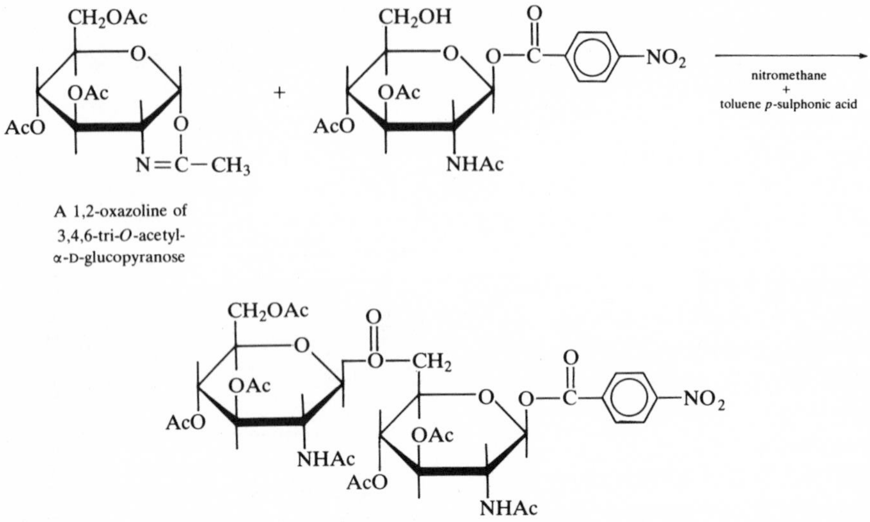
CH2OAc
OAc
AcO
N=C−CH3
A 1,2-oxazoline of 3,4,6-tri-O-acetyl-α-D-glucopyranose
CH2OH
NHAc
NO2
nitromethane
+
toluene p-sulphonic acid
CH2
NHAc

FIGURE 6.43. Use of oxazolines in oligosaccharide synthesis

Other reactions have now been developed for oligomer formation, e.g. the reaction of 1,2-orthoesters. Again, inversion of configuration takes place at the anomeric carbon atoms (see Figure 6.42(a)). A variation of this method, using a tricyclic orthoester, has been used to synthesize a synthetic arabinan (see Figure 6.42(b)).

Oxazoline derivatives of monosaccharides can be used in a similar way to give oligomers containing amino sugars (see Figure 6.43).

6.8 INDUSTRIAL UTILIZATION

Large industries are now based on the utilization of polysaccharides, particularly starch and cellulose.

Starch, which occurs naturally in our foodstuffs, is an important part of the diet, as a supply of glucose for energy. In addition, starch or modified starch is used widely in the food industry as a thickener or gelling agent for products such as soups, mayonnaise, and many desserts. *Starch derivatives* are used in the textile, paper, and laundry industries. In the manufacture of textiles, starch products are used as a size to strengthen the warp yarns and improve their resistance to abrasion. Further, during textile printing, starch derivatives hold dyes in the correct position on the cloth and prevent diffusion of the colours—after printing these starch derivatives are removed. Starches can also be used as finishes to give a glaze and polish to sewing thread.

Starches are, in addition, widely used to make adhesives, and in the laundry industry to stiffen and give a good finish to clothing and some household linens.

Modified starches are often made by partial acid hydrolysis, or partial oxidation with reagents such as alkaline hypochlorite and periodate. Such products have a lower viscosity in solution than native starches, and are used particularly in the textile and paper industries. Other starch derivatives are also manufactured e.g. hydroxyethyl starch (made by treating starch with ethylene oxide in the presence of basic catalysts), and its use as a blood volume plasma extender has been investigated.

Dextrose (α-D-glucopyranose monohydrate) and glucose syrups are the products of the *extensive* degradation of starch by acid and/or enzymes, and they have assumed considerable commercial importance. Dextrose results from the complete hydrolysis of starch, usually by an initial acid hydrolysis step which is followed by enzymic treatment. This product has many applications in food, pharmacy, and medicine. A glucose syrup is formed at intermediate stages of hydrolysis and consists of a mixture of molecules varying from glucose to oligomers of (say) 20 glucose units. Depending on the starch and the hydrolysis conditions, glucose syrups of various compositions are formed, and are defined in terms of their so-called 'dextrose equivalent', i.e. the apparent conversion into glucose as measured by the apparent reducing power. These products are used extensively in the food and confectionery industries.

Starch can also be converted enzymatically into a mixture of glucose and fructose according to the following scheme:

$$\text{Starch} \xrightarrow[\text{amyloglucosidase}]{\alpha\text{-amylase}} \text{D-glucose} \xrightarrow[\text{isomerase}]{\text{glucose}} \text{D-glucose} + \text{D-fructose}$$

This product, which is similar in sweetness to sucrose, can be economically cheaper than the disaccharide, and has many uses in the food industry.

There are a large number of minor outlets for starch, e.g. it is used as a flotation agent in ore refining, and is added to drilling mud when oil wells are sunk to seal off the pores in the surrounding hole wall.

The production of beer and spirits is based, of course, on the enzymic conversion of starch (from rye, barley, potatoes, etc.) first into glucose and then into ethyl alcohol.

In the paper industry, starch is of minor importance compared to *cellulose*, for it is the entangled fibres of cellulose which form the basis of paper. Wood chips are heated in alkali, or with sulphur dioxide in bisulphite solution, to remove lignin, gums, waxes, and most of the hemicelluloses, leaving the wood pulp used for paper manufacture. This pulp consists mainly of cellulose fibres with a little associated hemicellulose. The fibres are brought into aqueous suspension, and rubbed or crushed mechanically to increase tensile strength, smoothness, and hardness. At the same time, however, tearing resistance is decreased, and starch derivatives may be added at this stage to improve and strengthen the product. The fibres are then diluted and cleaned, and spread out on a wire belt where most of the water is lost by filtration. Presses squeeze out more water, and the paper passes through a dryer section where there is usually a size press; this applies a surface treatment to improve writing properties and strength. Modified starch is an important surface-sizing agent—a dilute starch solution cements surface fibres to the body of the paper, thus increasing the strength of the surface. Lastly, starch products can be used in a coating on the paper, a process giving better finish and printability.

In addition to paper manufacture, other industries are based on the use of cellulose fibres. Thus the seed hairs of the cotton plant, after removal of waxes and pectic substances, are almost pure cellulose and are used to make cotton cloth. Other plants are rich in cellulose fibres—for example, flax, jute, and sisal—and from these textiles and rope are made.

A number of cellulose derivatives are now very important industrially. Regenerated or viscose cellulose is obtained from wood pulp by forming a soluble *cellulose xanthate* derivative, and then reprecipitating the cellulose:

$$\underset{\text{Insoluble cellulose}}{R{-}OH} \xrightarrow[CS_2]{NaOH} \underset{\text{Soluble cellulose xanthate}}{R{-}O{-}\underset{\underset{S}{\|}}{C}{-}S{-}Na} \xrightarrow{H^+} \underset{\text{Insoluble viscose cellulose}}{R{-}OH}$$

The xanthate solution can be spun into an acid bath to form a fibre, viscose rayon, or can be made into a film and regenerated by acid to form cellophane. Regenerated cellulose may be acetylated or nitrated. *Cellulose acetates* can be used for plastics, sheeting, and rayon, and as a photographic film base. *Cellulose nitrates* are used as explosives, but with a lower nitrate content are used in plastics and lacquers.

Methyl- and *ethyl-celluloses* are also prepared industrially: the first finds use as a defoamer, thickener, and stabilizer, the second in protective coatings and plastics.

Carboxymethylcellulose, which is not hydrolysed in the human digestive tract, is used in foodstuffs and pharmaceuticals. In laboratories, this cellulose derivative is important as an ion-exchanger, and is widely utilized in the purification of many natural polymers (see Section 2.3.1.4, p. 19). Azides may be formed from the carboxymethyl groups, and these in turn can react with, say, a protein to give an enzyme, or an antibody, immobilized on an inert support. This modified cellulose can then be used for affinity chromatography (see Section 2.3.1.5, p. 20). Other cellulose derivatives, are used in laboratories for ion-exchange chromatography, e.g. *dimethylaminoethyl-(DEAE)cellulose* or for preparing purified ribonucleic acids, e.g. *naphtholyated DEAE-cellulose*.

Dextrans have been used as blood plasma volume extenders, but are also widely used in pharmaceuticals because of their high vicosities and as emulsion stabilizers in foodstuffs such as icecream. In laboratories, cross-linked dextrans are extremely important as gels for gel-permeation chromatography (see Section 2.3.1.3, p. 15). The most well-known of these, 'Sephadex', can be prepared by the action of epichlorohydrin on partially degraded dextran:

$$\underset{\text{dextran}}{R{-}OH} + \underset{\text{(epoxide ring, C1–C2 via O)}}{CH_2{-}CH{-}CH_2{-}Cl} \xrightarrow{OH^-} R{-}O{-}CH_2{-}\underset{\underset{OH}{|}}{CH}{-}CH_2Cl \xrightarrow{OH^-} R{-}O{-}CH_2{-}CH{-}CH_2 \text{ (epoxide ring via O)}$$

$$R{-}O{-}CH_2{-}CH{-}CH_2 \text{ (epoxide ring via O)} + R{-}OH \xrightarrow{OH^-} R{-}O{-}CH_2{-}\underset{\underset{OH}{|}}{CH}{-}CH_2{-}OR$$

Cross-linked dextran

Derivatives of 'Sephadex', e.g. DEAE-Sephadex, can be prepared and are used as ion-exchangers.

Pentosans such as xylans and arabinans yield furfural when treated with acid. Furfural itself is important industrially, and is used directly as a solvent, as well as in the manufacture of plastics and varnishes.

Many plant polysaccharides, because of their high viscosities and/or gelling capacities, are known as industrial gums. Most of these are non-toxic to humans, and so are used for thickening foodstuffs or producing edible gels. Many are also useful as textile and paper sizes. Thus galactomannans, gums arabic and tragacanth, carrageenans, agar and alginates are used singly or in combination as food additives in gelled desserts, icecream, salad dressings, pie fillings, etc.

The same polysaccharides, except agar, are widely used in the pharmaceutical industry—some as emulsion stabilizers, and others such as galactomannans and gum arabic as encapsulating materials for pills.

Pectic acid is extremely important in the jam and jelly industry, because of its ability to form gels with sucrose. Fruits with a high pectic acid content gel most easily when boiled with sucrose and water. Pectic acid is also used in making jellied confectionary.

In addition to its importance as a food additive, agar in the form of a gel is used in laboratories as a culture medium for bacteria. Further, agarose, one component of agar (see Section 6.5.2.3, p. 303) is used in scientific research as a support for electrophoresis and for gel-permeation chromatography. Proteins can be covalently attached to agarose, and if these proteins have a special affinity for other molecules, the protein–agarose complex can be used for affinity chromatography (see Section 2.3.1.5, p. 20).

6.9. ADDITIONAL READING

The Carbohydrates (Eds. W. Pigman and D. Horton), 2nd ed., Academic Press, London and New York, 1970.

Polysaccharides G. O. Aspinall, Pergamon Press, Oxford and New York, 1970.

MTP International Review of Science: Biochemistry of Lipids, *Biochemistry Series One*, Vol. 4 (Eds. H. L. Kornberg and D. C. Phillips), Butterworths, London, 1974.

Biochemistry D. E. Metzler, Academic Press, London and New York, 1977.

Industrial Uses of Starch and its Derivatives (Ed. J. A. Radley), Applied Science Publishers, Ltd., London, 1976.

Industrial Gums (Ed. R. L. Whistler), 2nd ed., Academic Press, London and New York, 1973.

Phytochemistry (Ed. L. P. Miller), Vol. 1, Van Nostrand Rheinhold Co., New York and London, 1973.

Starch and its Components W. Banks and C. T. Greenwood, Edinburgh University Press, Edinburgh, 1975.

Comprehensive Biochemistry (Eds. M. Florkin and E. H. Stotz), Elsevier, Amsterdam, London and New York; particularly vol. 5 (1963), vol. 17 (1969), and vol. 26 (1968).

'Newer observations on the synthesis of *O*-glycosides' R. J. Ferrier, in *Fortschritte der Chemischen Forschung*, Vol. 14, Springer-Verlag, Berlin, 1970, p. 389.

Annual Reviews:

Methods in Carbohydrate Chemistry (Ed. by R. L. Whistler), Academic Press, London and New York.

Advances in Carbohydrate Chemistry and Biochemistry (Ed. R. S. Tipson), Academic Press, London and New York.

Annual Reviews of Biochemistry (Ed. E. E. Snell), Annual Reviews Inc., Palo Alto, California.

Chapter 7

Rubber and Lignin

7.1. INTRODUCTION

In this chapter, we shall discuss rubber and lignin, two additional interesting and contrasting plant polymers. *Rubber*—and the related *gutta percha*—are of great industrial importance, but relatively little is known of their function in plants. On the other hand, although the biological function of *lignin* is more easily appreciated, it has relatively few industrial uses. The mode of biosynthesis of both these macromolecules is, however, quite well understood.

7.2. RUBBER AND GUTTA PERCHA

These hydrocarbon polymers are produced by many tropical vines, shrubs and trees, although some temperate plants, such as dandelion, also contain a little rubber. In general, few plants produce both rubber and gutta percha. Rubber is obtained commercially from *Hevea brasiliensis*, a tree native to Brazil, but now grown extensively in the Malay Peninsula and the East Indies. Gutta and balata are prepared from Malayan and South American trees, respectively; the difference in names is purely geographical—both describe the same hydrocarbon polymer.

Rubber and gutta are synthesized in a network of interconnected cells, the latex cells, located just below the bark. Latex is a slightly viscous juice made up of a suspension of the hydrocarbon polymer particles in a medium containing the enzymes, nucleic acids, and coenzymes necessary for the biosynthesis. Most rubber-synthesizing plants have a low concentration of the hydrocarbon in the latex, but, in the commercially-important *Hevea brasiliensis*, 30% of the latex may be rubber.

As the latex vessels of a tree are interconnected, the latex itself may be obtained by tapping, i.e. a cut is made in the bark through which latex exudes, forced out by the hydrostatic pressure of the vessel contents. Latex flows from a tapping-cut for minutes, or hours, depending on the plant involved, until it coagulates. The latex vessels then take up water, and the synthesis of the hydrocarbon polymer starts again. In *Hevea brasiliensis*, the regeneration of rubber is rapid, and a tree may be tapped on alternate days for 30 or more years to give the same, or an increasing amount of rubber. Indeed, commercial yields in Malaysia are about 1500 lb of rubber each year from one acre.

Gutta latex coagulates very rapidly in air, and so the polymer cannot be obtained commercially by tapping; gutta can, however, be prepared from the leaves of certain trees.

The biological function of both of these macromolecules in the plant is not yet understood.

7.2.1. The monomer and interunit linkages

Rubber and gutta percha are formed from the same monomer, isopentenyl pyrophosphate, i.e.:

$$CH_2{=}C(CH_3){-}CH_2{-}CH_2{-}O{-}\overset{O}{\overset{\|}{P}}(O^-){-}O{-}\overset{O}{\overset{\|}{P}}(O^-){-}O^-$$

During biosynthesis of the polymers, these monomers are covalently linked together and pyrophosphate is eliminated. The condensation reaction is accompanied by a translation of the position of the double bond, and two

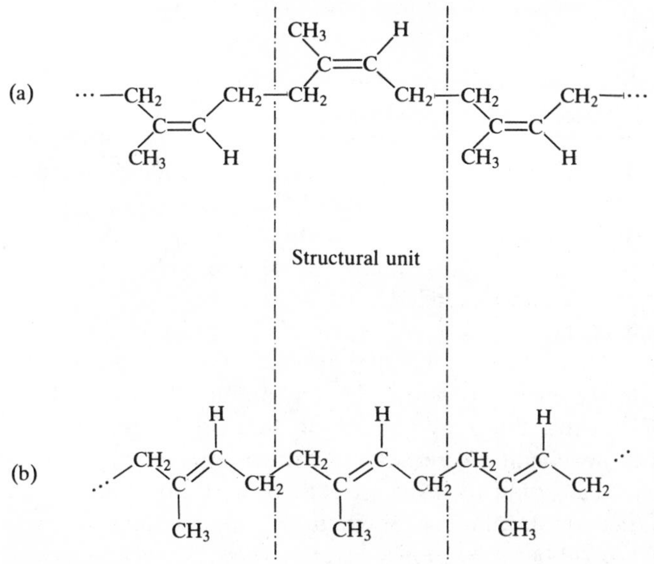

(a) ···—CH_2 $C=C$ CH_3 H CH_2—CH_2 $C=C$ CH_3 H CH_2—CH_2 $C=C$ CH_3 H CH_2—···

Structural unit

(b) ··· CH_2 C H C CH_3 CH_2 CH_2 C H C CH_3 CH_2 CH_2 C H C CH_3 CH_2 ···

FIGURE 7.1. The structure of (a) rubber, and (b) gutta percha and balata chains

stereoisomeric hydrocarbon polymers are formed—rubber where the carbon–carbon bonds of the polymer backbone are *cis* to the double bond, and gutta where these carbon–carbon bonds lie *trans* to the double bond (see Figure 7.1). The enzyme system of the plant controls the stereochemistry of the polymerization to produce either rubber or the gutta hydrocarbon (see Section 7.2.3). The structural unit of both polymers is an isoprene unit, and hence rubber and gutta are *polyisoprenes*.

7.2.2. Molecular size and structure; the phenomenon of elasticity

Rubber and gutta percha are both linear polymers, but X-ray diffraction patterns have shown that the polyisoprene chains are non-planar. Both materials occur in Nature with a wide distribution of molecular weights, the average molecular weight depending on the plant source. Thus, in freshly extracted *Hevea brasiliensis* latex the weight-average molecular weight of the rubber molecules may vary from 100,000 to 4,000,000; rubber from other sources usually has a smaller molecular weight. Gutta chains are on average shorter than those of *Hevea* rubber, and their molecular weights vary from 40,000 to 200,000. For both types of polyisoprene, purification often causes some degradation, with a concurrent decrease in the molecular weights.

The structure of rubber and gutta, particularly the configurations about the double bonds, are all-important in determining the physical properties of the polymers.

The most characteristic property of rubber is its elasticity—raw rubber can be extended 1000%, and will spring back to its original length on release of the tension (provided that the stress is applied for a short time).

The *cis* configuration at the double bonds and the fairly free rotation round single bonds allow the rubber molecules to coil-up, and become tangled with one another, in a more-or-less random manner. Entropy considerations favour the coiled state of the macromolecules. There are an immense number of possible conformations for the coiled state, but only *one* fully extended conformation, and so the polymer has the highest probability of being in the coiled state.

When the rubber is stretched, parts of the polymer molecules partially uncoil and tend to line up parallel to one another, this alignment being restricted by the extent of entanglement among the macromolecules (see Figure 7.2). Thus the randomness of the orientation (and hence the entropy) of the chains decreases on stretching; indeed, X-ray diffraction patterns can be obtained from stretched, but not unstretched, rubber, indicating that molecular order increases under tension. The ordered regions of the rubber are known as *crystallites*, and parts of one polymer molecule may lie in tangled regions, whilst other sections of the same molecule lie in crystallites. A small number of crystallites exist in unstretched rubber at room temperature, but these are not sufficient to give an X-ray diffraction pattern, and, in general, unstretched rubber exists in an amorphous state. When the tension on stretched rubber is released,

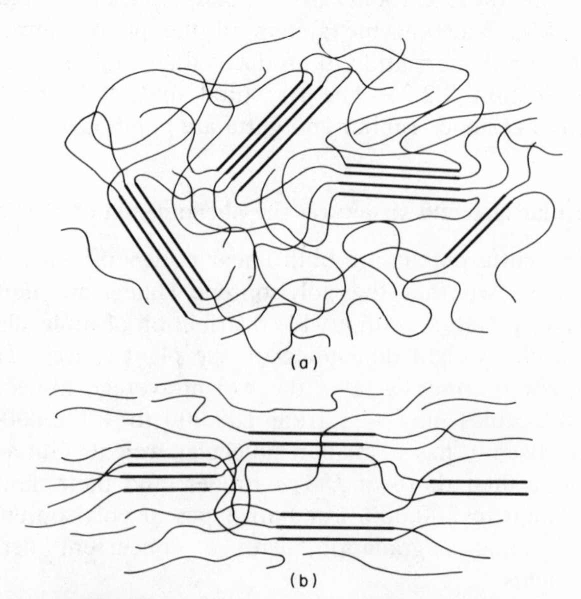

FIGURE 7.2. Arrangement of rubber molecules in (a) unstretched, and (b) stretched state. (Crystallites are shown as thick lines)

the original, randomly coiled conformation is favoured by the increase in entropy, and the rubber springs back to its original length.

When rubber is warmed, the increased thermal motion of the molecules reduces the number of crystallites, and increases the freedom of rotation about single bonds, and so the material becomes more elastic and plastic. On cooling to a temperature a little below 20°C, the converse behaviour occurs, and the rubber becomes more crystalline, and hence less elastic. At very low temperatures (less than −70°C), the polymer becomes brittle or inextensible and behaves as a glass because the thermal motion of the molecules is so reduced that even rotation about single bonds can no longer occur.

In contrast, the *trans* configuration at the double bonds in gutta percha restricts the coiling of the polymer chains, and the degree of order of gutta chains is much greater than for rubber. Indeed, gutta at 20°C exhibits an X-ray diffraction pattern similar to that for stretched rubber. As chain-folding is restricted, gutta is much less elastic than rubber, but at higher temperatures the gutta chains are able to contort more easily, and the polymer becomes plastic and more elastic.

Chemical synthesis has confirmed that the stereochemistry about the double bonds of rubber and gutta determines the physical properties of the polymers. Synthetic *cis*-1,4-polyisoprene with the properties of natural rubber can be synthesized from isoprene itself, as can a *trans*-1,4-polyisoprene resembling the gutta hydrocarbon (see Section 7.2.4, p. 336).

7.2.3. Biosynthesis

The isopentenyl pyrophosphate molecules, which polymerize to form rubber or gutta, are themselves synthesized from acetate. Radioactive tracer studies have shown, in fact, the distribution of the carbon atoms of

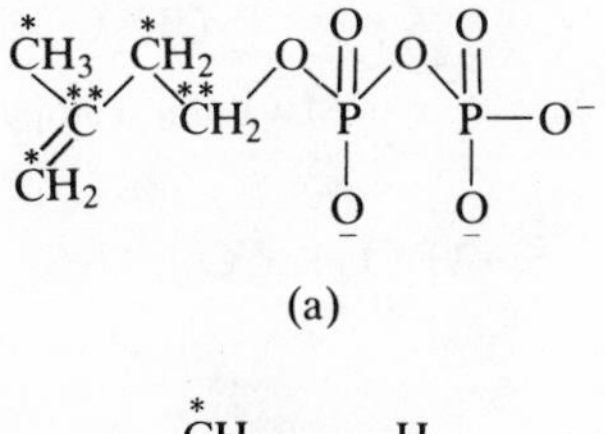

(a)

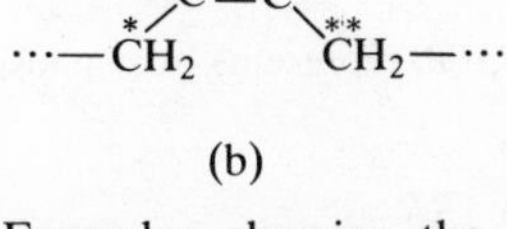

(b)

FIGURE 7.3. Formulae showing the carbon atoms of (a) isopentenyl pyrophosphate, and (b) isoprene units of rubber derived from carbonyl (C**) and methyl (C*) carbon atoms of acetate, $C^{*}H_3C^{**}OO^-$

the acetate in the products, isopentenyl pyrophosphate and the isoprene units of rubber (see Figure 7.3).

Conversion of acetate to isopentenyl pyrophosphate involves the enzyme-catalysed linking of three molecules of acetate to form a derivative of hydroxymethyl glutaric acid, i.e. hydroxymethyl glutaryl coenzyme A, which is then reduced to give the important intermediate, mevalonic acid:

$$3\ CH_3COOH \xrightarrow[+ATP\ (\text{see Figure 5.2}b)]{\text{HS-coenzyme A}} CH_3{-}\underset{\underset{CH_2COOH}{|}}{\overset{\overset{OH}{|}}{C}}{-}CH_2{-}CO{-}S{-}\text{Coenzyme A}$$

Hydroxymethyl glutaryl coenzyme A

$$\Big\downarrow \text{reduction}$$

$$CH_3{-}\underset{\underset{CH_2COOH}{|}}{\overset{\overset{OH}{|}}{C}}{-}CH_2CH_2OH$$

Mevalonic acid

After phosphorylation, the mevalonic acid undergoes enzymic decarboxylation and dehydration to form isopentenyl pyrophosphate:

$$CH_3{-}\underset{\underset{CH_2COOH}{|}}{\overset{\overset{OH}{|}}{C}}{-}CH_2CH_2OH \xrightarrow{ATP} CH_3{-}\underset{\underset{CH_2COOH}{|}}{\overset{\overset{OH}{|}}{C}}{-}CH_2CH_2OPP_i \longrightarrow$$

Mevalonic acid pyrophosphate

$$CH_3{-}\underset{\underset{CH_2}{||}}{C}{-}CH_2CH_2OPP_i + H_2O + CO_2$$

Isopentenyl pyrophosphate

(In this and later equations, $-OPP_i$ represents pyrosphosphate, $-O{-}\underset{\underset{O^-}{|}}{\overset{\overset{O}{||}}{P}}{-}O{-}\underset{\underset{O^-}{|}}{\overset{\overset{O}{||}}{P}}{-}O^-$.)

Studies of rubber biosynthesis, involving radioactive isopentenyl pyrophosphate, have shown, however, that the radioactivity is incorporated into already existing rubber molecules. Thus, isopentenyl pyrophosphate is used for chain extension, and not for the initiation of new hydrocarbon chains, i.e.:

$$H-(CH_2-\overset{\displaystyle CH_3}{\overset{|}{C}}=CH-CH_2)_n-OPP_i + CH_2=\overset{\displaystyle CH_3}{\overset{|}{C}}-CH_2-CH_2-OPP_i$$

$$\downarrow$$

$$H-(CH_2-\overset{\displaystyle CH_3}{\overset{|}{C}}=CH-CH_2)_n-CH_2-\overset{\displaystyle CH_3}{\overset{|}{C}}=CH-CH_2-OPP_i + HOPP_i$$

New rubber chains are formed only when dimethyl allyl pyrophosphate, an isomer of isopentenyl pyrophosphate, is present. An enzyme in the latex catalyses the isomerization of isopentenyl pyrophosphate to dimethyl allyl pyrophosphate. Chain initiation occurs as follows:

$$(CH_3)_2C=CH-CH_2-OPP_i + CH_2=C(CH_3)-CH_2-CH_2-OPP_i$$

Dimethyl allyl pyrophosphate — Isopentenyl pyrophosphate

$$\downarrow$$

$$(CH_3)_2CH=CH-CH_2-CH_2-C(CH_3)=CH-CH_2-OPP_i + HOPP_i,$$

Nerol pyrophosphate

or

$$(CH_3)_2CH=CH-CH_2-CH_2-C(CH_3)=CH-CH_2-OPP_i + HOPP_i$$

Geraniol pyrophosphate

The configuration of the bonds on either side of the double bond, (indicated by the dashed lines in the dimers), can be *cis*, i.e. to form nerol pyrophosphate, or *trans*, i.e. to form geraniol pyrophosphate, depending on the stereospecificity of the enzyme.

Further stereospecific condensation reactions of the dimer with more isopentenyl pyrophosphate then give either the polymeric rubber (from nerol pyrophosphate) or gutta percha (from geraniol pyrophosphate). The strict, steric control exerted by the biosynthetic enzymes has been shown using tritiated mevalonic acid of known configuration (see Section 7.5, p. 350).

7.2.4. Chemical synthesis of polyisoprenes

Synthetic rubber or gutta can be obtained by the stereospecific polymerization of isoprene; synthetic *cis*-1,4-polyisoprene is prepared on the industrial scale, and has the properties of natural rubber, whilst the *trans*-1,4-polyisoprene (synthetic gutta) has little industrial importance.

cis-1,4-Polyisoprene can be synthesized from isoprene dissolved in a hydrocarbon solvent (usually *n*-pentane) under moderate pressure at 50°C, in the presence of a catalyst such as butyllithium, or a Ziegler-type catalyst, e.g. titanium tetrachloride plus triisobutylaluminium.

Catalysis by butyllithium brings about anionic polymerization, where a negative ion may be formed from the monomer, isoprene, to initiate chain propagation. In this case, however, where a non-polar solvent is used, initiation and polymerization probably involve π-complex formation between the hydrocarbon and the lithium (see Figure 7.4).

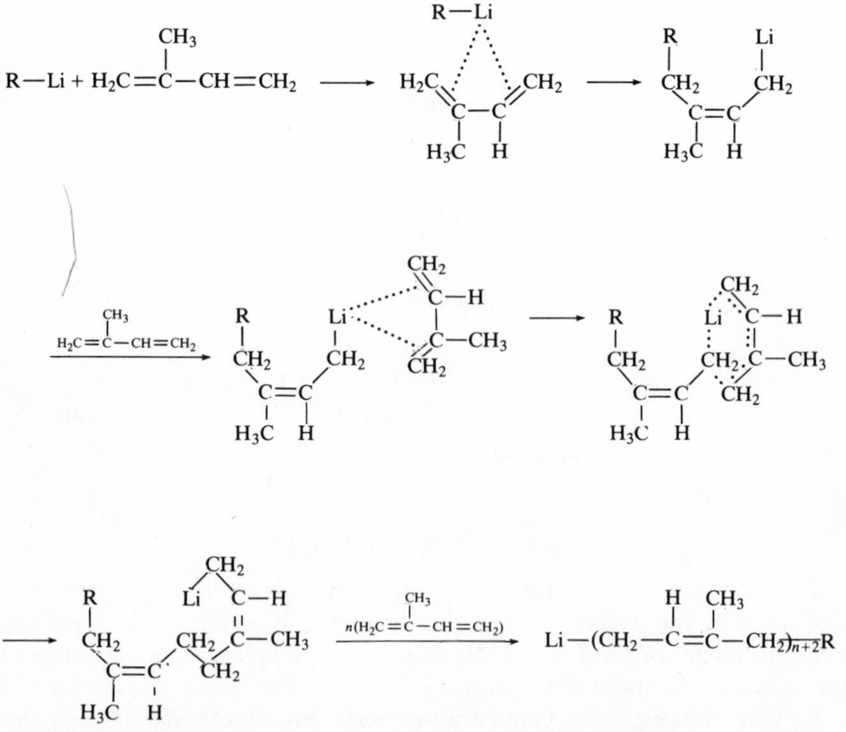

$$R{-}Li + H_2C{=}C(CH_3){-}CH{=}CH_2 \longrightarrow \cdots \xrightarrow{n(H_2C=C(CH_3)-CH=CH_2)} Li{-}(CH_2{-}C(CH_3){=}CH{-}CH_2)_{n+2}R$$

FIGURE 7.4. Synthesis of *cis*-1,4-polyisoprene on an alkyllithium catalyst

π-Complex formation between lithium and the monomer takes place with the same orientation of substituents on the isoprene each time, so that stereospecific polymerization occurs.

The two components of a Ziegler catalyst, e.g. titanium tetrachloride and triisobutylaluminium, react together to form complexes of the type

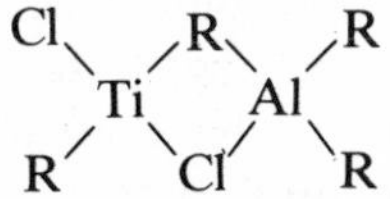

This complex in turn probably forms a π-complex with the monomer isoprene, so that chain initiation and propagation can take place (see Figure 7.5). The polymer grows from the catalyst surface by insertion of new monomer.

The mechanism by which stereospecific polymerization takes place is not clear, but it is believed that polymerization occurs at localized sites on the catalyst surface. Presumably the monomer must approach the surface *from the same side each time* to give opening of the double bonds in the same way, and hence give a stereospecific polymer. This type of polymerization is known as *coordination polymerization*, and the polymer produced depends on the nature of the catalyst. Thus synthetic rubber, the *cis*-1,4-polyisoprene, is obtained if the catalyst is a mixture of titanium tetrachloride and triisobutylaluminium or triethylaluminium, provided that the aluminium to titanium ratio is greater than one. Using

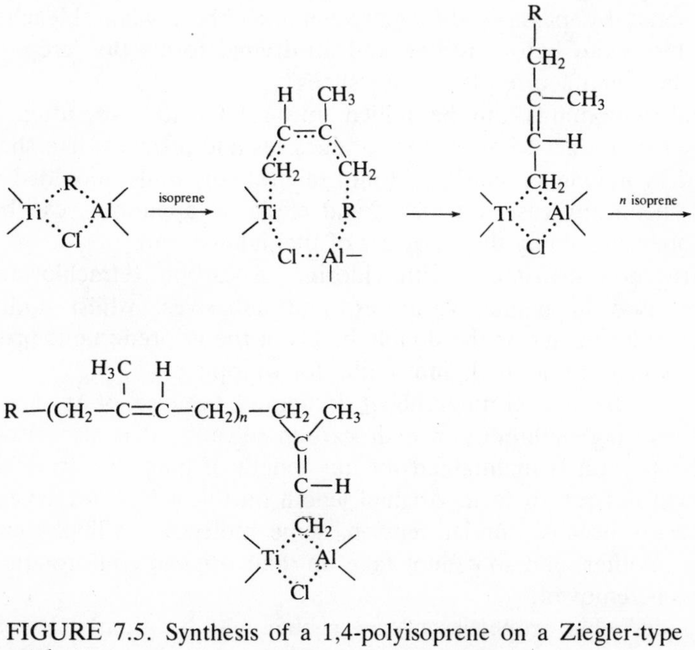

$$R-(CH_2-C(CH_3)=C(H)-CH_2)_n-CH_2-C(CH_3)=CH-CH_2-[Ti\cdots Al]$$

FIGURE 7.5. Synthesis of a 1,4-polyisoprene on a Ziegler-type catalyst

triethylaluminium and titanium tetrachloride at an Al/Ti ratio of less than one results in the *trans*-1,4-polyisoprene, as does triethylaluminium mixed with vanadium trichloride.

The synthetic *cis*-1,4-polyisoprene must compete economically with other synthetic rubbers such as polyurethanes, silicone rubbers, polysulphides, polychloroprenes, and polymers of butadience with styrene, as well as with natural rubber itself. Improved agricultural techniques and chemical stimulation of rubber biosynthesis have greatly increased the output from rubber plantations, and so natural rubber is assuming greater industrial importance as the cost of producing synthetic rubbers from petroleum-based monomers increases.

7.2.5. Processing and uses

Both rubber and gutta have considerable industrial importance, but, of the two polymers, rubber has the greater commercial value and competes successfully with many synthetic rubbers because of its excellence as an elastic polymer.

7.2.5.1. Rubber

In latex, rubber exists as negatively-charged particles about one μm in diameter. These particles may be coagulated by the addition of acid, and the resulting coagulum, which has a dough-like consistency, can be formed into a sheet by passage through rollers. Rubber, when bleached with sodium bisulphite before rolling and air-drying, forms the ‘crepe rubber’ used in the manufacture of soles for shoes.

Several compounds can be milled into rubber to give, after heating, many useful products. For example, adhesives and products like shellac are prepared by milling sulphuric acid into raw rubber, whilst moulded articles, such as chemically resistant dishes and electrical apparatus, can be made from rubber containing the chlorides of titanium or iron.

The reaction of rubber with chlorine in carbon tetrachloride yields products used in paints, varnishes, and adhesives, whilst addition of hydrogen chloride across the double bonds of the isoprene units produces a material which can be made into a film for wrapping.

However, the use of raw rubber is limited because of its low tensile strength and high solubility in hydrocarbon solvents. It is also viscoelastic, i.e. if the tension is maintained for any length of time, the stretched, raw rubber will not return to its original length on release of the tension. This effect occurs because, under tension, some molecules will physically slip past one another, and so cannot take up their original conformations once the stress is removed.

The undesirable properties of raw rubber may be greatly decreased by the process known as *vulcanization*, which involves the formation of new chemical cross-links in the rubber structure. Although vulcanization can be brought about by heating with reagents which produce free radicals (e.g. peroxides, nitro or azo compounds), or several elements (e.g. sulphur, selenium, or tellurium), sulphur vulcanization is at present the most important commercial process.

The solubility and viscoelasticity of rubber are both greatly reduced on such vulcanization, for the new sulphur cross-links prevent the polymer molecules slipping past one another on stretching. In addition, the tensile strength increases with the amount of sulphur incorporated into the rubber, although at the same time the elasticity decreases. The elastic properties of rubber depend on the possibility of easy rotation around single bonds, and this potential is greatly reduced when the vulcanized rubber contains a high proportion ($>30\%$) of sulphur, and consequently a very large number of cross-links. At lower sulphur contents ($<5\%$), the elastic properties are maintained, because rotation about single bonds and coiling of the polymer chains is possible between the cross-links; these segments of the chains can then uncoil on stretching, but can return to their coiled conformation (to give an entropy increase) on release of tension.

Sulphur vulcanization involves an ionic mechanism, where species of the type RS_x^+ add on to the double bond. The reaction results in the formation of sulphur cross-links, e.g.:

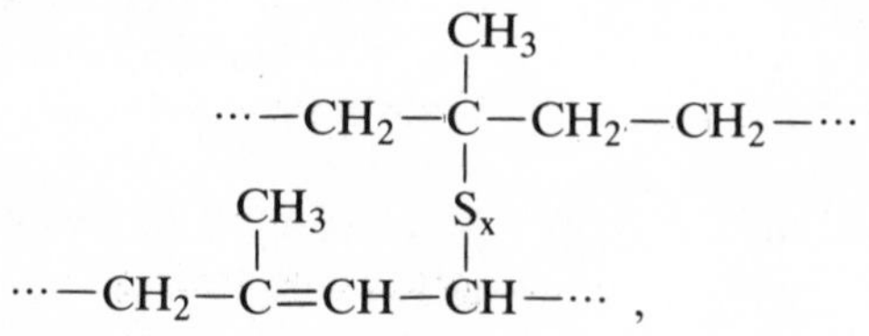

$$
\begin{array}{ccccccccccc}
 & & & & & CH_3 & & & & & \\
 & & & & & | & & & & & \\
 & & \cdots-CH_2 & - & & C & - & CH_2 & - & CH_2-\cdots & \\
 & & & & & | & & & & & \\
 & CH_3 & & & & S_x & & & & & \\
 & | & & & & | & & & & & \\
\cdots-CH_2- & C & = & CH & - & CH & -\cdots\ , & & & &
\end{array}
$$

and also cyclic sulphides, e.g.

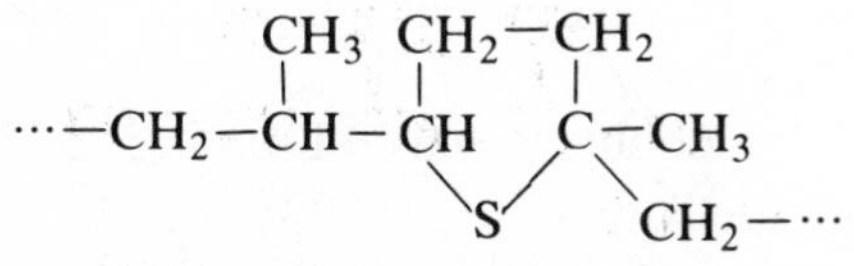

$$
\begin{array}{cccccc}
 & CH_3 & CH_2 & - & CH_2 & \\
 & | & | & & | & \\
\cdots-CH_2- & CH- & CH & & C & -CH_3 \\
 & & & \diagdown S \diagup & & \diagdown \\
 & & & & & CH_2-\cdots
\end{array}
$$

Vulcanization of rubber by heating with sulphur alone is a relatively slow process, but it may be speeded up considerably by the incorporation of an accelerator into the vulcanizate. Organic compounds such as thioureas, thiophenols, mercaptans, xanthates, ureas, and guanidines can all act as *accelerators*. Many accelerators function best in the presence of an activator, like zinc oxide. Because the solubility of the accelerator and activator is important, a rubber-soluble soap, e.g. the zinc salt of a long-chain fatty acid, is usually incorporated into the mixture. The accelerator, zinc oxide, fatty acid (R^1COOH), and sulphur are believed to form a soluble zinc mercaptide species $(XS)_2ZnL_2$, where X is a substituent from the accelerator capable of electron withdrawal, and L (formally R^1COO^-) renders the complex soluble. This species interacts with molecular sulphur to form a sulphurating reagent, $(XS.S_n)_2ZnL_2$, which can replace a hydrogen of the rubber (R^2H) with a sulphur chain bearing an accelerator moiety, e.g.:

$$R^2H + (XS.S_n)_2\ ZnL_2 \rightarrow R^2 - S_x.\ SX + ZnS + XS.S_zH + 2\ L$$

The product, $R^2 - S_x.SX$, can react with another rubber chain, or itself, to give disubstituted polysulphides, such as:

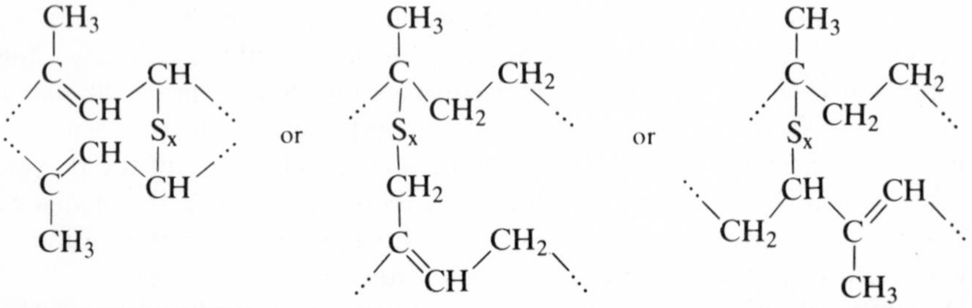

```
   CH3                     CH3                          CH3
   |                       |                            |
   C      CH               C       CH2                  C        CH2
 ./ \\CH/  |  \.         ./|\CH2/    \.               ./|\CH2/     \.
.          Sx   .    or     Sx                 or       Sx
 .\ //CH\  |  /.            |                           |
   C       CH               CH2                  .\    CH      CH
   |                        |                      CH2/  \C//    \.
   CH3                      C       CH2                   |
                         ./ \\CH/     \.                  CH3
```

These polysulphides in turn can lose sulphur to give *disulphide* and *monosulphide cross-links* ($R^2S_xR^2 \rightarrow R^2\text{-S-S-}R^2$ or $R^2\text{-S-}R^2 + S$), or can give cyclic sulphides (see above) with loss of cross-links. With a large accelerator: sulphur ratio most of the polysulphides are transformed into monosulphide cross-links, and the undesirable cyclic sulphide formation is suppressed.

The resulting vulcanized rubber is relatively soft, pliable and extensible, provided that the sulphur content does not exceed 5%. This product is used for rubber tubing, rubber bands, and gloves. When the sulphur content is high (30–50%), the rubber is no longer elastic, and a rigid polymer called ebonite is obtained. Ebonite is commercially important, for it is an electric and thermal insulator, has a high corrosion resistance, and is chemically inert.

Although some of the double bonds of rubber are saturated during vulcanization (e.g. by cyclic sulphide formation), many remain intact and, as a result, vulcanized rubber is sensitive to heat, light, and oxygen. Thus it can 'age', and lose many of its useful properties, for oxidation brings about the scission of carbon–carbon bonds in the polymer. The first stage in oxidation is believed to be the formation of a hydroperoxide, which then decomposes to give free radicals:

$$2\ R''OOH \rightarrow R''OO\cdot + R''O\cdot + H_2O$$

The peroxy radical can then react with a rubber chain to form a product which is easily oxidized further, and eventually chain scission results (see Figure 7.6).

To inhibit such oxidation, an antioxidant such as phenyl-α-naphthylamine

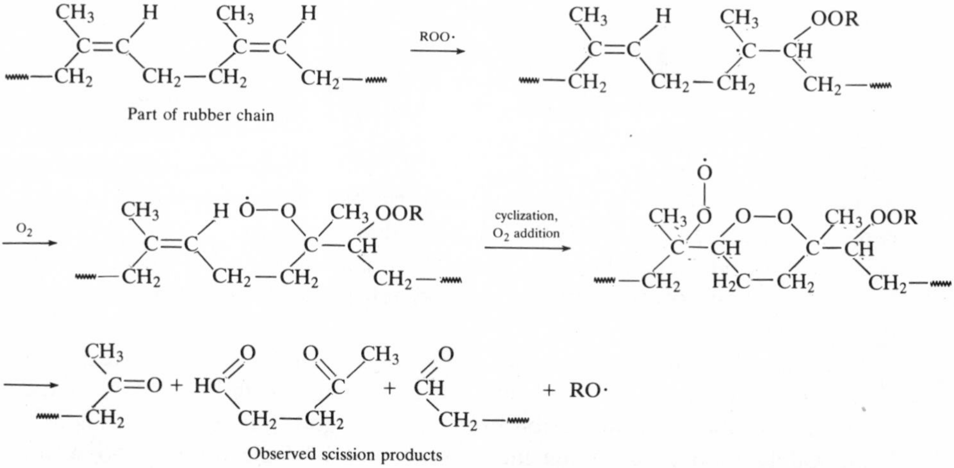

Part of rubber chain

$\xrightarrow{ROO\cdot}$

$\xrightarrow{O_2}$

cyclization, O_2 addition

$+ RO\cdot$

Observed scission products

FIGURE 7.6. Oxidative degradation of rubber chain by peroxy radical ROO·

is usually incorporated into the vulcanizing mixture. Antioxidants act by decomposing the initial hydroperoxide, or by deactivating the peroxy free radical, ROO·.

The vulcanizates of rubber (described above) do not have great commercial importance, because their tear and abrasion resistance is fairly low. These properties can be greatly improved by incorporating a '*filler*' into the rubber before vulcanization—'Carbon Black' is the most widely used filler. The reinforced rubber then has increased tensile strength (see Table 7.1) and greatly improved abrasion resistance, without much loss of elasticity.

It is believed that some type of cross-linking bond occurs between the Carbon Black particles and the rubber, and that these bonds are responsible for the reinforcement. Reinforced rubber is a product of great commercial importance, and is used for the manufacture of tyres, particularly for aircraft and large trucks, and engineering components —suspensions, mountings and couplings—which make heavy demands on physical performance.

TABLE 7.1. Mechanical properties of rubbers

	Type of rubber		
Property	Raw rubber	Vulcanized rubber	Reinforced vulcanized rubber
Tensile strength (psi)	300	3000	4500
% Elongation at break-point	1200	700–800	600

7.2.5.2. Gutta percha

Gutta percha is commercially much less important than rubber, but can nevertheless be processed to give some useful products. Although purified gutta is readily oxidized in air, and becomes brittle, a resin in unpurified gutta provides protection. As discussed earlier, at ordinary temperatures, gutta is much harder and less elastic than rubber, but on heating it softens, can easily be moulded at 100°C and hardens again on cooling. As it is acid-resistant, gutta can be used to make containers for hydrogen fluoride. Its resistance to solvents such as carbon tetrachloride and chloroform is, however, very low.

Gutta can be vulcanized in the same way as rubber, but the resultant mechanical properties are much inferior to vulcanized rubber. When mixed with a little rubber, gutta can be used to make golf ball covers. It has also been widely used as a submarine cable cover, because of its good water resistance. In the production of such cable, the gutta is warmed and extruded, through a die whilst still hot, over the metal conducting core. The cable is then cooled in cold water, and rolled onto drums.

Finally, it may be noted that chicle, a mixture of low molecular weight gutta and rubber with terpene alcohols, obtained from a tropical tree, may be used in the manufacture of chewing gum!

7.3. LIGNIN

Lignin is a biochemically inert polymer which functions as the structural support material in plants. During synthesis of plant cell-walls, polysaccharides such as cellulose and hemicelluloses (see Chapter 6), are laid down first, and then lignin fills the spaces between the polysaccharide fibres, cementing them together. This lignification process causes a stiffening of cell-walls, and the carbohydrate is protected from chemical and physical damage. Indeed, the lignified fibre plays no active metabolic rôle in the life of a plant, and acts merely as an inert support.

Although lignin constitutes, on average, 25% of wood and is therefore an extremely abundant polymer, less is known about its structure than many other natural polymers.

As lignin is inert and insoluble in the plant, it is not easily isolated in a pure and undegraded state. Lignin may be solubilized in strong alkali, or calcium bisulphite solution, but extensive modification of the polymer takes place, and the solutions are dark brown, or black, due to the formation of free radicals, quinones, and phenoxide ions. A portion of wood lignin can also be extracted in ethanol or dioxane; these solutions are light-coloured and contain less modified polymers which have been used for structural investigations.

7.3.1. The monomers and interunit linkages

The structure of lignin is not yet fully known, but catalytic hydrogenation of the polymer, with copper chromite as the catalyst, yields derivatives of cyclohexylpropane(I) whilst derivatives of phenylpropane(II) result when Raney nickel is the catalyst. Oxidation of coniferous wood with nitrobenzene in alkali yields vanillin (III), whilst oxidation of deciduous wood gives vanillin and syringaldehyde (IV).

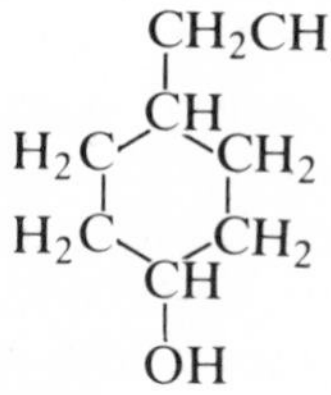

$CH_2CH_2CH_3$ — CH — H_2C CH_2 — H_2C CH_2 — CH — OH

1-(4′Hydroxycyclohexyl)propane

I

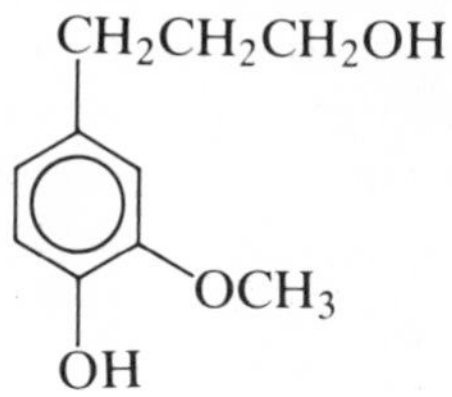

$CH_2CH_2CH_2OH$ — OCH_3 — OH

3-(3′-Methoxy-4′-hydroxy phenyl)-propanol

II

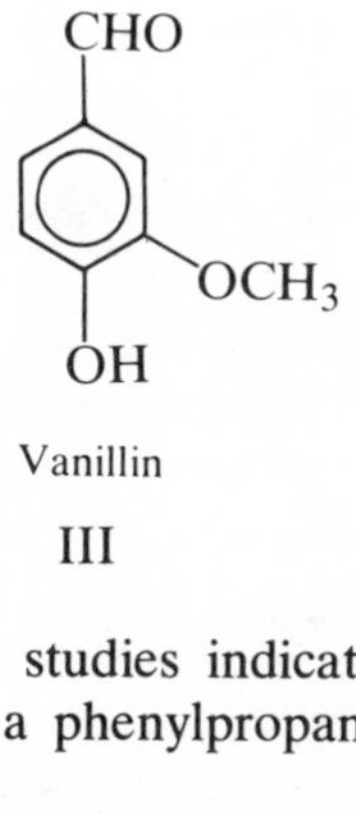

CHO — OCH_3 — OH

Vanillin

III

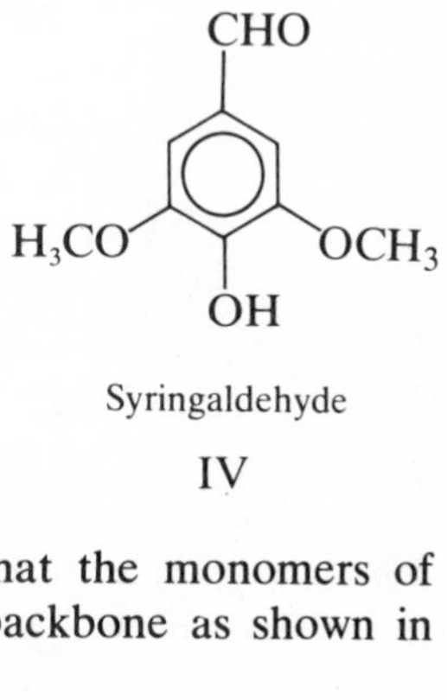

CHO — H_3CO — OCH_3 — OH

Syringaldehyde

IV

The results of these and other studies indicate that the monomers of lignin are aromatic alcohols with a phenylpropane backbone as shown in Figure 7.7.

The proportions of the different alcohols polymerizing to form lignin depend on the species of plant, e.g. coniferous lignin contains mainly

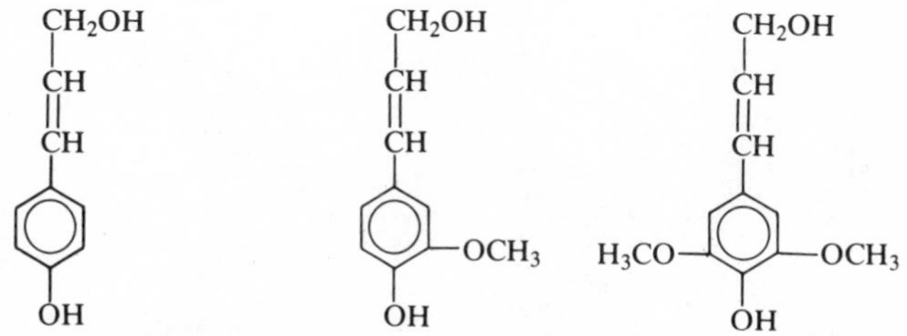

CH_2OH — CH — CH — OH

p-Hydroxycinnamyl alcohol (*p*-coumaryl alcohol)

CH_2OH — CH — CH — OCH_3 — OH

Coniferyl alcohol

CH_2OH — CH — CH — H_3CO — OCH_3 — OH

Sinapyl alcohol

FIGURE 7.7. Monomers of lignin

coniferyl alcohol derivatives, whilst deciduous lignin also contains sinapyl derivatives. Monocotyledons, such as grasses, contain a significant proportion of products from *p*-hydroxycinnamyl alcohol.

Lignin is believed to be formed by oxidative polymerization of these phenylpropane units to give large, cross-linked molecules containing carbon–carbon and ether linkages between the monomers. New carbon–carbon bonds may be formed between two propane side-chains, or between a side-chain and a benzene ring, or between the aromatic rings of two monomers. The ether links are most commonly formed between the phenolic hydroxyl group of one monomer and the propane side-chain of another, although ether formation between two side-chains is also possible. In addition, two subunits may be joined by multiple bonds.

As the monomers can be linked in several ways, there are many different structural units in lignin, and degradation of the polymer gives a wide variety of products; some typical dimers are shown in Figure 7.8.

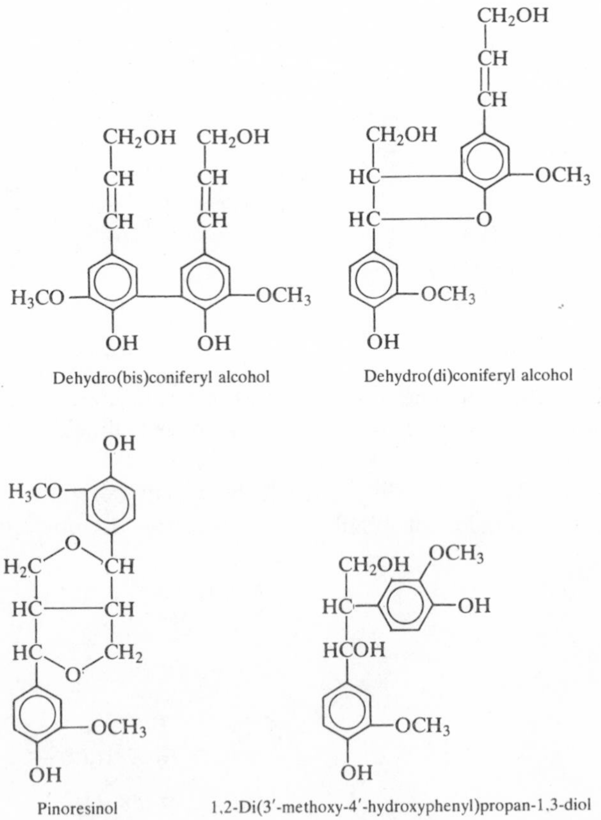

Dehydro(bis)coniferyl alcohol

Dehydro(di)coniferyl alcohol

Pinoresinol

1,2-Di(3′-methoxy-4′-hydroxyphenyl)propan-1,3-diol

FIGURE 7.8 Interunit linkages in coniferous wood lignin

7.3.2. Molecular size, structure, and properties

It is difficult to obtain soluble, undegraded lignin, and the size and properties of soluble lignin depend, therefore, on the method of extraction as well as on the species of plant. Lignin molecules probably have a wide distribution of molecular weight. The weight-average molecular weight of soluble lignin is usually small (about 10,000), although values of up to 48×10^6 have been reported for alkali-extracted fractions.

It is not known whether lignin molecules in the native state are of a finite size, or form an infinite, cross-linked network extending throughout a plant. Solid lignin is thought to be amorphous, and no one structural formula can be written for a lignin molecule, in view of the diversity of ways in which the monomers can be combined. A proposed, partial structure for a lignin molecule is shown in Figure 7.9.

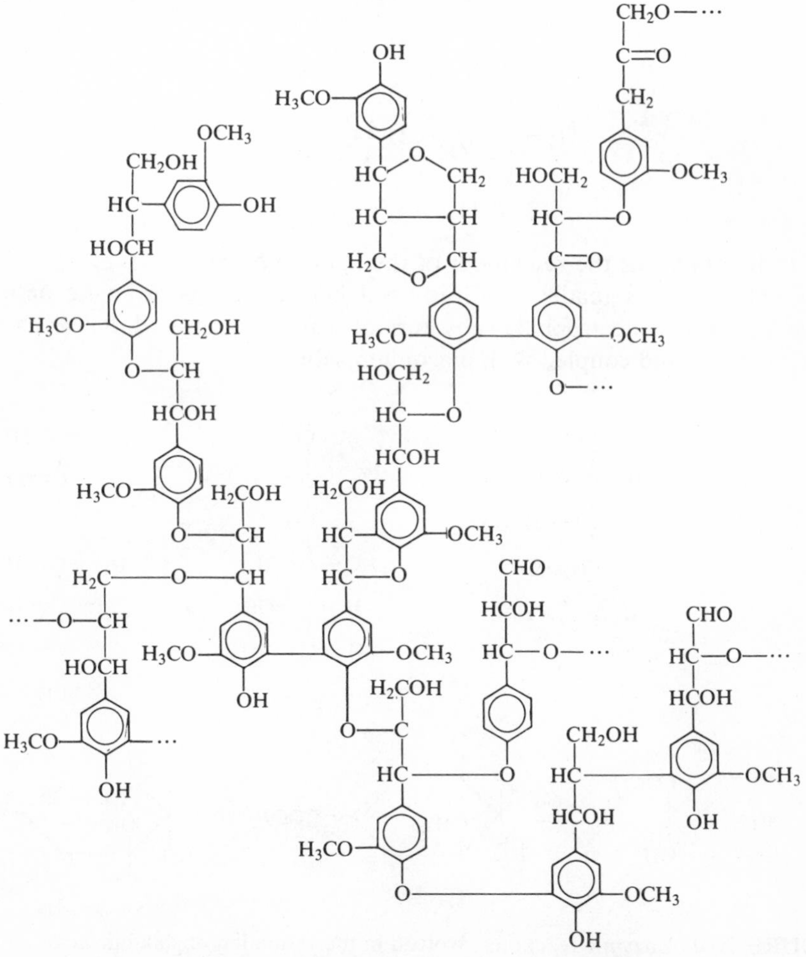

FIGURE 7.9. Partial structure of coniferous wood lignin

It is also not yet known whether lignin is covalently bonded to the polysaccharide components of the plant cell-wall, or whether it occurs in simple, physical association. The suggestion has been made that carbohydrate could be linked to the propyl side-chain of a phenylpropane unit, e.g.:

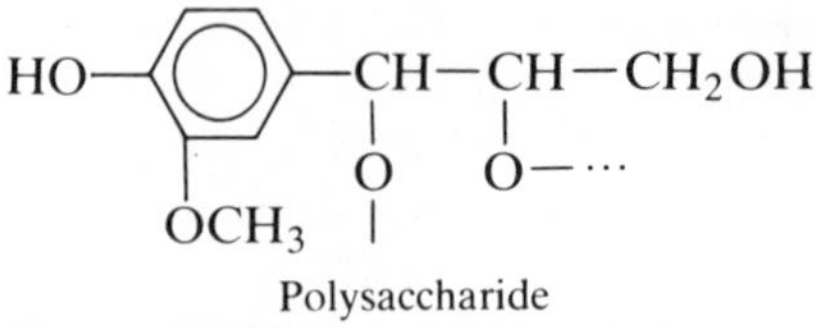

HO—(C₆H₃)(OCH₃)—CH(O—Polysaccharide)—CH(O—···)—CH₂OH

Polysaccharide

Electron spin resonance studies have shown that a small number of stable free radicals exist in coniferous wood lignin; a slightly larger number are found in deciduous wood lignin. The concentration of free radicals increases after chemical and enzymic degradation. These radicals may be of the type

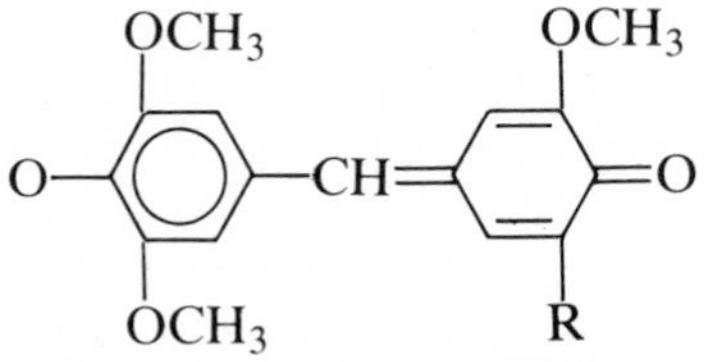

$O^{\cdot}$—(C₆H₂)(OCH_3)₂—CH=(C₆H₂)(OCH_3)(R)=O

where R represents the remainder of the lignin molecule.

Soluble lignin is readily oxidized, and because of its aromatic nature, easily undergoes electrophilic substitution; it may be nitrated, halogenated, hydroxylated, and coupled with diazonium salts.

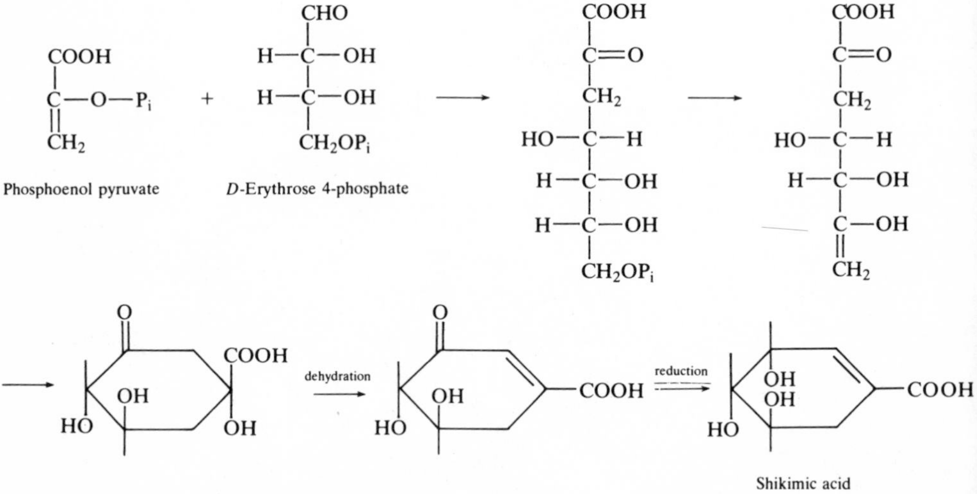

Phosphoenol pyruvate + *D*-Erythrose 4-phosphate → (dehydration) → (reduction) → Shikimic acid

FIGURE 7.10. Enzymic reactions involved in the formation of shikimic acid; $-OP_i$ represents phosphate $-(OPO_3)^{2-}$

7.3.3. Biosynthesis

The biosynthesis of lignin has been studied by feeding plants with radioactive substances which are thought to be the precursors, and studying the resultant activity appearing in the lignin. As the biochemical pathways have become known, the enzymes involved have been studied individually.

In some experiments, a polymer with properties similar to coniferous wood lignin has been prepared *in vitro* by the action of an oxidase enzyme on coniferyl alcohol. These studies indicate that the phenylpropane monomers are synthesized from photosynthetic carbohydrate, e.g. glucose. This hexose is converted enzymically to phosphoenol pyruvate and erythrose-4-phosphate, which react together to give—after a series of steps—shikimic acid (see Figure 7.10). Shikimic acid itself is then phosphorylated, and reacts with phosphoenol pyruvate to give an aromatic ring system, phenylpyruvic acid, and the amino acid, phenylalanine. This series of reactions is shown in Figure 7.11. Phenylalanine then undergoes deamination, hydroxylation, and methylation to give acids corresponding to the alcohols shown in Figure 7.7, i.e.:

Phenylalanine —deamination→ Cinnamic acid —hydroxylation→ *p*-Coumaric acid (V) —hydroxylation→

—methylation→ Ferulic acid (VI) —hydroxylation and methylation→ Sinapic acid (VII)

The acids V, VI and VII can then be reduced to the corresponding alcohols.

The first step in the polymerization of these alcohols involves a dehydrogenation, catalysed most probably by a peroxidase in the presence

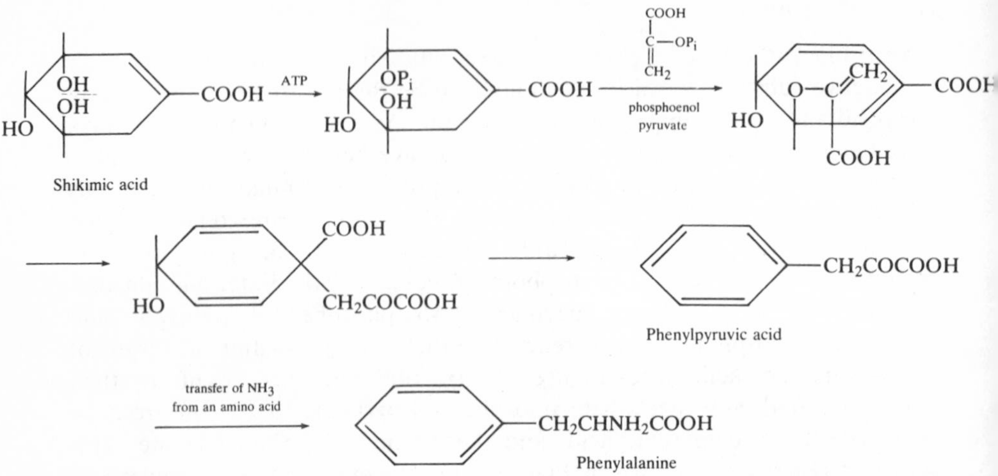

FIGURE 7.11. The conversion of shikimic acid to phenylalanine; $-OP_i$ represents phosphate $-(OPO_3)^{2-}$

of hydrogen peroxide. The proposed intermediates in the synthesis of coniferous lignin are free radicals of the type:

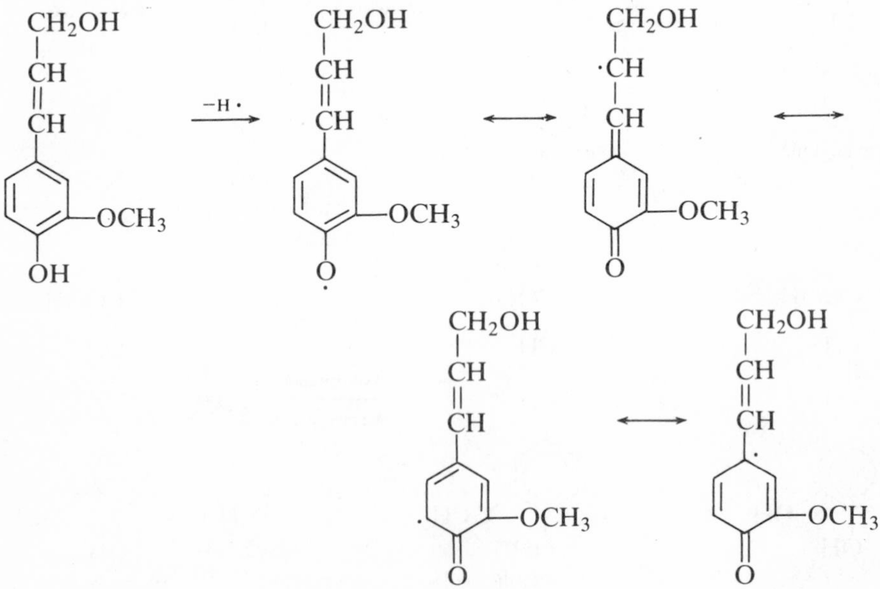

(VIII)

These radicals can combine in various ways to give carbon–carbon or ether linkages, and so form dimers such as those shown in Figure 7.8. Further dehydrogenation gives new radicals which can in turn produce trimers, tetramers, etc., until a large cross-linked polymer is formed. This mechanism is unusual for the biosynthesis of a polymer, because it is believed that, after initial production of the free radicals, the polymerization proceeds without enzymic control.

In vitro experiments using coniferyl alcohol and a mushroom oxidase have given a product with properties similar to coniferous lignin. Although sinapyl alcohol by itself could not be polymerized by the enzyme (presumably because the extra methoxyl group renders difficult formation of radicals of type VIII above), a lignin-type product containing sinapyl residues was formed when sinapyl alcohol was added to coniferyl alcohol in the presence of the oxidase.

7.3.4. Uses

Degraded lignin is usually obtained in solution as a by-product in the preparation of cellulose for paper-making (see Chapter 6). Wood may be pulped by heating, under pressure, with sulphur dioxide and calcium bisulphite solution, or with strong alkali in combination with sodium sulphide. In each case, the lignin is dissolved, leaving the insoluble cellulose fibres.

The lignin sulphonates resulting from the bisulphite pulping process are used as dispersants and wetting agents—they are important additions in the preparation of oil-well drilling-muds, and this outlet is at present one of the main uses of lignin products. Lignin sulphonates are also used in adhesives, road binders, cement products, industrial cleaners, and in leather tanning. In addition, vanillin (an important flavouring in the food industry) is produced when calcium lignin sulphonate is heated with alkali. Vanillic acid can be obtained also, and is now used as a monomer for the synthesis of textile polyesters. In addition, ethyl vanillate is toxic to some microorganisms but not to human beings, and so finds a use in medicine and as a food preservative.

Lignin obtained by alkali extraction can be used as a dispersant for dyes, as an additive for concrete, and a filler for natural rubber. It also finds uses in ore flotation, adhesives, as a component of resins, and as a stabilizer and emulsifying agent. Insufficient uses have been found, however, for the large quantities of lignin extracted by the paper-pulping industry. With the rising costs of energy sources, it is nevertheless useful to burn much of the lignin at pulping plants as fuel. In future, it may even become economical to use lignin and wood as the starting material for many organic chemicals now obtained from petroleum.

7.4. ADDITIONAL READING

Rubber and Gutta Percha

J. Bonner in '*Biogenesis of Natural Compounds*' (Ed. P. Bernfeld), Pergamon Press, London, 1967, p. 941.

Articles on rubber and rubber processing in *Encyclopedia of Polymer Science and Technology*, **12**, 178–355. Wiley–Interscience, New York, 1970

Textbook of Polymer Science, F. W. Billmeyer, 2nd Edition, Wiley–Interscience, New York, 1971.

B. L. Archer and B. G. Audley, *Advances in Enzymology,* **29**, 221 (1967).

M. Porter in *The Chemistry of Sulphides* (Ed. A. V. Tobolsky), Wiley–Interscience, New York, 1968, p. 165.

Phytochemistry, Vol. II (Ed. L. P. Miller), Van Nostrand Rheinhold Co., New York and London, 1973.

Introduction to Polymer Science L. R. G. Treloar and W. F. Archenhold, Wykeham Publications Ltd., London, 1974.

Polymer Chemistry, an Introduction M. P. Stevens, Addison–Wesley Publishing Co., Reading, Massachusetts and London, 1975.

E. D. Beytia and J. W. Porter, *Annual Reviews of Biochemistry,* **45**, 113 (1976).

Lignin

F. F. Nord and W. J. Schubert in *Biogenesis of Natural Compounds* (Ed. P. Bernfeld), Pergamon Press, London, 1967, p. 903.

W. J. Schubert in *Encyclopedia of Polymer Science and Technology*, Vol. 8, Wiley–Interscience, New York and London, 1968, p. 233.

The Constitution and Biosynthesis of Lignin K. Freudenberg and A. C. Neish, Springer–Verlag, New York and Berlin, 1968.

T. Higuchi, *Advances in Enzymology,* **34**, 207 (1971).

Lignins, Occurrence, Formation, Structure and Reactions (Eds. K. V. Sarkanen and C. H. Ludwig), Wiley–Interscience, New York, 1971.

Phytochemistry, Vol. III (Ed. L. P. Miller), Van Nostrand Rheinhold Co., New York and London, 1973.

G. H. N. Towers in 'MTP International Review of Science', *Biochemistry Series One*, Vol. 11 (Ed. D. H. Northcote), Butterworths, London, 1974, p. 247.

The Structure, Biosynthesis and Degradation of Wood (Eds. F. A. Loewus and V. C. Runeckles), Vol. 11 of *Recent Advances in Phytochemistry*, Plenum Press, New York and London, 1977.

7.5. APPENDIX: STEREOCHEMICAL CONTROL BY THE ENZYMES INVOLVED IN THE BIOSYNTHESIS OF RUBBER AND GUTTA PERCHA

In discussing the biosynthesis of rubber and gutta percha (see page 333), it was stressed that the enzymes responsible for the polymerization step exerted strict stereochemical control over the reaction. This fact has been established by using two tritiated mevalonic acids of known stereochemical configuration as the precursor in the biosynthesis.

The stereochemistry of the reaction to form labelled and unlabelled polymers is as shown:

H_3C, OH, HOOC, C, CH_2OH, CH_2, C, T, H → CH_3, C, CH_2OPP_i, CH_2, C, T, H → CH_3, T, C=C, ···$-CH_2$, CH_2-···

4S-[4³H] Mevalonic acid

Tritiated rubber

+

Unlabelled gutta

CH_3, CH_2-···, C=C, ···$-CH_2$, H

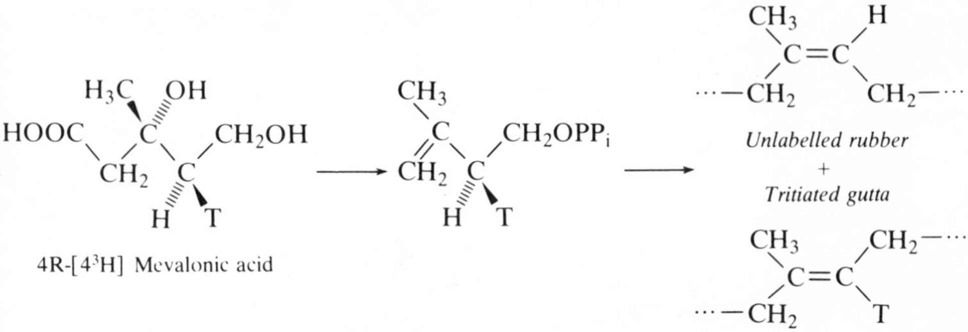

H_3C OH
HOOC C CH_2OH
CH_2 C
H T

4R-[4^3H] Mevalonic acid

CH_3
C CH_2OPP_i
CH_2 C
H T

CH_3 H
C=C
$\cdots-CH_2$ $CH_2-\cdots$

Unlabelled rubber
+
Tritiated gutta

CH_3 $CH_2-\cdots$
C=C
$\cdots-CH_2$ T

▼ indicates atom or group *above* plane of the paper.

⦙ indicates atom or group *below* plane of the paper.

Originally, chain initiation and growth (see Section 7.2.3, p. 333) were believed to take place by an S_N1 mechanism, involving the loss of pyrophosphate from the growing chain to give a carbonium ion. This ion was then thought to react with an incoming isopentenyl pyrophosphate molecule. However, such a mechanism would give rubber and gutta containing 50% of the tritium from reactions shown above, and could not account for the observed results. Instead, an S_N2 mechanism has been postulated where, in the synthesis of gutta, a *trans* addition across the isopentenyl pyrophosphate double bond is followed by a *trans* elimination:

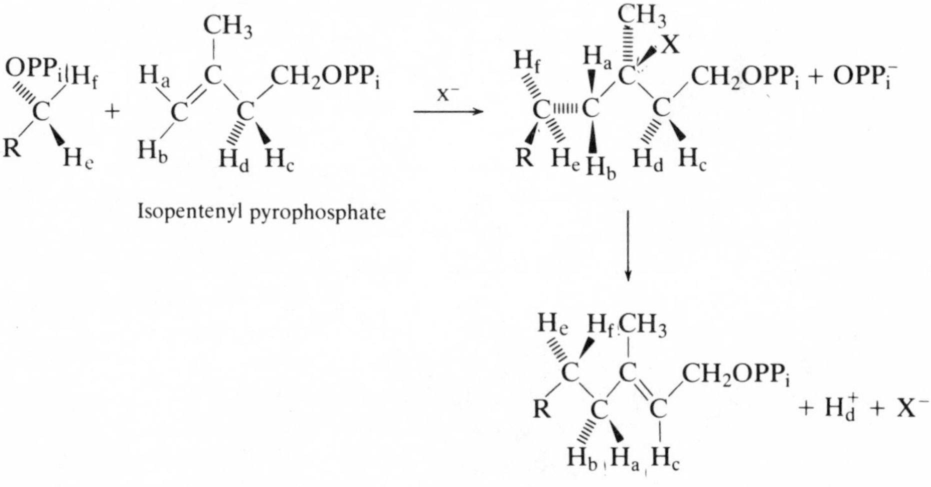

OPP_i H_f
C + H_a C CH_2OPP_i $\xrightarrow{X^-}$
R H_e H_b C H_d H_c

Isopentenyl pyrophosphate

CH_3
H_f H_a C X $CH_2OPP_i + OPP_i^-$
C C C
R H_e H_b H_d H_c

↓

H_e H_f CH_3
C C CH_2OPP_i
R C C $+ H_d^+ + X^-$
H_b H_a H_c

Gutta

R = Growing chain

X^- may come from the enzyme catalyzing the reaction

A similar mechanism with rotation round a single bond in the intermediate before the elimination step, could give rubber with elimination of H_c.

$$\text{(intermediate: } H_f,\ H_a,\ CH_3,\ X,\ H_d,\ R,\ H_e,\ H_b,\ H_c,\ CH_2OPP_i) \longrightarrow \text{(rubber: } H_e,\ H_f,\ CH_3,\ H_d,\ R,\ H_b,\ H_a,\ CH_2OPP_i) + X^- + H_c^+$$

Rubber

The steric requirements of the enzyme–substrate complex involved determine which of the above two reaction mechanisms takes place.

Cornforth was awarded the Nobel Prize in 1975 for his work on the stereochemistry of the biosynthesis of polyisoprenes.

Chapter 8

Inorganic Polymers

8.1. INTRODUCTION

The rocks, clays, soils, and sands of the Earth's surface are composed mainly of the *silicates*, some of which are polymers. Indeed, silicon and oxygen are the most abundant elements in the Earth's crust, and so the geological significance of such polymers is as great as the biochemical importance of the carbon-based macromolecules. Furthermore, the properties of the silicates are also often of great value in industry and agriculture.

Earlier chapters have discussed the biopolymers in terms of their functions and biosynthesis, but this approach is not possible with the inorganic polymers. Most of these materials were synthesized at extremely high temperatures and pressures in the interior of the Earth, and the details of such processes are unknown. Then later, these polymers were forced to the surface, where many were changed into different polymers by the chemical action of water, and the oxygen and carbon dioxide of the air.

Early attempts to classify silicates on the basis of chemical composition alone were unsuccessful, for on such a basis some classes contained minerals of widely divergent properties, whilst substances with obviously similar properties were often assigned to different classes. With the advent of X-ray diffraction methods of determining molecular structure (see Section 3.6, p. 65), it became apparent that the most satisfactory classification of silicates was that based on their structure. This system is adopted here, but before the more abundant silicates are considered, the simplest natural inorganic polymers—those formed from *carbon*—are discussed.

8.2. CARBON POLYMERS

The 'monomers' of the inorganic carbon polymers may be considered to be single carbon atoms. In the naturally occurring polymers, different spacial arrangements of the carbon atoms produce two substances with very different properties, *diamond* and *graphite*.

8.2.1. Diamond

The carbon atoms in diamond are arranged tetrahedrally in space, and each atom is linked to four others by normal single covalent bonds of length 1.54 Å (see Figure 8.1a). The result is an infinite three-dimensional network, and a diamond crystal is, in effect, one very large molecule of carbon. The crystal lattice in natural diamonds is not necessarily perfect, however, and interstitial carbon atoms may be found between the atoms shown in Figure 8.1(a).

The important properties of diamond are all related to this structure. As the four valency electrons are completely involved in bond formation, there are no free electrons and the structure does not conduct electricity. Also, the strong carbon–carbon bonds must be broken when the crystal is cleaved, and so diamond is extremely hard—the hardest natural material known. It is used industrially, therefore, for cutting, as an abrasive, in manufacturing machine tools, for rock-drilling in mining, and to make dies for the preparation of thin wires such as the tungsten filaments of incandescent lamps.

Diamond also has a high refractive index which enhances the internal reflection of light in the crystal. Although diamond is extremely hard, definite cleavage planes exist in the crystals, and the clear diamonds used as gem-stones are cut along these planes to maximize the internal

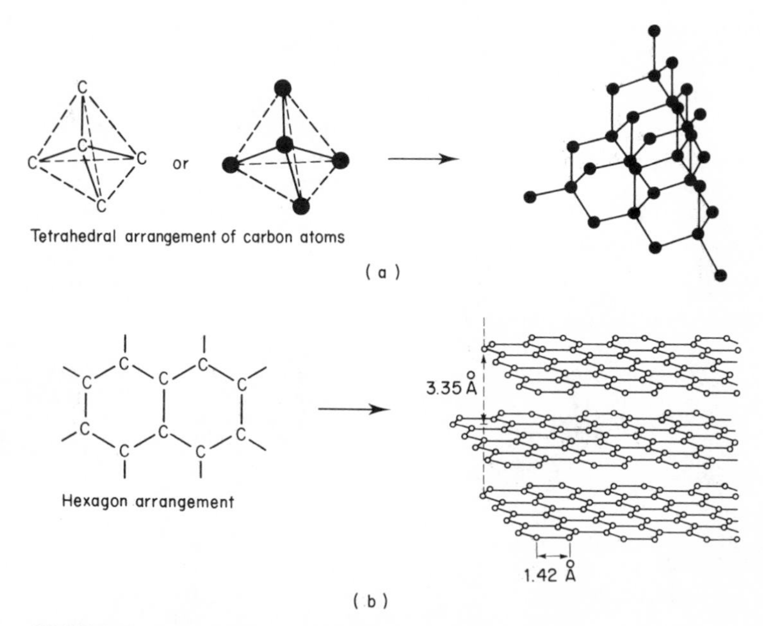

FIGURE 8.1. Polymers of carbon: spatial arrangements of carbon atoms in (a) diamond, and (b) graphite

reflection of light, thus producing the brilliance for which diamond gems are so well known.

Chemically, diamond is stable and inert. It can ignite in oxygen above 800°C to give carbon dioxide, and is very slowly attacked by sulphur vapour at 1000°C. As graphite is the stable high-temperature modification of carbon, diamond can be readily graphitized in a helium atmosphere at 1500°C.

Small, industrial diamonds can be synthesized by heating the second polymeric type of carbon, graphite, to 3000°C at a pressure of 10^5 atmospheres in the presence of a little manganese, iron, or cobalt.

Both natural and synthetic diamonds contain nitrogen as an impurity. In synthetic diamonds, this nitrogen is thought to occur as single nitrogen atoms replacing carbon atoms in the crystal lattice, and causing a yellow colour. Such coloured stones occur naturally, but are of little value as gems. Colourless, or blueish, gem diamonds contain more nitrogen in complex arrangements, often with N—N pairs replacing C—C in the polymer.

Studies have shown that when yellow diamonds are kept at high temperatures and pressures, the arrangement of nitrogen in the structure changes to resemble more closely the arrangement in colourless stones. It is thus likely that diamonds of gem-stone quality may be prepared synthetically, and further studies of this type will probably give information about the formation of diamonds in the Earth's crust.

8.2.2. Graphite

In graphite, the carbon atoms are arranged in coplanar, hexagonal groups: each atom is bonded to three other atoms in the same plane, and the parallel planes are stacked one above the other, so that half of the carbon atoms of one sheet lie directly above, and half directly below, atoms in adjacent sheets (see Figure 8.1b). Within each layer, the carbon–carbon bond lengths are 1.42 Å, whilst the layers themselves are 3.35 Å apart. The layers of graphite can be regarded as fused sheets of benzene rings, where each bond has one-third double-bond character. There is a σ- and π-bond between each pair of carbon atoms, and the whole structure involves a number of infinite two-dimensional, parallel molecules held together by residual forces involving the π-electrons. The π-electrons can readily change from one energy state to another by the absorption of visible light, and so graphite is black and opaque, even in thin slices. Graphite is thus used as a pigment in industrial paints.

Again the structure determines the important properties of graphite. The π-electrons of graphite are mobile within one plane, and so graphite can conduct electricity. Electrons do not normally cross from one plane to another, and the conductivity parallel to the layers is 10^5 greater than that in a perpendicular direction. Because of its conductivity, graphite is used to make brushes for electric motors and generators.

As the forces holding the parallel layers together are relatively weak, graphite is soft, and readily cleaves parallel to the layers, to form flakes. Thus graphite is used as a lubricant, and for this purpose is normally suspended in oil and water with tannin as a stabilizer. If powdered graphite is mixed with clay and water to give a paste, this may be extruded as thin threads for use as the 'lead' in pencils.

Graphite should have, in theory, high strength in the plane of the sheet of carbon atoms, but low resistance to the sliding of the sheets past each other. In ordinary graphite, sliding can occur in any direction, for the crystals have random orientation. It is possible now, however, to produce carbon fibres in which the sheets are all orientated in the direction of the fibre axis. This is done by careful carbonization of synthetic acrylic fibres. The resultant carbon fibres have strengths almost as high as that expected for single molecules where C—C bonds would have to be broken to break the fibres. Such fibres, which are stronger than steel, are often embedded in synthetic polymers to give increased strength.

Graphite is chemically inert, and so is used as electrodes for the electrolytic production of, for example, chlorine. Graphite rods are now widely used as moderators in nuclear reactors. This form of the polymer is also a good heat conductor, and is used to make crucibles for molten metals. In this instance, clay is mixed with the graphite in the manufacture of the crucibles, to bind the graphite together and prevent its oxidation. Graphite is also frequently used for foundry facings, as it is not wetted by most molten metals.

Graphite occurs naturally in shales, slates, and other metamorphic rocks (rocks which have been altered by great heat or pressure), but may be synthesized in an electric furnace by passing an alternating current through petroleum coke, or granular anthracite mixed with pitch, in the presence of a little silica sand and iron oxide as catalysts. (The silicon and iron form carbides, which decompose to give graphite and regenerate the elements silicon and iron.)

Although chemically inert, graphite will ignite in oxygen at 690°C to give carbon dioxide. When graphite is heated with many metals or their oxides above 1500°C, *metal carbides* are formed. With strong oxidizing agents, such as potassium chlorate in nitric acid, graphite oxide is formed; various C:O ratios have been reported, but these are usually less than two. The graphite swells in one direction on oxidation, and also takes up water between the layers. The interplanar distance in the oxide varies with the amount of water from 6.4 Å to a maximum of 11.3 Å, when a monomolecular layer of water exists on both sides of each layer. The bonding in the oxide is not completely understood but —C—C— (bridged by O), C=O, and —C—OH may exist. As the π-electrons of the graphite are involved in the bonding, the electrical conductivity is lost. The oxide decomposes at 200°C to give carbon monoxide, carbon dioxide, and carbon.

However, the relatively large interplanar spacing in graphite also enables other ions, atoms, or molecules to fit between the layers to form intercalation compounds, the graphite 'salts'; e.g. a blue complex, $C_{24}H_2SO_4.2\,H_2SO_4$, forms with H_2SO_4. Chlorides of multivalent metals in their highest oxidation state also form compounds with graphite, and the ease with which many paramagnetic compounds are intercalated suggests that electrons are transferred from the graphite to the metal. Other interlayer compounds can be formed with fluorine and bromine. The compound with fluorine is non-conducting, whilst that with bromine does conduct electricity. Conducting complexes are also formed with the alkali metals, with the exception of sodium and lithium. In all cases, the interplanar distance in graphite increases on compound formation. Thus graphite absorbs liquid potassium, and swells in a direction at right angles to the cleavage plane to form a bronze-coloured compound, KC_8, with potassium between every layer of the carbon atoms. On heating, KC_{24} is obtained with the potassium atoms between alternate carbon layers only. Unlike the mainly covalent bonding in the oxide, the bonding in the potassium compounds is essentially ionic, and KC_8 is a good electrical conductor.

8.3. SILICATE POLYMERS

The basic unit of the silicate polymers is a tetrahedron of four oxygen atoms surrounding a silicon atom (see Figure 8.2a). The Si—O bond is partly covalent, i.e. is intermediate between the following resonance forms:

$$\begin{array}{c}\bar{O}\\ |\\ {}^{-}O-Si-O^{-}\\ |\\ \underline{O}\end{array} \qquad \begin{array}{ccc} & O^{2-} & \\ \overset{2-}{O} & \overset{2+}{Si} & \overset{2-}{O}. \\ & \underset{2-}{O} & \end{array}$$

In the polymers, these tetrahedra are linked together at the corners to form either chains (Figure 8.2b), or layers (Figure 8.2c), or three-dimensional networks (see Sections 8.3.3 and 8.3.4, pp. 366, 368). An oxygen atom bound to only one silicon atom carries a negative charge, and various cations, e.g. Na^+, K^+, Mg^+, Ca^{2+}, Fe^{2+}, etc., are found in the structures of the naturally occurring silicates to maintain electrical neutrality. Many closely related silicates exist with the same silicon–oxygen arrangement, but differing only in the number and type of cations present. For example, a calcium ion may be substituted by one magnesium ion or

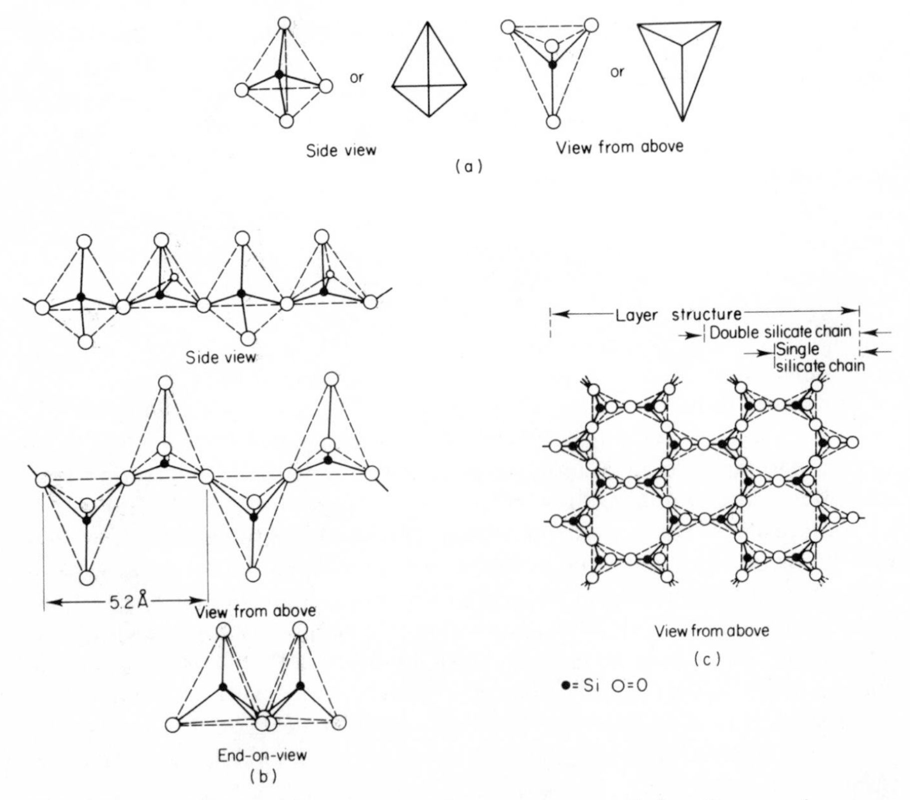

FIGURE 8.2. Silicate polymers: spatial arrangements of the silicon and oxygen atoms in (a) the basic tetrahedral unit, (b) one possible single chain, and (c) one possible double chain and a layer structure

two sodium ions, giving different electrically neutral polymers with the same silicate backbone.

Silicates often contain anions such as OH^- and F^-, and one of these ions may replace the other with no overall structural change in the polymer. This isomorphous replacement of ions in silicates made early attempts at chemical classification very difficult and unsatisfactory, and the relations between the different polymers were clarified only after structural determinations by X-ray diffraction. In addition, silicates can now be analysed and characterized by their optical and mineralogical properties, by infrared spectroscopy, and by differential thermal analysis.

Silicate structures are further complicated by the fact that aluminium can occur in a four- or six-coordinated state in these minerals. When aluminium is four-coordinated, it can replace silicon in the backbone of the polymer. This substitution can occur randomly, and causes the polymer framework to acquire an extra negative charge. A simultaneous increase in positive charge must occur to preserve electrical neutrality, and thus Ca^{2+} may be substituted for Na^+, or additional Na^+ ions may be found in a structure, where there is sufficient space for them. When the aluminium is six-coordinated it acts as a cation, Al^{3+}, to neutralize the charge on the silicate framework. Aluminium may exist in both states in one mineral, and information about the coordination of the aluminium can be gained by X-ray diffraction methods.

A few of the economically important silicate polymers will now be discussed in terms of their structures.

8.3.1. Chain structures

Both single and double chains (the simplest are shown in Figure 8.2b and c) are found in naturally occurring minerals in igneous or metamorphic rocks.

8.3.1.1. Pyroxenes

In the single chains, two of the four oxygen atoms of each tetrahedron are shared by other tetrahedra, giving $(SiO_3)^{2-}$ as the empirical formula of the polymeric anion. The important group of minerals, the pyroxenes, have structures of the type shown in Figure 8.2(b), and one of the simplest examples is *diopside*, $CaMg(SiO_3)_2$ (see Figure 8.3). In this structure, the vertices of different chains point alternatively in opposite directions, and the chains are held together by ionic bonding between the metal cations and oxygen atoms of the silicate backbone. (It should be noted that the chemical formulae given here for silicate minerals are idealized; because of isomorphous substitution, the formulae of naturally occurring mineral samples are usually much more complex.)

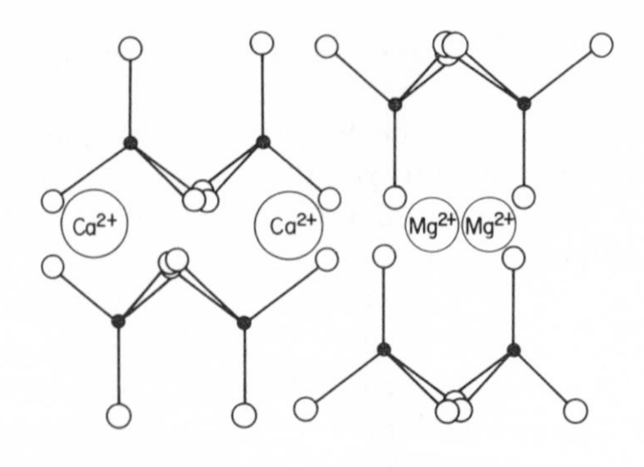

● = Si
○ = C

FIGURE 8.3. The structure of the pyroxene, diopside. The chains are shown end-on, and should be considered to extend indefinitely above and below the plane of the paper at right angles to it: only one pair of tetrahedra are visible in each chain, as each tetrahedron covers those behind it

8.3.1.2. Amphiboles

In the double chains of the amphibole minerals (Figure 8.2c), the silicon:oxygen ratio is 4:11, and so the empirical formula of the polymeric anion is $(Si_4O_{11})^{6-}$. The amphiboles differ further from the pyroxenes in containing hydroxyl and fluoride ions in their structures; these ions are bonded to the metal cations and not directly to the silicon. One of the simplest examples is *tremolite*, $Ca_2Mg_5(Si_4O_{11})_2(OH)_2$; again the vertices of the chains point alternately in opposite directions (compare Figure 8.3), and the chains are held together by the metal ions. In the amphiboles, some silicon may be replaced by aluminium, and the extra negative charge is balanced by the substitution of Al^{3+} for Mg^{2+}, or by the insertion of additional alkali, or alkaline earth, ions into the crystals, for there is sufficient space between the double chains to accommodate these extra ions.

Jade, which has been carved for centuries in the East to make ornaments and utensils, may be either of two minerals—one is a pyroxene, *jadeite*, $NaAl(SiO_3)_2$, whilst the other is an amphibole, *nephrite*, $Ca_2(Mg^{2+}Fe^{2+})_5(Si_4O_{11})_2(OH)_2$. High-resolution electron microscopy has shown the presence of isolated triple chains of silicate tetrahedra in minerals such as nephrite.

As the interchain binding is weaker than the Si—O backbone bonds, both pyroxenes and amphiboles can be cleaved relatively easily between

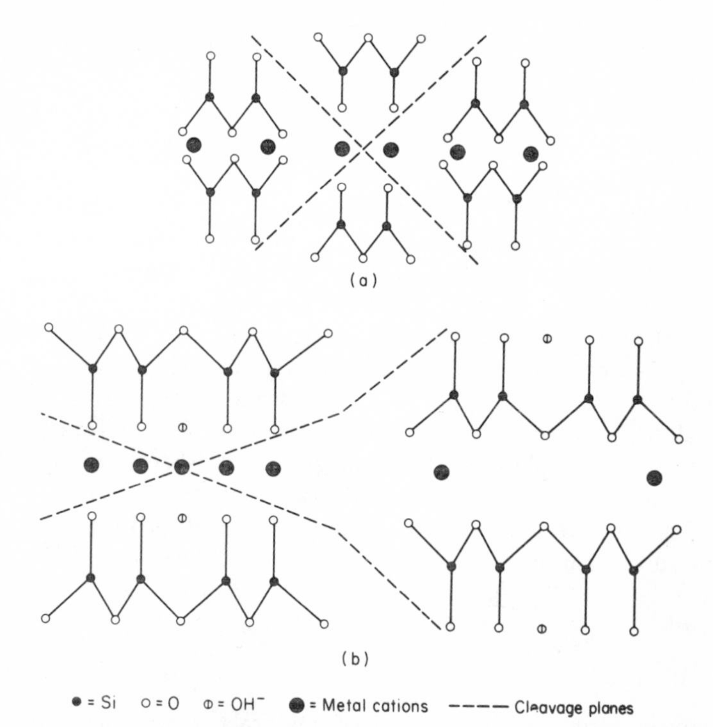

FIGURE 8.4. Cleavage planes of (a) pyroxenes, and (b) amphiboles. The chains are shown end-on

the chains, but not at right angles to them. The angles of cleavage are different for pyroxenes and amphiboles, and are used to distinguish between the two types of mineral (see Figure 8.4).

Several amphiboles are fibrous in character, and the fibres of some, e.g. tremolite and *crocidolite*, $Na_2Fe_3^{2+}Fe_2^{3+}(Si_4O_{11})_2OH_2$, can be processed to give the heat-insulating material, *asbestos*. Crocidolite fibres are extremely strong—stronger than carbon fibres—and have the highest strength at room temperature of any known mineral. These fibres are used for reinforcing building cement and thermosetting resins. The fibres are weakened at high temperatures, but can still be used for protective clothing. Tremolite fibres are weaker, and hence are of less industrial importance.

8.3.2. Layer structures

In the layer-type silicates (see Figure 8.2c), three of the oxygen atoms of each tetrahedron are shared by other tetrahedra, and the fourth oxygen atoms all lie on one side of the sheet; the empirical formula of the polymer ion is therefore $(Si_2O_5)^{2-}$. The layers are held together by metal ions, but this binding is weaker than the Si—O or Al—O binding within the layer, with the result that these materials cleave easily parallel to the layers and tend to be flaky and soft.

8.3.2.1. *Talc*

Talc, $Mg_3(Si_2O_5)_2(OH)_2$, is one of the softest minerals known, and consists of Si—O layers in double sheets with the unshared oxygen atoms pointing towards one another. The double layers are held together by magnesium ions, and contain hydroxyl ions for electrical neutrality (see Figure 8.5). Each magnesium ion is octahedrally coordinated by four oxygen atoms of the sheets and two hydroxyl ions. The hydroxyl ions lie at the centre of the hexagonal rings of unshared oxygen atoms. (These hexagonal rings are most easily seen in the view of the layers from above as shown in Figure 8.2c.) The bases of the double sheets are held together by weak van der Waals' forces, and this accounts for the extreme softness of talc. This property is utilized when talc is used as French chalk, as a lubricant and polish, in paper and leather manufacture, and in the cosmetic industry.

8.3.2.2. *Micas*

The micas form another group of minerals containing layer silicate structures. These layers are again arranged in double sheets as in talc, but in the micas, isomorphous replacement of one metal ion by another is very common, and so the formulae given are the ideal formulae, which do not often describe the minerals actually found in Nature. One quarter of the silicon atoms in micas are replaced by aluminium, and so the empirical formula of the polymer ion is $(AlSi_3O_{10})^{5-}$. To maintain electrical neutrality, potassium ions are present between the bases of the double sheets; there is one K^+ for every silicon replaced by aluminium, and each

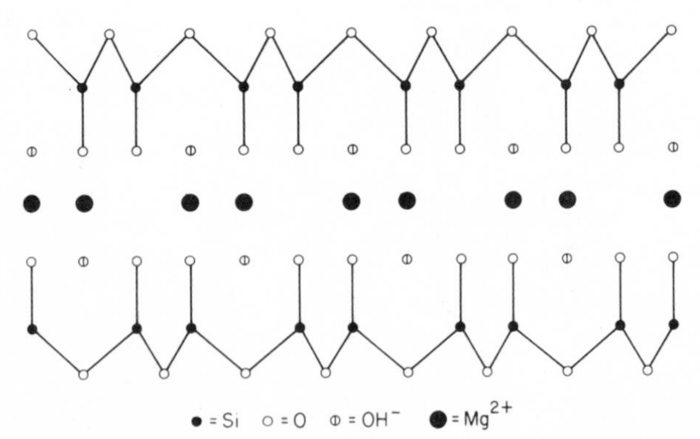

FIGURE 8.5. End-on view of layers of talc

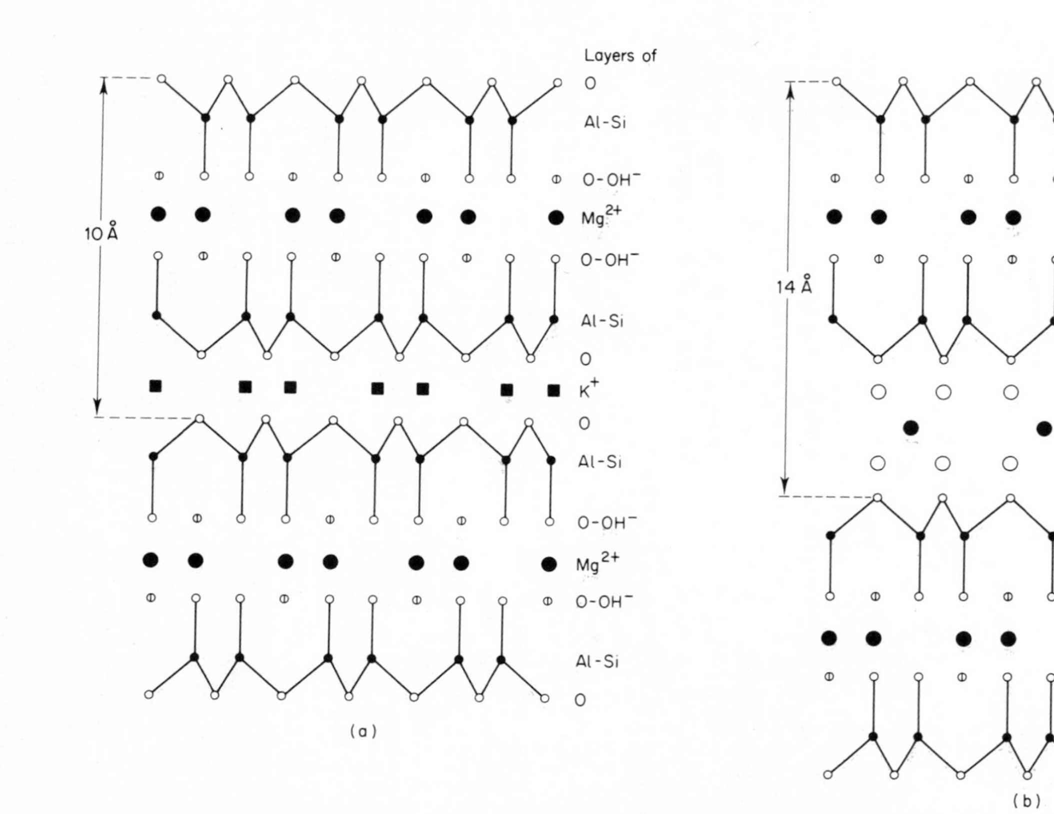

FIGURE 8.6 End-on view of layers of (a) biotite mica, and (b) a vermiculite

potassium is coordinated by 12 neighbouring oxygen atoms (see Figure 8.6a).

Phlogopite or *biotite mica*, $KMg_3(AlSi_3O_{10})(OH)_2$, has a structure similar to talc with the addition of the potassium ions, whilst *muscovite mica*, $KAl_2(AlSi_3O_{10})(OH)_2$, contains Al^{3+} instead of the magnesium. In this mineral, the aluminium exists in both the four- and six-coordinated states.

Micas are found in rocks such as granite, and are harder than talc because the bonding between the potassium ions and the bases of the layers is stronger than the van der Waals' forces of talc. However, micas still readily cleave parallel to the layers to give plates, which can be used as windows where glass could not withstand the thermal shock. Micas can also be used for electrical insulation, and are alkali- and acid-resistant.

An industrially important, synthetic mica, *fluorophlogopite*, $KMg_3(AlSi_3O_{10})F_2$, can be produced by melting a potassium feldspar (see Section 8.3.4.1, p. 368) with potassium fluorosilicate, alumina, magnesia and quartz sand (see Section 8.3.3, p. 366) at 1420°C using an arc resistance heat process, and then allowing the melt to crystallize.

Vermiculites are formed by the decomposition of mica, and contain layers of water molecules and magnesium ions in place of the potassium ions of mica (see Figure 8.6b). If vermiculite is subjected to temperatures of 800–1100°C for a short time, this water is converted to steam, which disrupts the structure. The vermiculite expands, and because of its low thermal conductivity and density, can be used as a thermal and sound insulator, and as an aggregate in lightweight concrete. It can also be used as a moisture-retaining soil conditioner.

8.3.2.3. Clays

The clays, a group of minerals formed by the weathering of volcanic rocks, also contain layer structures. However, the arrangement of the layers is different from that in talc and the micas, for in clays the sheets are not necessarily in pairs, but the unshared oxygen atoms of every sheet can point in the same direction. This arrangement means that clay minerals form large crystals with great difficulty.

Kaolinite, $Al_2(Si_2O_5)(OH)_4$, is the main constituent of the china clay used to make porcelain, and contains the silicate layer ions with hydroxyl ions lying in the hexagonal ring of unshared oxygens, as in talc. Above this sheet of unshared oxygen atoms and OH^- ions lies another sheet of OH^- ions; the Al^{3+} ions are situated between the two sheets of oxygen atoms (see Figure 8.7). Each aluminium ion is coordinated by two oxygen atoms of the silicate backbone and four hydroxyl ions. The whole 'sandwich' of aluminium between the (O—OH) and the OH layers is bound to the next 'sandwich' lying parallel to it by van der Waals' forces, and by hydrogen bonds between OH groups and the oxygen atoms attached to silicon. As these forces are weak, kaolinite, and indeed most clays, are soft and easily cleaved.

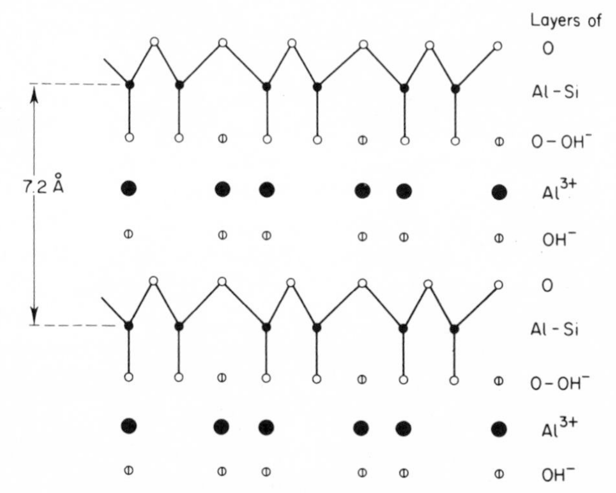

FIGURE 8.7. End-on view of layers of kaolinite

Another clay, *bentonite*, containing double-layer structures more like mica, is used in cosmetics, as a filler for soaps, as a plasticizer, and for stabilizing suspensions—hence its use in drilling-muds for oil-wells.

Most natural clays contain finely divided quartz (see Section 8.3.3, p. 366), feldspar (see Section 8.3.4.1, p. 368), and micas, as well as clay minerals, and hence they play an important part in determining soil quality—a fact which is of importance to agriculture. Clays rich in iron oxide are used to make ordinary pottery and terra cotta articles, whilst bricks and tiles are produced from clays containing sand in addition to iron oxide. Nowadays, bentonite and kaolinite clays are used (after treatment with sulphuric acid to generate acidic sites on the surface) as petroleum-cracking catalysts.

Clays with a high calcium and magnesium carbonate content, known as marls, are used in the cement industry. Even when the alkaline earth ion content is low, cement can be made from clay by mixing and heating with slaked lime. The main components of cement are dicalcium and tricalcium silicates, tricalcium aluminate, and calcium aluminoferrite. Extremely complex reactions take place after the addition of water during the setting of cement, and these probably involve both hydration and hydrolysis. The calcium silicates become hydrated, and these hydrates grow out from cement particles as fibrils which interlock to give a rigid structure. Thus solidification and the development of strength take place as the hydrates form an interlocking matrix, which replaces the water between the cement particles.

8.3.2.4. Chrysotile

The mineral of greatest economic importance as a source of asbestos is chrysotile, $Mg_3(Si_2O_5)(OH)_4$, which arises from the weathering of

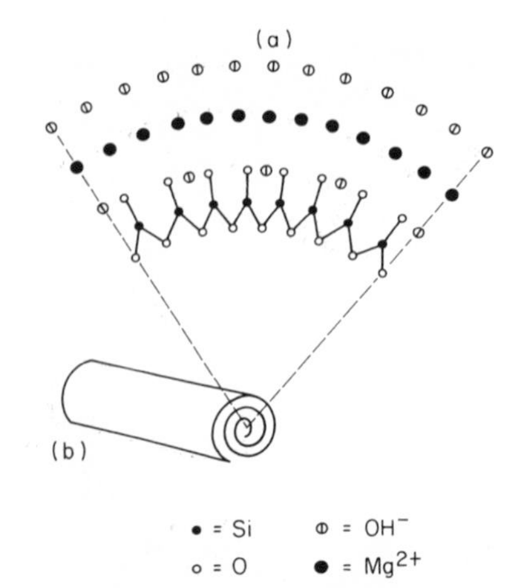

FIGURE 8.8. Structure of chrysotile: (a) curved layer; (b) fibrous tube formed from rolled-up sheet

magnesium-rich silicates in rocks such as granite. This fibrous mineral was believed to contain the double-chain backbone of the amphiboles, but the structure is now known to consist of rolled-up sheets, which form fibrous tubes. Chemically, the structure is related to kaolinite, but a layer of Mg^{2+} replaces the sheet of Al^{3+} of kaolinite (see Figure 8.8). The axes of the chrysotile scrolls are orientated along the fibre axis of the mineral, thus giving very strong fibres for reinforcing cement. The fibres also pose a health hazard, as airborne asbestos dust can give rise to lung cancer in humans.

8.3.3. Three-dimensional silica frameworks

In the three-dimensional silicate polymers each oxygen is shared by two tetrahedra, giving a silicon:oxygen ratio of 1:2. In the simplest case, this gives the electrically neutral *silica*, SiO_2. Silica exists in many forms in Nature. There are three main crystalline types, found mostly in granitic rocks. These are *quartz, cristobalite*, and *tridymite*, in all of which the Si—O distance is 1.60 Å. The three types differ in the arrangement of the SiO_4 tetrahedra in space, and in cristobalite, the silicon atoms are arranged like the carbon atoms of diamond, but with oxygen atoms between them. Like the diamond, a crystal of this type of silica may be considered as one large molecule, and Si—O bonds must be broken during cleavage of the crystal. However, the Si—O bond is not as strong as the C—C bond, and therefore the silica crystals are not as hard as diamond. In quartz and tridymite, there are helical arrangements of the SiO_4 tetrahedra. At room temperature, a type of quartz, α-*quartz*, is the only thermodynamically stable form of crystalline silica, but both cristobalite and tridymite occur naturally, as interconversion to α-quartz is extremely slow.

Crystalline silica is resistant to alkali at room temperature, and to most acids except HF. Fluorine readily attacks silica to give silicon tetrafluoride and oxygen, whilst silicon results from the reduction of silica with hydrogen or carbon above 1050°C.

Commercially, quartz is the most important of the three pure crystalline types, and quartz crystals are used as piezoelectric oscillators for frequency stabilization in radios, and for ultrasonic devices. Thin plates of quartz can be used as monochromators in X-ray diffraction. When pure quartz is melted in an electric vacuum furnace, the plastic mass can be worked into various shapes, which if allowed to cool under pressure, form clear quartz glass. This glass is used for the lenses and prisms of optical instruments working in the UV range of the electromagnetic spectrum, and in the manufacture of mercury vapour lamps. The observation windows of several Apollo spacecraft were made from silica glass.

Silica is found in quartzites and sandstones, both of which are used as building stones. Another source of silica is in the feathers of some birds, and the skeletons of certain marine organisms, e.g. sponges and diatoms. In the case of diatoms, the siliceous skeletons fall to the sea-floor as a fine-grained, porous silica, known as *kieselguhr*, which can be used as a polishing agent and adsorbent.

Several types of natural silica are used as decorative gem-stones. These include the crystalline but more impure form, *amethyst*, the crypto-crystalline types *carnelian*, *agate*, and *onyx*, and an amorphous form, *opal*. Opal is a dried and hardened gel formed by the action of hot water on silicate rocks. Precious opal consists of transparent spherical particles of amorphous silica, tightly packed together in a regular array. As a beam of light penetrates opal, the refractive index changes as the light passes from a silica sphere to one of the holes between the particles (the holes are filled with air, or water vapour, or liquid). These regularly arranged discontinuities act as light-scattering points and form a three-dimensional diffraction grating, for the spacing between the holes is of the same order of magnitude as the wavelength of visible light. Thus opals diffract visible light in a manner analagous to the diffraction of X-rays by the atoms of crystals (see Section 3.6, p. 65). When white light shines on an opal, the wavelength diffracted to an observer from the opal depends on the angle of incidence of the light, and so a range of colours can be seen as the stone is turned in the light.

Flint is a dark-grey, cryptocrystalline silica used by primitive man for making tools, and now used in glass manufacture and as an abrasive.

Silica can also be used in paints and as a wood filler, and with alumina is used to make fine clays to line furnaces. Sodium hydroxide dissolves silica to give a soluble silicate known as water-glass, which is used for fire- and water-proofing of textiles and timbers, as a filler in some cheap soaps, and as an adhesive.

Silica gel can be prepared by acidification of a solution of an alkali metal silicate. After drying, the gel can itself be used as a drying agent, as it can absorb half its own weight of water. A finely divided silica can be formed by acidification of silicate solutions, in the presence of salts to prevent gel formation. Such silica can be used as a filler in the rubber industry, as well as a filler and coating in the paper industry.

The weathering of quartz-bearing rocks can produce *silica sand*, which is widely used in making cement and concrete, and in the manufacture of various types of glass. Soft glass is made by heating silica sand with sodium and calcium carbonates; if potassium is substituted for the sodium, a harder glass is obtained, and is used for windows and bottles. Borosilicate glass is produced if some of the silica sand is replaced by boric oxide, and because of its low coefficient of expansion, resistance to chemical action and thermal shock, this glass can be used for laboratory apparatus, kitchen ware, insulators, and thermometers. A heavy, brilliant glass of high refractive index, used for cut-glass dishes and the lenses and prisms of optical instruments, is made by heating together silica sand, lead oxide, and potassium carbonate. The structure of glasses is believed to be a three-dimensional network of silicon and oxygen as in the crystalline silicas, but lacking a regular structure—they behave as a super-cooled liquid. A natural silicate glass, *obsidian*, is found in some areas, but is not industrially important.

8.3.4. Three-dimensional aluminosilicate frameworks

More complex three-dimensional silicates are formed when some of the silicon is replaced by aluminium. The resulting frameworks are negatively charged, and must contain metal ions to preserve electrical neutrality. In these polymeric ions, the Si/Al:O ratio is 1:2. Three main types of aluminosilicate frameworks will be considered here, i.e. the feldspars, zeolites and ultramarines.

8.3.4.1. Feldspars

The feldspars are very widely distributed minerals, and comprise two-thirds of all igneous rocks. In feldspars, the aluminosilicate tetrahedra are grouped in four-membered rings, with alternate pairs of vertices pointing in opposite directions, so that these rings form the links of a kinked chain (see Figure 8.9). The kinked chains are joined in layers through the vertices of the tetrahedra. Each oxygen atom is linked to two silicon (or aluminium) atoms giving an infinite three-dimensional network. The cations fit into the interstices in the structure. Typical feldspars are *orthoclase*, $K(AlSi_3)O_8$, and *albite*, $Na(AlSi_3)O_8$, where one silicon in every four is replaced by aluminium, and *anorthite* $Ca(Al_2Si_2)O_8$, where two silicon atoms are replaced by aluminium. As the ionic radii of Na^+ and

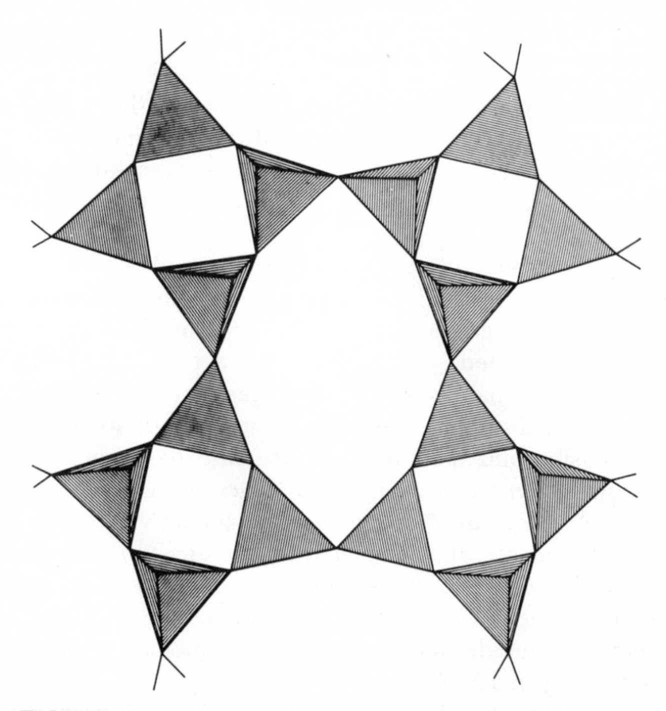

FIGURE 8.9. The structure of feldspar showing the aluminosilicate tetrahedra linked to form a very open three-dimensional structure

Ca^{2+} are very similar (0.95 and 0.99 Å, respectively), solid solutions form between albite and anorthite, and these minerals are found not only with the composition of albite or anorthite, but also with any intermediate composition. Good specimens of albite and orthoclase are known in jewellery as moonstones.

Feldspars (along with mica and quartz) are the main components of the building stone, granite. On weathering, the three-dimensional structure is broken down to layers, and so feldspars yield kaolinite and clay minerals which are very important in soil formation.

Mixed with quartz and kaolin, feldspars are used to make hard porcelain, for, during the heating, they produce a glass phase which gives the translucency of porcelain. They are also a constituent of the transparent glazes on porcelain, i.e. tableware after firing is dipped into a suspension of kaolin, clay, feldspar, and marble in water, and on heating, the transparent glaze is produced.

Synthetic feldspars can be formed by heating silica with an aqueous solution of a base at high pressures to temperatures above 400°C. Various three-dimensional aluminosilicate networks are produced in this way, e.g. zeolites (see next section) as well as feldspars.

8.3.4.2. Zeolites

Zeolites are found in cavities in rocks like basalt. They possess a much more open structure than the feldspars, and the aluminosilicate framework contains large channels and pores. The cations necessary to balance the negative charge on the polymeric ion are located in these channels, as are molecules of water.

An important property of zeolites is the ease with which the water can be removed (by heating) and taken up again, without destruction of the silicate framework. Consequently, zeolites are used as industrial drying agents. When dehydrated, the zeolites can act as adsorbents, and can reversibly take up many liquids and gases (e.g. alcohol, benzene, chloroform, carbon disulphide, mercury, ammonia, or iodine). As zeolites adsorb gases easily, and their internal structure has a high surface area, they show high catalytic activity in heterogeneous gas reactions. (Synthetic zeolites are now widely used as petroleum-cracking catalysts.) The size of the molecules which are able to penetrate the structure depends on each particular zeolite, and so zeolites may be used as a series of molecular sieves. Synthetic zeolites can be produced by treating a sodium silicate–sodium aluminate mixture with an aqueous solution of alkali, followed by heating to 100°C.

The channels in most zeolites are large enough to allow the free passage of ions, and the cations of a zeolite may be exchanged by soaking in a solution of other cations. Thus zeolites can act as ion-exchangers and may be used for water-softening. Unlike feldspars, where all the spaces in the structure are filled by cations, the zeolites contain more spaces than ions, and so one calcium ion may easily be replaced by two sodium ions.

Another type of synthetic zeolite, the *permutites*, are made by fusing mixtures of quartz sand, feldspar, kaolin and soda. Their structure is similar to that of natural zeolites, but is lacking in crystalline symmetry and regularity. Permutites are widely used for water-softening.

Many zeolites exist, with similar structures but differing in detail; both 'fibrous' and 'layer' zeolites occur naturally. The most extensively studied are those of the fibrous types, in which the aluminosilicate tetrahedra are

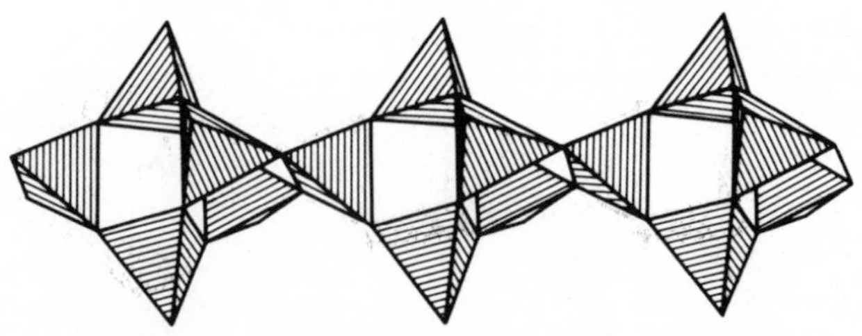

FIGURE 8.10. The arrangement of the aluminosilicate tetrahedra chain in a fibrous zeolite

joined to form chains with a repeating unit 6.6 Å long and containing five tetrahedra (see Figure 8.10). Thus the empirical formula of the polyanion of the fibrous zeolites is $(Al_nSi_{5-n}O_{10})^{n-}$ and one of the simplest examples is *natrolite*, $Na_2(Al_2Si_3)O_{10}.2\ H_2O$. These chains are linked side-by-side through oxygen atoms to give a three-dimensional network. Parallel to, and between, the chains are channels which can adsorb small molecules. In the fibrous zeolites, the oxygen cross-links are relatively few, and so the minerals can be cleaved more easily parallel to the fibres than at right angles to them.

8.3.4.3. *Ultramarines*

Ultramarines have the cage-type, aluminosilicate structure shown in Figure 8.11. The 'cages' are linked together through oxygen atoms to give an infinite three-dimensional framework. Ultramarines differ from feldspars and zeolites, because the large openings in the structures contain small anions as well as cations, but no water molecules. Thus ultramarine itself, which gives the bright blue colour to the semiprecious stone, lapis lazuli, has the idealized formula, $Na_8Al_6Si_6O_{24}{\cdot}S_2{\cdot}$. The blue colour is due to the presence of S_3^- ions, although a yellow ion S_2^- also exists in ultramarine. Other natural ultramarines contain chloride or sulphate ions.

Ultramarines can act as ion-exchangers for both cations and anions; thus different cations such as lithium, calcium, silver, and thallium can be introduced into the structures. When selenium or tellurium are incorporated into ultramarines, the products possess colours varying from red to yellow, and are used, in powdered form, as pigments.

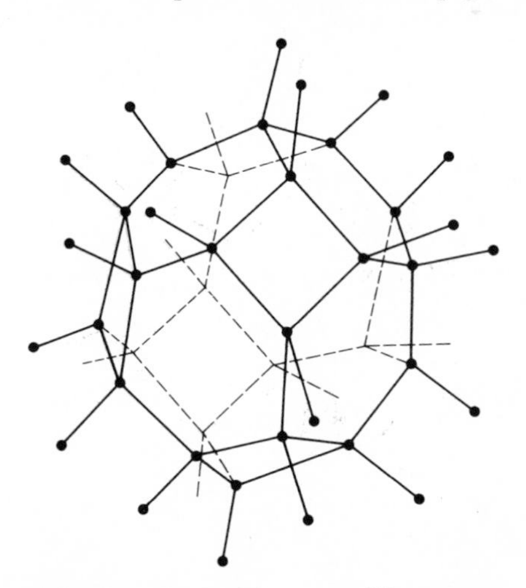

FIGURE 8.11. The cage-like structure of ultramarine: only the positions of the silicon or aluminium atoms are shown

Ultramarine blue (with the same chemical formula as lapis lazuli) is used as a pigment in oil paints and printing ink, as a 'blueing' agent in laundering, and to mask yellow tints in paper.

Synthetic ultramarines can be prepared by fusing a mixture of clay, anhydrous sodium carbonate, and sulphur, and are efficient hydrocarbon-cracking catalysts.

8.4. ADDITIONAL READING

Silicate Science, Vol. 1: *Silicate Structures* W. Eitel, Academic Press, London and New York, 1964.

Articles on carbon in *Kirk–Othmer Encyclopedia of Chemical Technology*, 2nd Ed., Vol. 4, Wiley–Interscience, London and New York, 1964, pp. 283–335.

Articles on silica in *Kirk–Othmer Encyclopedia of Chemical Technology*, 2nd Ed., Vol. 18, Wiley–Interscience, London and New York, 1969, pp. 46–111.

'Silica and silicates' A. E. R. Westman and M. K. Murthy in *Encyclopedia of Polymer Science and Technology*, Vol. 12, Wiley–Interscience, London and New York, 1970, p. 441.

An Introduction to Crystal Chemistry R. C. Evans, Cambridge University Press, 1964.

Inorganic Solids (D. M. Adams), John Wiley and Sons, London and New York, 1974.

Ionic Polymers (Ed. L. Holliday), Halsted Press, London and New York, 1975.

Inorganic Chemistry R. B. Helsop and K. Jones, Elsevier, Amsterdam, Oxford and New York, 1976.

Silicon Chemistry and Applications C. A. Pearce, *Chemical Society Monographs for Teachers*, No. 20, The Chemical Society, London, 1972.

Chapter 9

Epilogue

It is perhaps unnecessary to stress the importance of natural high polymers to man; they control the very process of life itself and, in addition, perform essential functions like constituting the basis of his food and clothing, and meeting many other everyday needs. But too often our knowledge of these macromolecules is still incomplete.

For all types of biopolymers there is a need to develop more sensitive methods of structural determination, which can be used efficiently on the sub-microscale. The development of more generally-applicable methods for determining the molecular weight distribution would also be advantageous. Hopefully, advances will be made in the determination of three-dimensional structure so that the time and effort required to establish the tertiary structure of a protein chain will be much reduced. In addition, more detailed investigations of helix formation in nucleic acids and polysaccharides are required in order to resolve some of the controversies regarding these groups of macromolecules.

Much research is necessary to appreciate the structure of 'supermolecular assemblies'. This area is the subject of a lot of current research activity, but—with the exception of simple viruses—much knowledge remains to be gained.

A wide variety of questions regarding the *proteins* remain unanswered. Of these, we might mention that it is not known how the insoluble complexes involving structural proteins are laid down. More information about the detailed structure of this type of protein is required, as well as details of their interaction with other biopolymers. A better understanding of the mechanisms of enzyme action is also necessary. More enzymes will be studied structurally and their modes of action will be revealed. There is then the possibility that enzymes might be designed and synthesized to carry out specific tasks. At the molecular level, more knowledge is required of the action of hormones and of antibody production. The rôle of histones in chromosomes is not understood, and it is also not known what function the oligosaccharides play in the behaviour of the glycoproteins—particularly on cell surfaces.

The solving of problems associated with *nucleic acids* will be of great importance to medicine and biology. For instance, it is not known in detail how the cells of higher animals can differentiate (i.e. produce cells as different as a muscle or red blood cell from one original cell). Here we are obviously concerned with understanding the control of nucleic acid activity. Another problem concerning the control mechanisms of nucleic acid synthesis is the loss of growth regulation so characteristic of cancerous cells. One of the great developments of the future will be the use of 'genetic engineering' (hopefully to benefit mankind), i.e. recombinant DNA technology may enable large quantities of currently rare substances to be produced for use in medicine and scientific research.

In the case of *polysaccharides*, several fundamental problems remain. We do not fully understand the biological function of many of these biopolymers. Structural investigations on plant gums and a large number of bacterial polysaccharides are incomplete. These latter investigations should yield information regarding the chemical nature of the bacterial surface responsible for the extremely important antigen–antibody interactions. Many questions remain with regard to the structural polysaccharides, e.g. the way in which the plant cell wall is synthesized, its exact structure, and the interaction of these polysaccharides with lignin. The architecture of the starch granule is not understood, and neither is the mode of biosynthesis of this unique insoluble cell component. For all complex polysaccharides there is a fundamental difficulty in that it is not known whether we are dealing with a unique polymer or a *range of related polymeric materials*. Often the latter appears to be the situation, and the development of some new technique for fractionating polysaccharides on the basis of fine structure would be an enormous advance.

The structure of *lignin* has yet to be solved completely, and finding a new economic use for this copious waste product of the cellulose industry would be beneficial. We have already stressed that the function of *rubber* in plants is not understood.

In spite of their great economic importance, research in the field of natural *inorganic polymers* is more limited. Further detailed investigations of structures are desirable, and advances in synthetic inorganic polymers are likely—particularly in the case of zeolites—which may improve their catalytic/adsorptive properties for industrial use.

It can be seen that the immense problems in all fields of natural high polymers present a great challenge to the chemist and biochemist!

Index

Definitions are indexed in bold type

UWIST